Applications of
Mathematics

22

Edited by   A.V. Balakrishnan
I. Karatzas
M. Yor

Applications of Mathematics

1. Fleming/Rishel, **Deterministic and Stochastic Optimal Control** (1975)
2. Marchuk, **Methods of Numerical Mathematics,** Second Ed. (1982)
3. Balakrishnan, **Applied Functional Analysis,** Second Ed. (1981)
4. Borovkov, **Stochastic Processes in Queueing Theory** (1976)
5. Liptser/Shiryayev, **Statistics of Random Processes I: General Theory** (1977)
6. Liptser/Shiryayev, **Statistics of Random Processes II: Applications** (1978)
7. Vorob'ev, **Game Theory: Lectures for Economists and Systems Scientists** (1977)
8. Shiryayev, **Optimal Stopping Rules** (1978)
9. Ibragimov/Rozanov, **Gaussian Random Processes** (1978)
10. Wonham, **Linear Multivariable Control: A Geometric Approach,** Third Ed. (1985)
11. Hida, **Brownian Motion** (1980)
12. Hestenes, **Conjugate Direction Methods in Optimization** (1980)
13. Kallianpur, **Stochastic Filtering Theory** (1980)
14. Krylov, **Controlled Diffusion Processes** (1980)
15. Prabhu, **Stochastic Storage Processes: Queues, Insurance Risk, and Dams** (1980)
16. Ibragimov/Has'minskii, **Statistical Estimation: Asymptotic Theory** (1981)
17. Cesari, **Optimization: Theory and Applications** (1982)
18. Elliott, **Stochastic Calculus and Applications** (1982)
19. Marchuk/Shaidourov, **Difference Methods and Their Extrapolations** (1983)
20. Hijab, **Stabilization of Control Systems** (1986)
21. Protter, **Stochastic Integration and Differential Equations** (1990)
22. Benveniste/Métivier/Priouret, **Adaptive Algorithms and Stochastic Approximations** (1990)

Albert Benveniste   Michel Métivier
Pierre Priouret

Adaptive Algorithms and Stochastic Approximations

Translated from the French by Stephen S. Wilson

With 24 Figures

Springer-Verlag

Berlin Heidelberg New York
London Paris Tokyo
Hong Kong Barcelona

Albert Benveniste
IRISA-INRIA
Campus de Beaulieu
35042 RENNES Cedex
France

Michel Métivier †

Pierre Priouret
Laboratoire de Probabilités
Université Pierre et Marie Curie
4 Place Jussieu
75230 PARIS Cedex
France

Managing Editors

A.V. Balakrishnan
Systems Science Department
University of California
Los Angeles, CA 90024
USA

I. Karatzas
Department of Statistics
Columbia University
New York, NY 10027
USA

M. Yor
Laboratoire de Probabilités
Université Pierre et Marie Curie
4 Place Jussieu, Tour 56
75230 PARIS Cedex
France

Title of the Original French edition:
Algorithmes adaptatifs et approximations stochastiques
© Masson, Paris, 1987

Mathematics Subject Classification (1980): 62-XX, 62L20, 93-XX, 93C40, 93E12, 93E10

ISBN 3-540-52894-6 Springer-Verlag Berlin Heidelberg New York
ISBN 0-387-52894-6 Springer-Verlag New York Berlin Heidelberg

This work is subject to copyright. All rights are reserved, whether the whole or part of the material is concerned, specifically the rights of translation, reprinting, reuse of illustrations, recitation, broadcasting, reproduction on microfilms or in other ways, and storage in data banks. Duplication of this publication or parts thereof is only permitted under the provisions of the German Copyright Law of September 9, 1965, in its current version, and a copyright fee must always be paid. Violations fall under the prosecution act of the German Copyright Law.

© Springer-Verlag Berlin Heidelberg 1990
Printed in the United States of America

2141/3140-543210 - Printed on acid-free paper

A notre ami Michel

Albert, Pierre

Preface to the English Edition

The comments which we have received on the original French edition of this book, and advances in our own work since the book was published, have led us to make several modifications to the text prior to the publication of the English edition. These modifications concern both the fields of application and the presentation of the mathematical results.

As far as the fields of application are concerned, it seems that our claim to cover the whole domain of pattern recognition was somewhat exaggerated, given the examples chosen to illustrate the theory. We would now like to put this to rights, without making the text too cumbersome. Thus we have decided to introduce two new and very different categories of applications, both of which are generally recognised as being relevant to pattern recognition. These applications are introduced through long exercises in which the reader is strictly directed to the solutions. The two new examples are borrowed, respectively, from the domain of *machine learning using neural networks* and from the domain of *Gibbs fields* or networks of random automata.

As far as the presentation of the mathematical results is concerned, we have added an appendix containing details of a.s. convergence theorems for stochastic approximations under Robbins–Monro type hypotheses. The new appendix is intended to present results which are *easily proved* (using only basic limit theorems about supermartingales) and which are *brief*, without over-restrictive assumptions. The appendix is thus specifically written for reference, unlike the more technical body of Part II of the book. We have, in addition, corrected several minor errors in the original, and expanded the bibliography to cover a broader area of research.

Finally, for this English version, we would like to thank Hans Walk for his interesting suggestions which we have used to construct our list of references, and Dr. Stephen S. Wilson for his outstanding work in translating and editing this edition.

April 1990

Preface to the Original French Edition

The Story of a Wager

When, some three years ago, urged on by Didier Dacunha-Castelle and Robert Azencott, we decided to write this book, our motives were, to say the least, both simple and naive. Number 1 (in alphabetical order) dreamt of a corpus of solid theorems to justify the practical everyday engineering usage of adaptive algorithms and to act as an engineer's handbook. Numbers 2 and 3 wanted to show that the term "applied probability" should not necessarily refer to probability with regard to applications, but rather to probability in support of applications.

The unfolding dream produced a game rule, which we initially found quite amusing: Number 1 has the material (examples of major applications) and the specification (the theorems of the dream), Numbers 2 and 3 have the tools (martingales, ...), and the problem is to achieve the specification. We were overwhelmed by this long and curious collaboration, which at the same time brought home several harsh realities: not all the theorems of our dreams are necessarily true, and the most elegant tools cannot necessarily be adapted to the toughest applications.

The book owes a great deal to the highly active adaptive processing community: Michèle Basseville, Bob Bitmead, Peter Kokotovic, Lennart Ljung, Odile Macchi, Igor Nikiforov, Gabriel Ruget and Alan Willsky, to name but a few. It also owes much to the ideas and publications of Harold Kushner and his co-workers D.S.Clark, Hai Huang and Adam Shwartz. Proof reading amongst authors is a little like being surrounded by familiar objects: it blunts the critical spirit. We would thus like to thank Michèle Basseville, Bernard Delyon and Georges Moustakides for their patient reading of the first drafts.

Since this book was bound to evolve as it was written, we saw the need to use a computer-based text-processing system; we were offered a promising new package, MINT, which we adopted. The generous environment of IRISA, much perseverance by Dominique Blaise, Philippe Louarn's great ingenuity in tempering the quirks of the software, and Number 1's stamina of a long-distance runner in implementing the many successive corrections, all contributed to the eventual birth of this book.

January 1987

Contents

Introduction   1

Part I. Adaptive Algorithms: Applications   7

1. General Adaptive Algorithm Form   9
1.1 Introduction 9
1.2 Two Basic Examples and Their Variants 10
1.3 General Adaptive Algorithm Form and Main Assumptions 23
1.4 Problems Arising 29
1.5 Summary of the Adaptive Algorithm Form: Assumptions (A) 31
1.6 Conclusion 33
1.7 Exercises 34
1.8 Comments on the Literature 38

2. Convergence: the ODE Method   40
2.1 Introduction 40
2.2 Mathematical Tools: Informal Introduction 41
2.3 Guide to the Analysis of Adaptive Algorithms 48
2.4 Guide to Adaptive Algorithm Design 55
2.5 The Transient Regime 75
2.6 Conclusion 76
2.7 Exercises 76
2.8 Comments on the Literature 100

3. Rate of Convergence   103
3.1 Mathematical Tools: Informal Description 103
3.2 Applications to the Design of Adaptive Algorithms with
    Decreasing Gain 110
3.3 Conclusions from Section 3.2 116
3.4 Exercises 116
3.5 Comments on the Literature 118

4. Tracking Non-Stationary Parameters · 120
4.1 Tracking Ability of Algorithms with Constant Gain 120
4.2 Multistep Algorithms 142
4.3 Conclusions 158
4.4 Exercises 158
4.5 Comments on the Literature 163

5. Sequential Detection; Model Validation · 165
5.1 Introduction and Description of the Problem 166
5.2 Two Elementary Problems and their Solution 171
5.3 Central Limit Theorem and the Asymptotic Local Viewpoint 176
5.4 Local Methods of Change Detection 180
5.5 Model Validation by Local Methods 185
5.6 Conclusion 188
5.7 Annex: Proofs of Theorems 1 and 2 188
5.8 Exercises 191
5.9 Comments on the Literature 197

6. Appendices to Part I · 199
6.1 Rudiments of Systems Theory 199
6.2 Second Order Stationary Processes 205
6.3 Kalman Filters 208

Part II. Stochastic Approximations: Theory · 211

1. O.D.E. and Convergence A.S. for an Algorithm with Locally Bounded Moments · 213
1.1 Introduction of the General Algorithm 213
1.2 Assumptions Peculiar to Chapter 1 219
1.3 Decomposition of the General Algorithm 220
1.4 L^2 Estimates 223
1.5 Approximation of the Algorithm by the Solution of the O.D.E. ...230
1.6 Asymptotic Analysis of the Algorithm 233
1.7 An Extension of the Previous Results 236
1.8 Alternative Formulation of the Convergence Theorem 238
1.9 A Global Convergence Theorem 239
1.10 Rate of L^2 Convergence of Some Algorithms 243
1.11 Comments on the Literature 249

2. Application to the Examples of Part I 251
2.1 Geometric Ergodicity of Certain Markov Chains 251
2.2 Markov Chains Dependent on a Parameter θ 259
2.3 Linear Dynamical Processes 265
2.4 Examples 270
2.5 Decision-Feedback Algorithms with Quantisation 276
2.6 Comments on the Literature 288

3. Analysis of the Algorithm in the General Case 289
3.1 New Assumptions and Control of the Moments 289
3.2 L^q Estimates 293
3.3 Convergence towards the Mean Trajectory 298
3.4 Asymptotic Analysis of the Algorithm 301
3.5 "Tube of Confidence" for an Infinite Horizon 305
3.6 Final Remark. Connections with the Results of Chapter 1 306
3.7 Comments on the Literature 306

4. Gaussian Approximations to the Algorithms 307
4.1 Process Distributions and their Weak Convergence 308
4.2 Diffusions. Gaussian Diffusions 312
4.3 The Process $U^\gamma(t)$ for an Algorithm with Constant Step Size 314
4.4 Gaussian Approximation of the Processes $U^\gamma(t)$ 321
4.5 Gaussian Approximation for Algorithms with Decreasing
    Step Size 327
4.6 Gaussian Approximation and Asymptotic Behaviour
    of Algorithms with Constant Steps 334
4.7 Remark on Weak Convergence Techniques 341
4.8 Comments on the Literature 341

5. Appendix to Part II: A Simple Theorem in the "Robbins–Monro" Case 343
5.1 The Algorithm, the Assumptions and the Theorem 343
5.2 Proof of the Theorem 344
5.3 Variants 345

Bibliography 349

Subject Index to Part I 361

Subject Index to Part II 364

Introduction

Why "adaptive algorithms and stochastic approximations"?

The use of adaptive algorithms is now very widespread across such varied applications as system identification, adaptive control, transmission systems, adaptive filtering for signal processing, and several aspects of pattern recognition. Numerous, very different examples of applications are given in the text. The success of adaptive algorithms has inspired an abundance of literature, and more recently a number of significant works such as the books of Ljung and Söderström (1983) and of Goodwin and Sin (1984).

In general, these works consider primarily the notion of an *adaptive system*, which is composed of:

1. The object upon which processing is carried out: control system, modelling system, transmission system,....
2. The so-called estimation process.

In so doing, they implicitly address the modelling of the system as a whole. This approach has naturally led to the introduction of boundaries between

- System identification from the control point of view.
- Signal modelling.
- Adaptive filtering.
- The myriad of applications to pattern recognition: adaptive quantisation,....

These boundaries echo the *classes of models* which conveniently describe each corresponding system. For example multivariable linear systems certainly have an important role to play in system identification, although they are scarcely ever met in adaptive filtering, and never appear in most pattern recognition applications. On the other hand, the latter applications call for models which have no relevance to the linear systems widely used in automatic control theory. It would therefore be foolish to try to present a *general theory of adaptive systems* which created a framework sufficiently broad to encompass all models and algorithms simultaneously.

However, in our opinion and experience, these problems have a major common component: namely the use (once all the modelling problems have been resolved) of *adaptive algorithms*. This topic, which we shall now study more specifically, is the counterpart of the notion of *stochastic approximation* as found in statistical literature. The juxtaposition of these two expressions in the title is an exact statement of our ambition to produce a reference work, both for engineers who use these algorithms and for probabilists or statisticians who would like to study stochastic approximations in terms of problems arising from real applications.

Adaptive algorithms.

The function of these algorithms is to adjust a *parameter vector*, which we shall denote generically by θ, with a view to an objective specified by the user: system control, identification, adjustment,.... This vector θ is the user's only interface with the system and its definition requires an initial *modelling* phase.

In order to tune this parameter θ, the user must be able to monitor the system. Monitoring is effected via a so-called *state vector*, which we shall denote by X_n, where n refers to the time of observation of the system. This state vector might be:

- The set consisting of the regression vector and an error signal, in the classical case of system identification, as for example presented in (Ljung and Soderström 1983) or in numerous adaptive filtering problems.

- The sample signal observed at the instant n, in the case of adaptive quantisation,....

In all these cases, the rule used to update θ will typically be of the form

$$\theta_n = \theta_{n-1} + \gamma_n H(\theta_{n-1}, X_n)$$

where γ_n is a sequence of small gains and $H(\theta, X)$ is a function whose specific determination is one of the main aims of this book.

Aims of the book.

These are twofold:

1. To provide the user of adaptive algorithms with a *guide to their analysis and design*, which is as clear and as comprehensive as possible.

2. To accompany this guide with a *presentation of the fundamental underlying mathematics*.

In seeking to reach these objectives, we come up against two contradictory demands. On the one hand, adaptive algorithms must, generally speaking, be easy to use and accessible to a large class of engineers: this requires the guide to use a minimal technical arsenal. On the other hand, an honest assessment

Introduction

of practices currently found in adaptive algorithm applications demands that we obtain fine results using assumptions which, in order to be realistic, are perforce complicated. This remark has led many authors to put forward the case for a similar guide, modestly restricted to the application areas of interest to themselves.

We have preferred to resolve this difficulty in another way, and it is this prejudice which lends originality to the book, which is, accordingly, divided into two parts, each of a very different character.

Part II presents the mathematical foundations of adaptive systems theory from a modern point of view, without shying away from the difficulty of the questions to be resolved: in it we shall make great use of the basic notions of conditioning, Markov chains and martingales. Assumptions will be stated in detail and proofs will be given in full. Part II contains:

1. "Law of large numbers type" convergence results where, so as not to make the proofs too cumbersome, the assumptions include minor constraints on the temporal properties of the state vector X_n and on the regularity of the function $H(\theta, X)$, and quite severe restrictions upon the moments of X_n (Chapter 1).

2. An illustration of the previous results, first with classical examples, then with a typical, reputedly difficult, example (Chapter 2).

3. A refinement of the results of Chapter 1 with weaker assumptions on the moments (Chapter 3).

4. The introduction of diffusion approximations ("central limit theorem type" results) which allow a detailed evaluation of the asymptotic behaviour of adaptive algorithms (Chapter 4).

Many of the results and proofs in Part II are original. They cover the case of algorithms with decreasing gain, as well as that of algorithms with constant gain, the latter being the most widely use in practice.

Part I concentrates on the presentation of the guide and on its illustration by various examples. Whilst not totally elementary in a mathematical sense, Part I is not encumbered with technical assumptions, and thus it is able to highlight the essential mathematical difficulties which must be faced if one is to make good use of adaptive algorithms. On the other hand, we wanted the guide to provide as full an introduction as possible to good usage of adaptive algorithms. Thus we discuss:

1. The convergence of adaptive algorithms (in the sense of the law of large numbers) and the consequence of this on algorithm analysis and design (Chapters 1 and 2).

2. The asymptotic behaviour of algorithms in the "ideal" case where the phenomenon upon which the user wishes to operate is time invariant (Chapter 3).

3. The behaviour of the algorithms when the true system evolves slowly in time and the consequences of this on algorithm design (Chapter 4).

4. The monitoring of abrupt changes in the true system, or the non-conformity of the true system to the model in use (Chapter 5).

The final two points are central to the study of adaptive algorithms (these algorithms arose because true systems are time-varying), yet, to the best of our knowledge they have never been systematically discussed in any text on adaptive algorithms.

Whilst the two parts of the book overlap to a certain extent, they take complementary views of the areas of overlap. In each case, we cross-reference the informal results of Part I with the corresponding theorems of Part II, and the examples of Part I with their mathematical treatment in Part II.

How to read this book.

The diagram below shows the organisation of the various chapters of the book and their mutual interaction.

Each chapter of Part I contains a number of exercises which form a useful complement to the material presented in that chapter. The exercises are either direct applications or non-trivial extensions of the chapter. Part I also includes three appendices which describe the rudiments of systems theory and Kalman filtering for mathematicians who wish to read Part I. Part II is technically difficult, although it demands little knowledge of probability: basic concepts, Markov chains, basic martingale concepts; other principles are introduced as required. As for Part I, the first two chapters only require the routine knowledge of probability theory of an engineer working in signal processing or control theory, whilst the final three chapters are of increasing difficulty.

The book may be read in several different ways, for example :

- *Engineer's introductory course on adaptive algorithms and their uses*: Chapters 1 and 2 of Part I;

- *Engineer's technical course on adaptive algorithms and their use*: all of Part I, the first two sections of Chapter 4 of Part II;

- *Mathematician's technical course on adaptive algorithms and their use*: Part II, Chapters 1, 2, 4 and a rapid pass through Part I.

Introduction

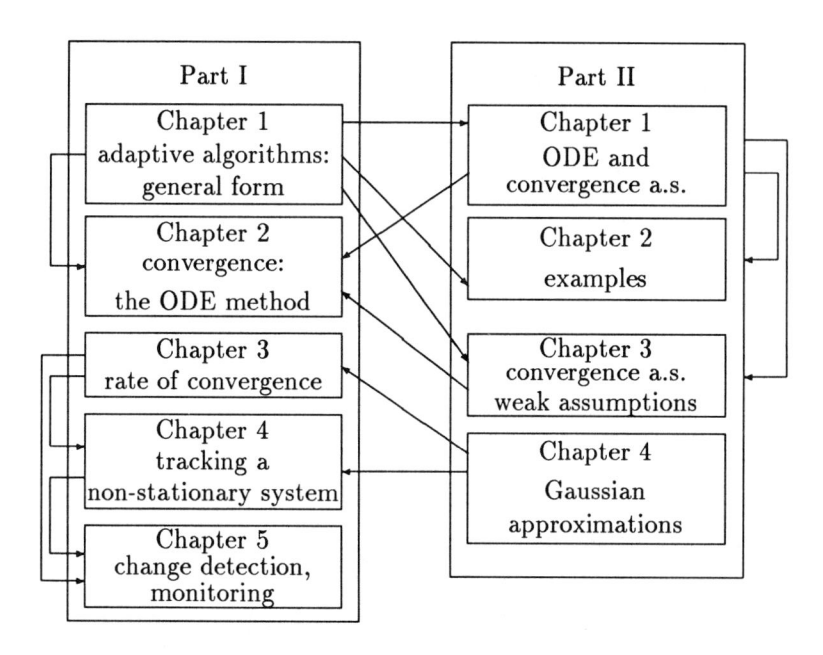

```
┌─────────────────────────┐   ┌─────────────────────────┐
│        Part I           │   │        Part II          │
│  ┌───────────────────┐  │   │  ┌───────────────────┐  │
│  │    Chapter 1      │──┼───┼─▶│    Chapter 1      │  │
│  │ adaptive algorithms:│ │   │  │    ODE and        │  │
│  │   general form    │  │   │  │ convergence a.s.  │  │
│  ├───────────────────┤  │   │  ├───────────────────┤  │
│  │    Chapter 2      │  │   │  │    Chapter 2      │  │
│  │   convergence:    │  │   │  │    examples       │  │
│  │  the ODE method   │  │   │  │                   │  │
│  ├───────────────────┤  │   │  ├───────────────────┤  │
│  │    Chapter 3      │  │   │  │    Chapter 3      │  │
│  │ rate of convergence│ │   │  │ convergence a.s.  │  │
│  │                   │  │   │  │ weak assumptions  │  │
│  ├───────────────────┤  │   │  ├───────────────────┤  │
│  │    Chapter 4      │  │   │  │    Chapter 4      │  │
│  │    tracking a     │  │   │  │    Gaussian       │  │
│  │non-stationary system│ │   │  │  approximations   │  │
│  ├───────────────────┤  │   │  └───────────────────┘  │
│  │    Chapter 5      │  │   │                         │
│  │  change detection,│  │   │                         │
│  │    monitoring     │  │   │                         │
│  └───────────────────┘  │   │                         │
└─────────────────────────┘   └─────────────────────────┘
```

Application domains.

As the title indicates, the adaptive algorithms are principally applied to system identification (one of the most important areas of control theory), signal processing and pattern recognition.

As far as system identification is concerned, comparison of the numerous examples of AR and ARMA system identification with (Ljung & Soderström 1983) highlights the importance of this area; of course this much is already well known. On the other hand, the two adaptive control exercises will serve to show the attentive reader that the *stability* of adaptive control schemes is one essential problem which is not resolved by the theoretical tools presented here.

The relevance of adaptive algorithms to signal processing is also well known, as the large number of examples from this area indicates. We would however highlight the exercise concerning the ALOHA protocol for satellite communications as an atypical example in telecommunications.

Applications to pattern recognition are slightly more unusual. Certainly the more obvious areas of pattern recognition, such as speech recognition, use techniques largely based on adaptive signal processing (LPC, Burg and recursive methods ...). The two exercises on adaptive quantisation are more characteristic: in fact they are a typical illustration of the difficulties and the techniques of pattern recognition; such methods, involving a learning phase, are used in speech and image processing. Without wishing to overload our already long list of examples, we note that the recursive estimators of motion in image sequences used in numerical encoding of television images are also adaptive algorithms.

Part I

Adaptive Algorithms: Applications

Chapter 1
General Adaptive Algorithm Form

1.1 Introduction

In which an example is used to determine a general adaptive algorithm form and to illustrate some of the problems associated with adaptive algorithms.

The aim of this chapter is to derive a form for adaptive algorithms which is sufficiently general to cover almost all known applications, and which at the same time lends itself well to the theoretical study which is undertaken in parallel in Part II of the book.

The general form is the following:

$$\theta_n = \theta_{n-1} + \gamma_n H(\theta_{n-1}, X_n) + \gamma_n{}^2 \varepsilon_n(\theta_{n-1}, X_n) \qquad (1.1.1)$$

where:

$(\theta_n)_{n \geq 0}$	is the sequence of vectors to be recursively updated;
$(X_n)_{n \geq 1}$	is a sequence of random vectors representing the on-line observations of the system in the form of a state vector;
$(\gamma_n)_{n \geq 1}$	is a sequence of "small" scalar gains;
$H(\theta, X)$	is the function which essentially defines how the parameter θ is updated as a function of new observations;
$\varepsilon_n(\theta, X)$	defines a small perturbation (whose role we shall see later) on the algorithm (the most common form of adaptive algorithm corresponds to $\varepsilon_n \equiv 0$).

In this chapter, we shall determine, by studying significant examples, the required properties of the state vector (X_n), of the function H, of the gain (γ_n) and of the residual perturbation ε_n. Furthermore, we shall examine the nature of the problems which may be addressed using an algorithm of type (1.1.1); we shall illustrate some of the difficulties which arise when studying such algorithms.

1.2 Two Basic Examples and Their Variants

These two examples are related to telecommunications transmission systems; the first example concerns transmission channel equalisation by amplitude modulation, the second concerns transmission carrier recovery by phase modulation. In order to set the scene for the rest of the book, we shall describe these applications in detail. We shall begin with a description of the physical system, then we shall examine the modelling problem, a preparatory phase indispensable to algorithm design, and finally we shall give a brief overview of the so-called algorithm design phase, ending with an introduction to the algorithms themselves.

1.2.1 Adaptive Equalisation in Transmission of Numerical Data

Amplitude Modulation.
Linear (or amplitude) modulation of a carrier wave of frequency f_c (e.g. $f_c = 1800$ Hz) by a data message is generally used to transmit data at high speed across a telephone line L. We recall that a telephone line has a passband of 3000 Hz (300–3300 Hz) and that the maximum bit rate commonly achieved on such lines is 9600 bits per second.

The simplest type of linear modulation is the modulation of a single carrier $\cos(2\pi f_c t - \Phi)$ where Φ is a random phase angle from the uniform distribution on $[0, 2\pi]$. Figure 1 illustrates this type of transmission.

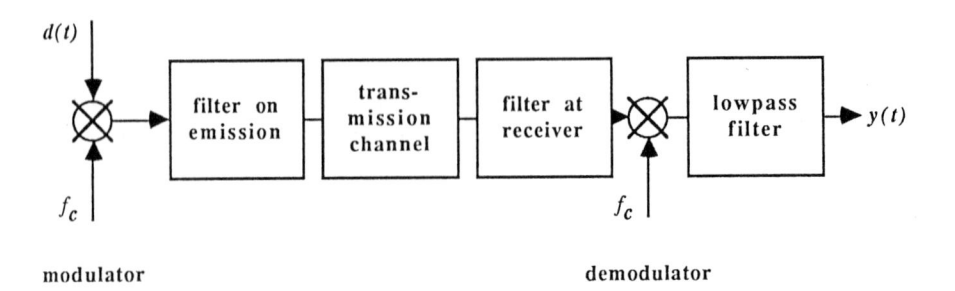

Figure 1. Data transmission by linear modulation of a carrier.

1.2 Two Basic Examples and Their Variants

The message $d(t)$ to be emitted is of the form

$$d(t) = \sum_{k=-\infty}^{+\infty} a_k \, g_\Delta(t - k\Delta) \tag{1.2.1}$$

The symbols a_k, which are the data items to be transmitted, take discrete values in a finite set, for example $\{\pm 1, \pm 3\}$. They are transmitted at regular intervals Δ.

The function $g_\Delta(t)$ is, for example, the indicator function of the interval $[0, \Delta]$. Figure 2 shows a message to be emitted.

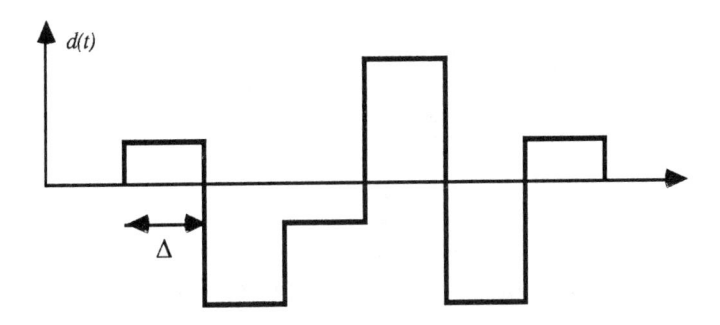

Figure 2. Example of a message.

Using the fact that a rectangular impulse of width Δ is the response of a particular linear filter whose input is a Dirac impulse, it can be shown that the emitter–line–receiver system of Fig. 1 is equivalent to the system in Fig. 3.

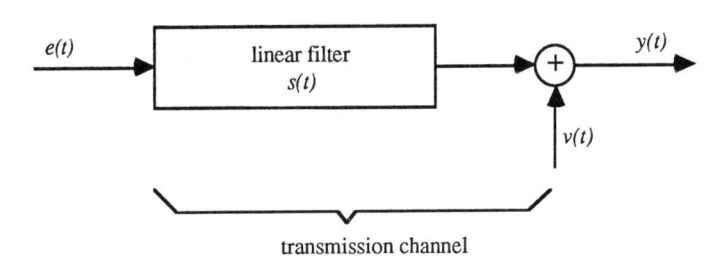

Figure 3. Baseband system.

In this system, which is called an equivalent baseband system, the emitted signal is

$$e(t) = \sum_{k=-\infty}^{+\infty} a_k \, \delta(t - k\Delta) \tag{1.2.2}$$

and the received signal may be expressed as:

$$y(t) = \sum_{k=-\infty}^{+\infty} a_k \, s(t - k\Delta) + \nu(t) \tag{1.2.3}$$

where $\nu(t)$ is an additive noise and $s(t)$ is the impulse response of a linear filter. Figure 4 gives an example of a response $s(t)$.

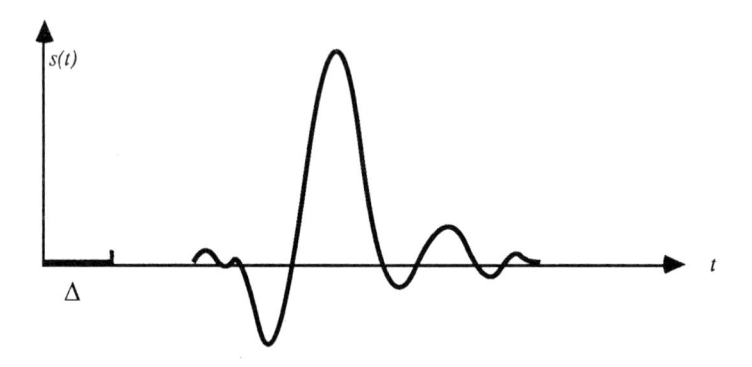

Figure 4. Example of an impulse response $s(t)$.

In practice, it is desirable to choose the interval Δ to be as small as possible so as to increase the data rate: $s(t)$ may have a duration of the order of 10 to 20 seconds. This causes an overlap of successive impulses, or intersymbol interference, and leads to problems in reconstituting the emitted data sequence from the received signal. We shall return to this topic later.

If the received signal $x(t)$ is sampled with period Δ, and if we set

$$y_n = y(n\Delta + t_0) \quad s_n = s(n\Delta + t_0) \quad \nu_n = \nu(n\Delta + t_0) \tag{1.2.4}$$

where t_0 is the chosen sample time origin, (1.2.3) may be rewritten in the form

$$y_n = \sum_{k=-\infty}^{n} a_k s_{n-k} + \nu_n$$

or in the form

$$y_n = \sum_{k=0}^{+\infty} s_k a_{n-k} + \nu_n \tag{1.2.5}$$

1.2 Two Basic Examples and Their Variants

The *equalisation* problem is then the following: in Fig. 5 below, what is the best way to tune the filter θ so that the output of the quantiser $\hat{a}_n$ is equal to a_n with least possible error rate?

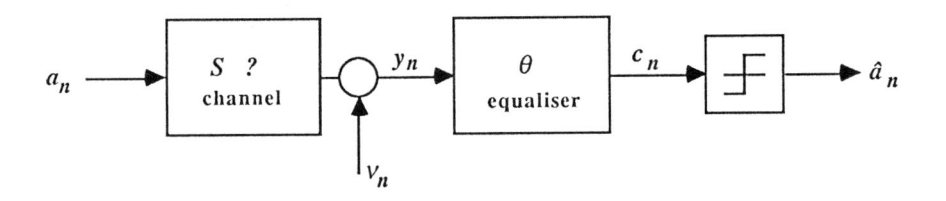

Figure 5. Equaliser, general diagram.

Reasons for Equalisation.

Note that if the effect of the additive noise ν_n is negligible compared with that of the channel (s_k), the adaptive filter θ must invert as closely as possible the transformation applied to the message (a_n) by the channel (s_k). We shall later give a precise definition of this objective which we shall denote symbolically by

$$\theta_* \approx S^{-1} \qquad (1.2.6)$$

The tasks of the equaliser then fall into three categories:

(*i*) *Learning the channel.*
Since the channel S is initially unknown, a learning phase is necessary prior to any emission proper. For this, a training sequence (a_n) known to all (and which is even the subject of an international CCITT convention) is used to tune the equaliser θ to approximate to the desired value S^{-1}.

(*ii*) *Tracking channel variations.*
In certain cases, following task (*i*), the equaliser is satisfactorily tuned and may then be fixed for transmission proper: this in particular is the case in *packet* transmission mode (cf. the well known TRANSPAC system) where the learning phase precedes the emission of a fixed-length packet of messages. In other cases, where the message length has no a priori limits, the channel may be subject to significant temporal variations: this in particular is the case for atmospheric transmissions (radio link channels) where the existence of transient multiple paths causes significant variations over a period of a second or so. A second equalisation phase, simultaneous with the message transmission, is then needed to maintain the desired condition (1.2.6). This is *self-adaptive equalisation*.

(iii) Blind equalisation.
In the case of a broadcast link (one emitter for several receivers) the channel learning phase (*i*) cannot be carried out, since it would necessitate the interruption of transmission whenever a new receiver entered service. In this case, the channel must be learnt directly from the data stream: learning and decoding go together right away. This is *blind* equalisation.

Of the three problems mentioned above, it is chiefly in the second (tracking the channel) that ongoing action is required. Such action naturally takes the form of a regular update of the filter θ as new data is received. We have seen a first illustration of one of the fundamental messages which we wish to put across in this book.

First message: the main reason for using adaptive algorithms is to track temporal system variations.

1.2.1.1 Modelling

Until now we have denoted the equaliser in an informal way by the letter θ. The *modelling* of dynamical systems (and a filter is a special case of a dynamical system) comprises

1. a given *model structure* which is capable of describing the dynamic input/output behaviour which interests us;
2. the specification of the *parameters* in the model structure which remain to be determined to complete the definition of the dynamical system; in general, these parameters will be represented by a vector denoted by θ, knowledge of which will suffice to determine the complete model;
3. the mathematical model of the behaviour of signals entering the dynamical system.

We shall apply this procedure to the equaliser example.

Choice of Model Structure.

We shall call upon two types of structure most frequently used to synthesise a filter: the transversal form (or "all zeroes") and the recursive form ("poles-zeroes").

Transversal or all zeroes form.
The output (c_n) of the equaliser is given as a function of the input (y_n) by

$$(i) \qquad c_n = \sum_{k=-N}^{+N} \theta(k) y_{n-k} \quad = Y_n^T \theta$$

1.2 Two Basic Examples and Their Variants

$$(ii) \quad Y_n^T := (y_{n+N}, \ldots, y_n, \ldots, y_{n-N})$$
$$(iii) \quad \theta^T := (\theta(-N), \ldots, \theta(0), \ldots, \theta(N)) \qquad (1.2.7)$$

The fact that c_n depends on y_{n+N} is here unimportant, it is simply a choice of numbering convention for samples which is justified in practice by the fact that in general, accurate tuning of θ corresponds to the presence of a preponderance of coefficients $\theta(k)$ with values of the index k close to 0. Note that if the noise ν_n is neglected, and if the channel $(s_k)_{k\geq 0}$ is described by a recursion (Auto-Regressive or all-poles form)

$$\sum_{k=0}^{2N+1} h_k y_{n-k} = a_{n-\Delta} \qquad (1.2.8)$$

where $\Delta \geq 0$ is a delay which takes into account the propagation delays; then if we set

$$\theta(k) = h_{k+N} \qquad (1.2.9)$$

we obtain $c_n = a_{n-\Delta-N}$ and the global message may be reconstituted exactly. It is customary to say that, in this case, the model describes the system (represented here by the channel S) exactly. Of course the system will not in general be described by (1.2.8), so that the chosen model will only give an approximate representation of the true system. Equally clearly, the choice of a model which is too removed from the reality of the system will affect the performance of the equaliser.

Recursive or poles-zeroes form.
The output (c_n) of the equaliser is given as a function of the input (y_n) by

$$(i) \quad c_n = \sum_{k=-N}^{+N} \theta'(k) y_{n-k} + \sum_{l=1}^{P} \theta''(l) \alpha_{n-l} = \Phi_n^T \theta$$
$$(ii) \quad \Phi_n^T := (y_{n+N}, \ldots, y_{n-N}; \alpha_{n-1}, \ldots, \alpha_{n-P})$$
$$(iii) \quad \theta^T := (\theta'(-N), \ldots, \theta'(N); \theta''(1), \ldots, \theta''(P)) \qquad (1.2.10)$$

In these formulae, the signal α_n may be chosen (respectively) to equal

$$(i) \quad \alpha_n = c_n$$
$$(ii) \quad \alpha_n = \hat{a}_n$$
$$(iii) \quad \alpha_n = a_n \qquad (1.2.11)$$

Option (1.2.11-i) is the main reason for the name "poles-zeroes" (see Section 6.1), this is simply a more general way than (1.2.7) of realising the equalisation filter. Option (1.2.11-ii) is directly derived from the previous case since $\hat{a}_n$ is the result of quantisation of the signal c_n, later we shall see this used in the self-adaptive equalisation phase. Lastly, option (1.2.11-iii) will be used in the learning phase. Of course, a larger class of channels S may be modelled exactly using (1.2.10) than using the transversal structure. In practice (where the modelling is never perfect), for θ of the same dimension in both cases, a better approximation may be obtained using a recursive rather than a transversal structure.

Signal Modelling.

In our case the problem reduces to modelling the behaviour of the signals (a_n) and (ν_n) in the data transmission system.

The noise (ν_n) is usually modelled as "white noise", that is to say as a sequence of independent, identically distributed random variables with mean zero and fixed variance (for example a Gaussian distribution $N(0, \sigma^2)$).

The case of the emitted message (a_n) is slightly more complicated, since (a_n) must not be the direct result of 2 bit encoding of the information to be transmitted. In fact the result of this encoding is first transformed by a **scrambler**, a non-linear dynamical system whose effect is to give its output (a_n) the statistical behaviour of a sequence of independent random variables having an identical uniform distribution over the set $\{\pm 1, \pm 3\}$; of course the inverse transformation, known to the receiver, is applied to the decoded signal $(\hat{a}_n)$ to re-obtain the message carrying the information. The importance of this procedure is that it permits a better spectral occupancy of the channel bandwidth.

In conclusion, (ν_n) is modelled as white noise (the assumption of independence is not indispensable, only the zero mean and stationarity are required), whilst (a_n) is a sequence of independent random variables, uniformly distributed over the set $\{\pm 1, \pm 3\}$. In particular, it follows from this that the received signal (y_n) is a stationary random signal with zero mean

1.2.1.2 Equalisation Algorithms: Some Variants

Once the modelling problem is settled, it remains to choose θ, which will then be updated by an adaptive algorithm in accordance with the data received. It is not proposed to discuss the design of such algorithms in this first chapter, since this will be the central theme of Chapter 2. Here we provide only a summary (and provisional) justification, since our main aim is to describe, on the basis of these examples, the characteristics of the algorithms to be studied.

These algorithms will be distinguished one from another in the following three ways:

- the nature of the task (learning the channel, self-adaptive equalisation or blind equalisation)
- the choice of the equaliser model (in this case, of the filter structure)
- the complexity chosen for the algorithm

1.2 Two Basic Examples and Their Variants

We shall begin by presenting the simplest case in detail, then we shall describe a number of variants.

Learning Phase, Transversal (All-Zeroes) Equaliser, Corresponding to Formulae (1.2.7).

This phase corresponds to task (i), in which the message (a_n) is known to the receiver. Since the goal is to tune the equaliser θ so that c_n is as close to a_n as possible, it is natural to seek to minimise the mean square error $E(c_n - a_n)^2$ with respect to θ.

Taking into account formulae (1.2.7), the desired value of θ is given by

$$\theta_* = [E(Y_n Y_n^T)]^{-1} E(Y_n a_n) \tag{1.2.12}$$

Formula (1.2.12) requires a knowledge of the joint statistics of Y_n and a_n, and therefore it may not be applied directly. In seeking to avoid the need to learn these statistics before calculating θ, some time ago engineers saw how to replace (1.2.12) by an iterative method of estimating θ; as each new data item is received, θ_n is adjusted in the direction opposite to the gradient of the instantaneous quadratic error $(c_n - a_n)^2$.

This gives the algorithm

(i) $\quad\quad\quad\quad\quad \theta_n = \theta_{n-1} + \gamma_n Y_n e_n$
(ii) $\quad\quad\quad\quad\quad e_n = a_n - c_n$
(iii) $\quad\quad\quad\quad\quad c_n = Y_n^T \cdot \theta_{n-1}$
(iv) $\quad\quad\quad\quad\quad Y_n^T := (y_{n+N}, \ldots, y_n, \ldots, y_{n-N})$
(v) $\quad\quad\quad\quad\quad \theta^T := (\theta(-N), \ldots, \theta(0), \ldots, \theta(N)) \tag{1.2.13}$

The gain γ_n may be fixed or variable; in signal processing literature, this algorithm is know as "Least Mean Squares" (LMS). Finally, we note that the simultaneous use of c_n and a_n is not a problem for the receiver, since (a_n) is permanently available there.

Learning Phase, Recursive Equaliser (Poles-Zeroes), Corresponding to Formulae (1.2.10, 1.2.11-iii).

We now consider another model of the equaliser. The algorithm, inspired by the previous one is:

(i) $\quad\quad\quad\quad\quad \theta_n = \theta_{n-1} + \gamma_n \Phi_n e_n$
(ii) $\quad\quad\quad\quad\quad e_n = a_n - c_n$
(iii) $\quad\quad\quad\quad\quad c_n = \Phi_n^T \theta_{n-1}$
(iv) $\quad\quad\quad\quad\quad \Phi_n^T := (y_{n+N}, \ldots, y_{n-N}; a_{n-1}, \ldots, a_{n-P})$
(v) $\quad\quad\quad\quad\quad \theta^T := (\theta'(-N), \ldots, \theta'(N); \theta''(1), \ldots, \theta''(P)) \tag{1.2.14}$

Learning Phase, Transversal Equaliser (Formulae (1.2.7)), Least Squares Algorithm.

This algorithm is obtained from (1.2.13) by replacing one step of the gradient algorithm to minimise e_n^2 by a step of a Newtonian method.

The algorithm is given by formulae (1.2.13-ii to v), whilst (1.2.13-i) is replaced by

$$\begin{aligned} \theta_n &= \theta_{n-1} + \gamma_n R_n^{-1} Y_n e_n \\ R_n &= R_{n-1} + \gamma_n (Y_n Y_n^T - R_{n-1}) \end{aligned} \quad (1.2.15)$$

It is well known that, using the "matrix inversion lemma" (Ljung and Soderström 1983), the inversion of R_n may be avoided by replacing the propagation of R_n by that of R_n^{-1} using a Riccati equation, but this does not concern us here.

Tracking Phase: Self-Adaptive Equalisers.

Here, the input message (a_n) is not known to the receiver: it may not be used in adjusting the filter. The tracking phase corresponding to task (ii) relies on the assumption that the adaptive filter θ always stays close to S^{-1}; this allows us to replace the true message a_n in the previous algorithms by the message $\hat{a}_n$ as reconstructed at the receiver. Self-adaptive equalisers are thus derived by simply replacing everywhere in (1.2.13, 1.2.14, 1.2.15) the message a_n by

$$\hat{a}_n = Q(c_n) \quad (1.2.16)$$

where Q is a quantiser matched to the (known) distribution of the message a_n. For example when a_n is uniformly distributed over the set $\{\pm 1, \pm 3\}$, $Q(c)$ is the point of that set closest to c.

To make the text more readable, we restate the version for the recursive equaliser (poles-zeroes), which is known as a *decision-feedback equaliser*

(i) $\quad \theta_n = \theta_{n-1} + \gamma_n \Phi_n e_n$
(ii) $\quad e_n = \hat{a}_n - c_n \quad \hat{a}_n = Q(c_n)$
(iii) $\quad c_n = \Phi_n^T \theta_{n-1}$
(iv) $\quad \Phi_n^T := (y_{n+N}, \ldots, y_{n-N}; \hat{a}_{n-1}, \ldots, \hat{a}_{n-P})$
(v) $\quad \theta^T := (\theta'(-N), \ldots, \theta'(+N); \theta''(1), \ldots, \theta''(P)) \quad (1.2.17)$

The approach to task (iii), so-called blind equalisation, will be discussed later, when we come to study the methodology of algorithm design in Chapter 2.

1.2.2 Phase-Locked Loop

Four State Phase Modulation.

This corresponds to the following diagram:

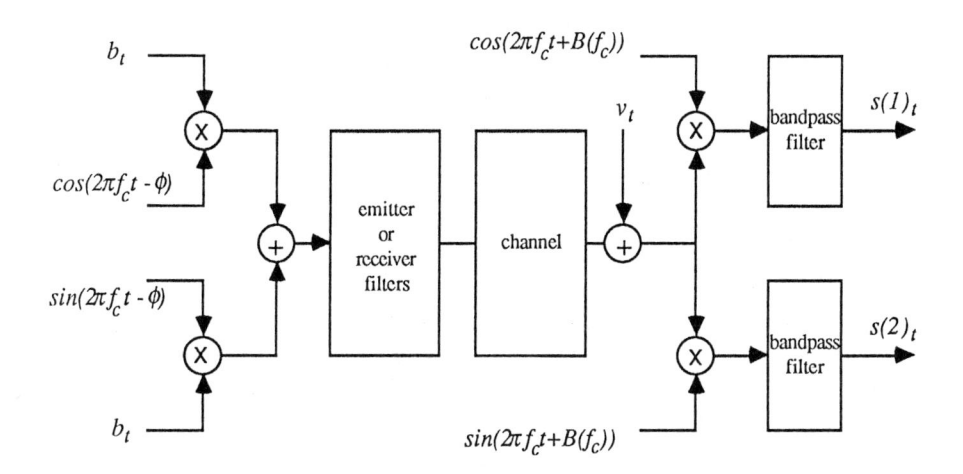

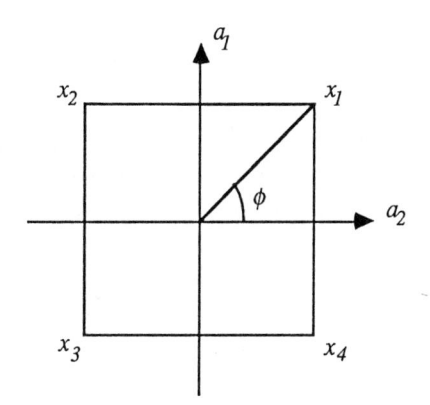

Figure 6. Four state phase modulation.

Instead of being coded by amplitude on the four levels $\{\pm 1, \pm 3\}$, as in the previous case, here, the message is encoded onto a 2-tuple $(b'_n, \tilde{b}'_n)$, each component of which is coded on 2 levels ± 1. Thus equation (1.2.1) is replaced here by

$$\begin{aligned} d(t) &= b(t)\cos(2\pi f_c t - \phi) - \tilde{b}(t)\sin(2\pi f_c t - \phi) \\ b(t) &= \sum_{k=-\infty}^{+\infty} b_k \, g_\Delta(t - k\Delta - t_0) \\ \tilde{b}(t) &= \sum_{k=-\infty}^{+\infty} \tilde{b}_k \, g_\Delta(t - k\Delta - t_0) \\ b'_n &= b_n - b_{n-1} \qquad \tilde{b}'_n = \tilde{b}_n - \tilde{b}_{n-1} \end{aligned} \qquad (1.2.18)$$

where
f_c is the carrier frequency,
g_Δ is an elementary waveform,
ϕ is an arbitrary, fixed phase angle, and
t_0 is the sample time origin.

Note that the message is **differentially** encoded (the message is $b'_n = b_n - b_{n-1}$) in order to eliminate the ambiguity modulo $\pi/2$ which appears when $d(t)$ is decoded in terms of the 2-tuple $(b, \tilde{b})$, where the phase angle ϕ is fixed, but unknown. Now we set

$$\begin{aligned} a_k &= b_k + i\tilde{b}_k \qquad i = \sqrt{-1} \\ a(t) &= b(t) + i\tilde{b}(t) \end{aligned} \qquad (1.2.19)$$

It follows that
$$d(t) = \mathrm{Re}(a(t) e^{i(2\pi f_c t - \phi)}) \qquad (1.2.20)$$

Thus the set {channel, emitter and receiver filters} acts on the emitted signal $d(t)$ as a continuous-time linear filter on which additive noise is superimposed. The effect of the channel may then (after an easy calculation) be represented as follows: the received signal is given by

$$\mathrm{Re}(y(t) \cdot e^{i 2\pi f_c t})$$
$$y(t) = \sum_{k=-\infty}^{+\infty} a_k s(t - k\Delta) + \nu(t) \qquad (1.2.21)$$

where, now $s(t)$ is a **complex** filter and $\nu(t)$ is a complex noise. It is now effectively possible for the receiver to reconstruct the signal $y(t)$: this requires a knowledge of the carrier frequency f_c and is known as **demodulation**. After sampling with period Δ, the so-called "baseband" model of the transmission system is obtained in the form:

$$y_n = \left(\sum_{k=-\infty}^{+\infty} s_k a_{n-k} + \nu_n\right) e^{i\phi_*} \qquad (1.2.22)$$

1.2 Two Basic Examples and Their Variants

where (s_k) is the complex channel **normalised** by the constraint $s_0 \in \mathbb{R}_+$, and ϕ_* is the channel **phase-shift**. Often, the "intersymbol interference" (s_k) is close to unity and the main distortion is that introduced by the channel phase shift ϕ_*.

The phase-locked loop estimates this phase shift in such a way that it may be compensated for prior to quantisation of the received signal. The signal forwarded to the quantiser is then given by

$$y_n e^{-i\phi} \qquad (1.2.23)$$

Aims and Difficulties.

We now encounter again some of the issues already discussed in relation to equalisation. The first goal is clearly the **learning** or **identification** of the channel phase shift ϕ_*. But the main difficulty is the **tracking of variations** in this phase shift. Such variations are very common in data transmission systems (phase drift due to a slight mismatch between the carrier frequency f_c at the emitter and at the receiver—cf. (1.2.21)—; phase jitter, i.e. sinusoidal variations, whose frequency on the French network is typically 50 Hz; arbitrary variations for atmospheric transmissions). Finally a new problem arises from the indeterminacy modulo $\pi/2$ of the phase shift ϕ_*: this is **cycle slipping**, which is described as follows. For a certain length of time the phase estimate ϕ_n remains close to a given estimate of the true phase ϕ_*, then a disconnection occurs, leading to a new phase of equilibrium around another estimate of ϕ_*: this change in the estimate will be translated into a packet of errors when the message is decoded at the receiver.

1.2.2.1 Modelling

Here things are very much easier: phase is simply modelled as a scalar which we shall denote by ϕ.

For reasons similar to those cited in the equaliser example, we use the following constructs to model the behaviour of signals entering the system:

- (b_n) and $(\tilde{b}_n)$ are two sequences of independent variables having an identical uniform distribution over the set $\{(\pm 1, \pm 1)\}$, the signals (b_n) and $(\tilde{b}_n)$ being additionally globally independent;
- (ν_n) is a complex white noise (sequence of complex, independent, identically distributed Gaussian variables with zero mean).

1.2.2.2 Two Phase-Locked Loops

These two loops do not use the true message (a_n) and thus a learning phase is not required. Since a_n is of the form

$$a_n = e^{i\Psi_n} \qquad \Psi \in \{\frac{\pi}{4} + k\frac{\pi}{2},\ k = 0, \ldots, 3\} \qquad (1.2.24)$$

there is an ambiguity modulo $\pi/2$ in the definition of the true phase shift ϕ_* (this justifies the differential coding of the information). This ambiguity is resolved in two different ways by the following two algorithms:

$$\phi_n = \phi_{n-1} + \gamma\varepsilon_n \qquad (1.2.25)$$

where γ is the loop gain, and where the loop error signal ε_n is given by

(i) $$\varepsilon_n = -\mathrm{Im}(y_n^4 e^{-i4\phi_n})$$

for the **Costas loop**, and by

(ii) $$\varepsilon_n = \mathrm{Im}(y_n e^{-i\phi_n} \overline{\hat{a}_n}) \qquad (1.2.26)$$

for the **decision-feedback loop**. In both cases the message is reconstructed according to the formula

$$\hat{a}_n = \mathrm{sgn}[\mathrm{Re}(y_n e^{-i\phi_n})] + i\,\mathrm{sgn}[\mathrm{Im}(y_n e^{-i\phi_n})] \qquad (1.2.27)$$

where $\bar{z}$ denotes the conjugate of z.

In Chapter 2 we shall see how these algorithms may be derived using a well-established method.

1.2.3 Conclusions from the Given Examples and a Comment on the Objectives of this Book

The conclusions are of two types.

On the Algorithm Design Process.

The two preceding examples show that it is possible to distinguish three major stages in the design of an adaptive algorithm.

A first stage consists of the *analysis of the physical system* of interest, and the *specification* in abstract terms of the *task to be accomplished* on this system. This part of the process is illustrated by the descriptions of amplitude modulation and of the aims of equalisation in Subsection 1.2.1, together with the descriptions of phase modulation and of the aims of the phase-locked loop in Subsection 1.2.2.

A second stage is that of *modelling*; this is the translation into precise mathematical terms of the task to be undertaken and of the objects on which it will be effected. This stage is illustrated in Subsection 1.1.3 for equalisation, and in Subsection 1.2.3 for the phase-locked loop.

Finally comes the so-called *design* of the adaptive algorithm: this is the definition of how the objects (systems, signals) are to be manipulated according to well-defined mathematical rules.

We must immediately stress that **this book will address only the algorithm design stage**: the modelling is assumed to be given. This is a different point of view from that classically adopted in works on system identification, of which (Ljung and Soderström 1983) is a typical example. The reason for this is that the methods which we shall describe are applicable to a much larger class of problems than the class of linear systems studied in classical works: the description of generic models and of estimation or identification problems for general systems is becoming a risky exercise which we have chosen to leave aside.

Reasons for Using Adaptive Algorithms.

As we have seen, the use of adaptive algorithms is largely motivated by the generally recognised ability of these algorithms to adapt to variations in the underlying systems. This is a major point which we shall stress in the remainder of Part I of the book. Lastly, the occurrence of unexpected phenomena, such as cycle slipping in phase-locked loops, will force us to pay great attention to the behaviour of adaptive algorithms over very long periods of time.

The following section is devoted to the derivation of a firm mathematical framework for the study of adaptive algorithms; the presentation relies heavily on an analysis of the previous examples.

1.3 General Adaptive Algorithm Form and Main Assumptions

The purpose of this section is to define precisely the mathematical conditions relating to the objects (in particular the gain γ_n, the function H, the state vector X_n, and the residual perturbation ε_n) which were informally introduced in formula (1.1.1). To this end we shall rely upon the properties of the two previous examples. In the first instance, we prove that the general form (1.1.1) is indeed adequate.

1.3.1 Expression of the Previous Examples in the General Adaptive Algorithm Form

We shall discuss two significant variants of the equaliser, other cases are left (with some considerable advice) for the reader to consider.

1.3.1.1 Recursive Decision-Feedback Equaliser

This is the algorithm described in formulae (1.2.17). It is readily expressed in the form (1.1.1) with $\varepsilon_n \equiv 0$ and with

$$\begin{aligned} X_n^T &= (y_{n+N}, \ldots, y_{n-N}; \hat{a}_n, \hat{a}_{n-1}, \ldots, \hat{a}_{n-P}) \\ H(\theta_{n-1}, X_n) &= \Phi_n(\hat{a}_n - \Phi_n^T \cdot \theta_{n-1}) \end{aligned} \quad (1.3.1)$$

where the vector Φ_n is obtained by omitting the coordinate $\hat{a}_n$ of the state vector X_n.

1.3.1.2 Transversal Equaliser, Least Squares Algorithm

This is the algorithm of formulae (1.2.15, 1.2.13-ii to v). Here, the procedure is slightly more complicated. Thanks to the first order approximation

$$\begin{aligned} R_n^{-1} &= R_{n-1}^{-1}(I + \gamma_n(Y_n Y_n^T R_{n-1}^{-1} - I))^{-1} \\ &\approx R_{n-1}^{-1} - \gamma_n R_{n-1}^{-1}(Y_n Y_n^T R_{n-1}^{-1} - I) \end{aligned}$$

(1.2.15-i) may be rewritten in the form

$$\begin{aligned} (i) \quad & \theta_n = \theta_{n-1} + \gamma_n R_{n-1}^{-1} Y_n e_n + \gamma_n^2 R_{n-1}^{-1}(I - Y_n Y_n^T R_{n-1}^{-1}) Y_n e_n \\ (ii) \quad & R_n = R_{n-1} + \gamma_n(Y_n Y_n^T - R_{n-1}) \end{aligned} \qquad (1.3.2)$$

If now we set

$$\Theta_n = \begin{bmatrix} \theta_n \\ \mathrm{col}(R_n) \end{bmatrix}$$

$$X_n = \begin{bmatrix} Y_n \\ a_n \end{bmatrix} \qquad (1.3.3)$$

where $\mathrm{col}(R)$ denotes the vector obtained by superposing the columns of the matrix R, then it is tedious but straightforward to write (1.3.2) in the form (1.1.1) where Θ replaces θ and the contribution following γ_n^2 in (1.3.2-i) provides the residual term ε_n. Naturally some assumptions will be needed to ensure that the term after γ_n^2 remains effectively bounded.

This type of coupled algorithm which introduces a form of **relaxation** (the solution of the second equation is fed directly back into the first) gives one reason for introducing the correction term ε_n. Another reason is for the analysis of algorithms with constraints, where in fact the parameter θ stays within a subvariety of $\mathbb{R}^d$ (see the description of the blind equaliser in Chapter 2).

1.3.2 The State Vector

1.3.2.1 The Recursive Decision-Feedback Equaliser

This example corresponds to the most complicated case met in practice. This is because the state X_n depends now on the previous values of θ. From 1977 onwards, Ljung was the first to take this into account as applied to systems with linear dynamics conditional on θ; in this case the situation is somewhat complicated by the non-linearities introduced by the quantiser Q in the construction of the signal $\hat{a}_n$.

If we assume that the unknown channel may be modelled by a rational, stable transfer function (see Section 6.1) we may write

$$y_n = \sum_{i=1}^{q} s_i y_{n-i} + \sum_{j=1}^{r} t_j a_{n-j} + \nu_n \qquad (1.3.4)$$

1.3 General Adaptive Algorithm Form and Main Assumptions

where (a_n) is the emitted message and (ν_n) is the additive channel noise. Using the results of Section 6.1, we may similarly replace (1.3.4) by the state variable representation

$$\begin{aligned} U_n &= AU_{n-1} + Ba_n \\ y_n &= CU_{n-1} + \nu_n \end{aligned} \tag{1.3.5}$$

where the transition matrix A has its eigenvalues strictly inside the complex unit circle, thus ensuring that (y_n) is asymptotically stationary.

Using formulae (1.2.17) and (1.3.5) it is possible to write

$$\begin{pmatrix} U_{n+N} \\ y_{n+N} \\ y_{n+N-1} \\ \vdots \\ y_{n-N} \\ \hat{a}_{n-1} \\ \vdots \\ \hat{a}_{n-P} \end{pmatrix} = \begin{pmatrix} A & 0 & \cdots & \cdots & \cdots & \cdots & \cdots & 0 \\ C & 0 & \cdots & \cdots & \cdots & \cdots & \cdots & 0 \\ 0 & 1 & 0 & & & & & \vdots \\ \vdots & & \ddots & \ddots & & 0 & & \vdots \\ 0 & \cdots & & 1 & 0 & & & \vdots \\ 0 & & \cdots & & 0 & 1 & & \vdots \\ \vdots & & & 0 & & \ddots & \ddots & \vdots \\ 0 & \cdots & \cdots & \cdots & \cdots & & 0 & 1 & 0 \end{pmatrix} \begin{pmatrix} U_{n+N-1} \\ y_{n+N-1} \\ y_{n+N-2} \\ \vdots \\ y_{n-N-1} \\ \hat{a}_{n-1} \\ \vdots \\ \hat{a}_{n-P-1} \end{pmatrix} + \begin{pmatrix} Ba_{n+N} \\ \nu_{n+N} \\ 0 \\ \vdots \\ \vdots \\ \vdots \\ 0 \end{pmatrix}$$

$$\begin{aligned} c_n &= (y_{n+N}, \ldots, y_{n-N}; \hat{a}_{n-1}, \ldots, \hat{a}_{n-P})\theta_{n-1} \\ \hat{a}_n &= Q(c_n) \end{aligned} \tag{1.3.6}$$

For increased clarity, we set (cf. formulae (1.3.1)):

$$\begin{aligned} X_n^T &:= (y_{n+N}, \ldots, y_{n-N}; \hat{a}_n, \hat{a}_{n-1}, \ldots, \hat{a}_{n-P}) \\ \xi_n^T &:= (U_{n+N}^T; X_n^T) \\ \eta_n^T &:= (U_{n+N}^T; y_{n+N}, \ldots, y_{n-N}; c_n, \hat{a}_{n-1}, \ldots, \hat{a}_{n-P}) \\ w_n^T &:= (a_{n+N}, \nu_{n+N}) \end{aligned} \tag{1.3.7}$$

This notation allows us to rewrite (1.3.6) in the form

(i) $$\eta_n = A(\theta_{n-1}) \cdot \xi_{n-1} + B \cdot w_n$$
(ii) $$\xi_n = Q(\eta_n) \tag{1.3.8}$$

where (1.3.8-i) and (1.3.8-ii) are respectively (1.3.6-i,ii) and (1.3.6-iii) rewritten for appropriate matrices $A(\theta), B$ and function $Q(\eta)$.

Since (w_n) is a stationary sequence of independent random variables, formulae (1.3.8) imply that (ξ_n) is a **Markov chain controlled by** θ_n, and so the conditional (on the past) distribution of ξ_n is of the form

$$P(\xi_n \in G | \xi_{n-1}, \xi_{n-2}, \ldots; \theta_{n-1}, \theta_{n-2}, \ldots) = \int_G \pi_{\theta_{n-1}}(\xi_{n-1}, dx) \tag{1.3.9}$$

where $\pi_\theta(\xi, dx)$ is the transition probability (with parameter θ) of a homogeneous Markov chain. Moreover, the state vector X_n is simply a function of ξ_n (X_n is obtained by extracting components of ξ_n).

By studying this example we have been able to determine the appropriate general form for the state vector (X_n); we next describe this form explicitly.

1.3.2.2 Form of the State Vector and Conditions to be Imposed

We have chosen to represent (X_n) by a **Markov chain controlled by the parameter θ**. Thus X_n is defined from an extended state ξ_n as follows:

$$P(\xi_n \in G | \xi_{n-1}, \xi_{n-2}, \ldots; \theta_{n-1}, \theta_{n-2}, \ldots) = \int_G \pi_{\theta_{n-1}}(\xi_{n-1}, dx)$$
$$X_n = f(\xi_n) \qquad (1.3.10)$$

where, for fixed θ, $\pi_\theta(\xi, dx)$ is the transition probability of a homogeneous Markov chain, and f is a function. We shall call ξ_n the extended state; in fact, we shall only use this notion when we wish to verify in detail that the theorems of Part II apply to a particular example.

The theorems of Part II of the book depend on assumptions on the regularity and on the asymptotic behaviour of the transition probabilities $\pi_\theta(\xi, dx)$. Such conditions on $\pi_\theta(\xi, dx)$ are described informally below.

Asymptotic behaviour.

For θ fixed in the effective domain of the algorithm, the Markov chain (ξ_n) must have a unique invariant probability μ_θ and the convergence of the conditional distributions

$$L_\theta(\xi_{n+k} | \xi_n, \xi_{n-1}, \ldots) = \pi_\theta^k(\xi_n, .) \to \mu_\theta \qquad (1.3.11)$$

is uniform in θ as k tends to infinity; these conditions are similar to very weak mixing conditions given in (Billingsley 1968).

Regularity.

The function

$$\theta \to \mu_\theta \qquad (1.3.12)$$

is suitably regular.

To summarise, **for θ fixed, the state (X_n) must be asymptotically stationary, and its limiting behaviour must be regular in θ.**

In the case of the recursive decision-feedback equaliser, which we use as a test case, the existence of a unique stationary asymptotic behaviour for all θ is easy to check; on the other hand, the condition of regularity with respect to θ is difficult to verify (cf. Section 2.5 of Part II).

The importance of the Markov representation is that properties (1.3.10, 1.3.11, 1.3.12) are preserved by numerous transformations, several useful examples of which are given below:

1.3 General Adaptive Algorithm Form and Main Assumptions

Proposition 1. Transformations preserving the Markov representation.
(i) If (X_n) has a Markov representation controlled by θ which satisfies (1.3.10, 1.3.11, 1.3.12) then so also do

$$Y_n = g(X_n) \qquad (1.3.13)$$

where g is a suitably regular function, and

$$Z_n = (X_n, \ldots, X_{n-p}) \qquad (1.3.14)$$

where p is a fixed integer.
(ii) If (X_n) and (a_n) have Markov representations controlled by θ, defined using the **same** extended state (ξ_n), then so also does

$$U_n = (X_n, a_n) \qquad (1.3.15)$$

Note in passing that the extended state of (Z_n) is $(\xi_n, \ldots, \xi_{n-p})$ which is of course again Markov. We now present some useful particular cases.

1.3.2.3 Stationary State with Parameter-Free Distribution

This case is quite common: all the examples using the transversal form of the equaliser fall into this category, as do phase-locked loops. By way of example, the least squares transversal equaliser falls into this category, as is shown by formula (1.3.3), since X_n has a Markov representation independent of θ. In this case, the necessary conditions on the state reduce to the convergence

$$L(\xi_{n+k}|\xi_n, \xi_{n-1}, \ldots) = \pi^k(\xi_n, .) \to \mu \qquad (1.3.16)$$

which is a very weak mixing property.

1.3.2.4 Algorithms with Conditionally Linear Dynamics

Introduced by Ljung (Ljung 1977a,b), these are of the form

$$\begin{aligned} \theta_n &= \theta_{n-1} + \gamma_n H(\theta_{n-1}, X_n) \\ X_n &= A(\theta_{n-1}) X_{n-1} + B(\theta_{n-1}) W_n \end{aligned} \qquad (1.3.17)$$

where (W_n) is a stationary sequence of independent variables. The existence of an invariant probability for the chain π_θ which defines directly the dynamics of X_n is tied to the stability of the matrix $A(\theta)$ (which must have all its eigenvalues strictly inside the complex unit circle). This type of dynamic behaviour is frequently found in the identification of linear systems, to which topic the book (Ljung and Soderström 1983) is entirely given over. Note that if the $\hat{a}$ are replaced everywhere by the corresponding c in (1.3.6-i), then behaviour of this type arises.

1.3.3 Study of the Vector Field: Introduction of the Ordinary Differential Equation (ODE) Associated with the Algorithm.

1.3.3.1 Case of the Phase-Locked Loop

Let us consider the case of the decision-feedback phase-locked loop, corresponding to formulae (1.2.25, 1.2.26-ii). This is readily written in the form (1.1.1) with $\varepsilon_n \equiv 0$, $X_{n+1} = y_n$ (assuming y_n has a Markov representation according to formula (1.3.5)) and

$$H(y, \phi) = \operatorname{Im}(y e^{-i\phi} \overline{\hat{a}}(y, \phi)) \qquad (1.3.18)$$

where $\hat{a}$ is given by formula (1.2.27). It is thus clear in this case that, for fixed y,

$$\phi \to H(y, \phi)$$

introduces a discontinuity at points ϕ with

$$\operatorname{Re}(y e^{-i\phi}) \cdot \operatorname{Im}(y e^{-i\phi}) = 0$$

The essential conclusion here is that we must allow conditions on H of the form

for fixed X, $\theta \to H(\theta, X)$ may have discontinuities. (1.3.19)

This last difficulty is effectively taken on board by Kushner, but it is completely ignored by the school of control scientists as represented in the book (Ljung and Söderström 1983).

1.3.3.2 Joint Conditions on the Vector Field and on the State; Introduction of the ODE.

Even if the function $\theta \to H(\theta, X)$ is allowed to be discontinuous, the 3-tuple (H, π_θ, f) must satisfy the following condition

$$h(\theta) := \lim_{n \to \infty} E_\theta(H(\theta, X_n)) = \int H(\theta, f(\xi)) \mu_\theta(d\xi) \qquad (1.3.20)$$

exists and is regular (locally Lipschitz), where E_θ denotes the expectation with respect to the distribution of X_n for a fixed value of the parameter θ. Recall that μ_θ is the invariant probability of the chain (ξ_n) with the transition matrix $\pi_\theta(\xi, dx)$ whose existence and uniqueness are assumed, so that the second equality of (1.3.20) is a consequence of our assumption that the state (X_n) is asymptotically stationary.

The existence and regularity of the **mean vector field** $h(\theta)$ allow us to introduce the ODE associated with the algorithm (1.1.1)

(i) $\qquad\qquad \dot{\theta} = h(\theta) \quad \theta(0) = z$

1.3 General Adaptive Algorithm Form and Main Assumptions

where $z = \theta_0$ is the initial value of the parameter θ_n. We shall denote the solution of this equation by

$$(ii) \qquad \theta(t) \quad \text{or} \quad \theta(z,t) \ t \geq 0 \qquad (1.3.21)$$

according as to whether or not it is useful to make the dependence on the initial conditions explicit.

The essential point here is that **condition (1.3.20) does not exclude functions $H(\theta, X)$ which are discontinuous in θ** (cf. self-adaptive equalisers and the decision-feedback phase-locked loop), since the regularity condition applies to $H(\theta, X)$ averaged over X, an operation which in all sensible known cases has a regularising effect.

1.3.4 Form of the Residual Perturbation

There is not much to say. The theorems require only controls on the size of ε_n, nothing more. As already mentioned, the flexibility introduced in this way will allow us to handle algorithms with variable gain matrices, and more generally algorithms with two components in the form of a relaxation (where one of the two iterations is carried out first and the result is fed back into the other) and also algorithms with constraints. The reader should refer to Exercise 1 of this chapter and to the study of the blind equaliser in Chapter 2.

The conditions imposed on the algorithm (1.1.1) will be restated at the end of Chapter 1. The nature of the gain γ_n will be examined in the next section, as we look at problems relating to adaptive algorithms.

1.4 Problems Arising

We have placed these in increasing order of refinement; thus the first problems are at once the most crucial and the easiest.

1.4.1 Convergence

The previous examples show that the primary purpose of an adaptive algorithm is **to learn** an unknown parameter θ_*, which characterises the system of interest. The first problem is therefore to study the convergence of algorithm (1.1.1) to this parameter θ_*.

The corresponding mathematical analysis will postulate a fixed θ_* and will formulate convergence results in more or less precise terms. Two types of results will be given, according as

1. the gain γ_n decreases towards 0
2. the gain γ_n is asymptotically equal to a constant $\gamma > 0$.

We shall speak in the first case of **algorithms of decreasing gain** and in the second of **algorithms of constant gain**. The former are the more commonly studied in the literature, whilst the latter are almost the only ones used in practice, we shall shortly see the reason for this.

One of the problems of concern to the user is the risk of **explosion** of the state X_n, and also of the algorithm, which may occur when X_n is Markov, controlled by θ.

1.4.2 The Transient Phase

Alas! As will be seen, there is very little to say.

1.4.3 Rate of Convergence and Tracking Ability

By rate of convergence we understand the following: given that θ_n converges towards θ_* (assumed fixed), how quickly does $\theta_n - \theta_*$ decrease towards zero? Asymptotic efficiency measures, as frequently used by statisticians, will be applied here. The results permit optimal design of adaptive algorithms for the identification of a fixed parameter.

What most interests the engineer is in fact the ability of the adaptive algorithm to track slow variations in the true system, as represented now by a **time-varying** parameter θ_*. The user is actually interested in the following two questions:

1. Given a certain a priori knowledge of the variations of the true system (for example via a dynamical model), what is the best way to tune the various parameters of the algorithm so as to improve the tracking performance? In particular, it is well known that whilst the use of a small gain (constant) decreases the magnitude of the fluctuations in θ_n (which is nice), it also decreases the ability to track the variations in the true system. The first question is how to quantify this compromise to obtain an optimal solution of the problem?

2. How can one evaluate directly the ability of the algorithm to track a non-stationary parameter, without a priori knowledge of the true system? This is best illustrated by the phase-locked loop, where the two algorithms (1.2.26-i) and (1.2.26-ii) are in competition: which is the better?

1.4.4 Detection of Abrupt Changes

As we shall see, adaptive algorithms behave passively with respect to temporal variations of the true system. The best possible description of this is the following.

The true system θ_* may be thought of as a moving target; the estimator θ_n is attached to θ_* by a piece of elastic and moves over a rough surface. The elastic allows θ_n to follow θ_*, whilst the rough surface causes the fluctuations of θ_n.

Extending the metaphor a little further, too abrupt a manoeuvre of θ_* may overstretch the elastic and even break it. Thus there is a lively procession; above all when an abrupt change in θ_* is detected. Such situations occur quite commonly with adaptive algorithms, although we have not described them in our examples.

1.4.5 Model Validation

Given a model which is said to represent a dynamical system, one is often led to question the true validity of the model as a description of the physical system under consideration; this is the model validation problem.

This issue may arise in the following two ways:

- the model is a model obtained from measurements of the system taken at a previous time: is the model still representative of the system at the moment in hand?
- the structure of the model itself may be inappropriate; the model validation must therefore aim to verify that the given model takes satisfactory account of the system behaviour.

These two points are in fact related, and we shall see that they themselves are associated with the problem of detecting abrupt changes.

1.4.6 Rare Events

This is a totally different problem; caused not by variations in the true system, but solely by the algorithm itself. A typical example is the cycle slip in a phase-locked loop; here, the estimator θ_n escapes from what should have been a domain of attraction for it centred around θ_* (supposed fixed). In Exercise 13 of Chapter 2 we shall see another example of this phenomenon, which we shall call a **rare event** or **large deviation**. The phrase rare event refers to the generally very long period of time before such an escape occurs.

1.5 Summary of the Adaptive Algorithm Form: Assumptions (A)

1.5.0.1 Form of the Algorithm

$$\theta_n = \theta_{n-1} + \gamma_n H(\theta_{n-1}, X_n) + \gamma_n^2 \varepsilon_n(\theta_{n-1}, X_n) \tag{1.5.1}$$

where θ lies in $\mathbb{R}^d$ or a subvariety of $\mathbb{R}^d$, and the state X_n lies in $\mathbb{R}^k$.

1.5.0.2 Nature of the Gain

We distinguish algorithms with **decreasing gain** satisfying

$$(i) \qquad \gamma_n \geq 0 \qquad \sum_n \gamma_n = +\infty \qquad \sum_n \gamma_n^\alpha < \infty \text{ for some } \alpha > 1$$

whose purpose is to estimate fixed parameters, from algorithms of **fixed gain** satisfying

$$(ii) \qquad \gamma_n \geq 0 \quad \gamma := \lim_{n \to \infty} \gamma_n > 0 \qquad (1.5.2)$$

(where the gain limit is "small"), whose purpose is to track slowly varying systems.

1.5.0.3 Nature of the State Vector

General Case: Semi-Markov Representation Controlled by θ.

$$(i) \qquad P(\xi_n \in d\xi | \xi_{n-1}, \xi_{n-2}, \ldots; \theta_{n-1}, \theta_{n-2}, \ldots) = \pi_{\theta_{n-1}}(\xi_{n-1}, d\xi)$$
$$X_n = f(\xi_n)$$

where, for fixed θ, the extended state (ξ_n) is a Markov chain with transition probability $\pi_\theta(\xi, d\xi)$ a function of θ. It is assumed that for all θ in the effective domain of the algorithm, the Markov chain (ξ_n) has a unique stationary asymptotic behaviour.

Special Case in which (X_n) is Stationary and Independent of θ.

This corresponds to the previous expression where $\pi_\theta(\xi, d\xi) \equiv \pi(\xi, d\xi)$ does not depend on θ

$$(ii) \qquad P(\xi_n \in d\xi | \xi_{n-1}, \ldots; \theta_{n-1}, \ldots) = \pi(\xi_{n-1}, d\xi)$$
$$X_n = f(\xi_n)$$

In practice, this amounts to the direct assumption that (ξ_n) (and so also (X_n)) has become stationary.

Special Case in which (X_n) is Conditionally Linear.

$$(iii) \qquad X_n = A(\theta_{n-1})X_{n-1} + B(\theta_{n-1})W_n \qquad (1.5.3)$$

where $A(\theta)$ and $B(\theta)$ are matrices and (W_n) is a sequence of independent identically distributed variables with mean zero. For θ fixed, (X_n) is asymptotically stationary if and only if the matrix $A(\theta)$ has all its eigenvalues strictly inside the complex unit circle.

1.5.0.4 Conditions on the Vector Field and Introduction of the ODE.

The function $H(\theta, X)$, may admit discontinuities, but we assume the existence and regularity of the **mean vector field** defined by

$$h(\theta) := \lim_{n \to \infty} E_\theta(H(\theta, X_n)) \qquad (1.5.4)$$

where P_θ denotes the distribution of the state $(X_n)_{n \geq 0}$ for a fixed value of θ. This allows us to introduce the ODE

$$\dot{\theta} = h(\theta) \quad \theta(0) = z \qquad (1.5.5)$$

whose unique solution we shall denote by $[\theta(t)]_{t \geq 0}$ or $[\theta(z, t)]_{t \geq 0}$.

1.5.0.5 Nature of the Complementary Term

The functions $\varepsilon_n(\theta, X)$ must be uniformly bounded for (θ, X) in some fixed compact set.

1.6 Conclusion

We have derived, and illustrated via important examples, an appropriate form which may be used to describe almost all adaptive algorithms met in practice. Its characteristic features are:

1. the Markov representation (controlled by θ) of the state X_n which models the randomness coming into play in the algorithm;

2. the possibility that the function $H(\theta, X)$ may be discontinuous; this permits the use of algorithms with quantised signals;

3. the presence of a complementary term of order γ_n^2 (where γ_n is the gain): this allows us to use algorithms with constraints and algorithms where the components of θ are updated successively rather than simultaneously.

This degree of generality is obtained only at a price. Firstly, the formal mathematics is quite involved. We have chosen to describe only the essential features of the mathematics here; readers interested in more formal details are referred to Part II of the book. Secondly, and more importantly, we assume the availability of an existing model which will allow us to describe the algorithm and to discuss its properties: this principle was illustrated in detail in the equaliser example. This is a fundamental difference from the classical works of control science on system identification, which consider only linear systems, but which examine the problems of selecting models of such systems.

We have also illustrated, via examples, certain questions concerning adaptive algorithms, which questions we shall examine in later chapters:

Convergence.

This is the study in the classical setting in which the true system is assumed fixed.

Rate of Convergence.

We retain the same setting.

Tracking Non-Stationary Parameters.

The study of the behaviour of the algorithm in the presence of slowly varying systems.

Detection of Abrupt Changes.

Model Validation.

Rare Events and Large Deviations.

These may apparently invalidate predictions of convergence.

1.7 Exercises

1.7.1 Algorithms with Cyclic Update

An adaptive algorithm is given

$$\theta_n = \theta_{n-1} + \gamma H(\theta_{n-1}, X_n) \tag{1.7.1}$$

which is assumed to satisfy Assumptions (A); in particular, it is assumed that, for each fixed θ, the state (X_n) is asymptotically stationary. In order to decrease the volume of calculation, it is decided not to update the components of θ according to formula (1.7.1), but to adjust them cyclically, one by one. This corresponds to the following algorithm, where $d = \dim(\theta)$ and where $\theta(i)$ and $H^{(i)}$ denote the i-th coordinate of the parameter θ and of the vector field H (respectively) :

$$\begin{aligned}
\theta_{m,0} &= \theta_{m-1} \quad \text{for } i = 1, \ldots, d \\
\theta_{m,i}(j) &= \theta_{m,i-1}(j) \quad \text{if } j \neq i \\
\theta_{m,i}(i) &= \theta_{m,i-1}(i) + \gamma H^{(i)}(\theta_{m,i-1}, X_{(m-1)d+i})
\end{aligned}$$

and

$$\theta_m = \theta_{m,d} \tag{1.7.2}$$

1. Show that $(\theta_m)_{m \geq 0}$ is given by an adaptive algorithm of the form

$$\theta_m = \theta_{m-1} + \gamma K(\theta_{m-1}, Y_m) + \gamma^2 \varepsilon_m(\theta_{m-1}, Y_m) \tag{1.7.3}$$

where
$$Y_m := (X_{md}, X_{md-1}, \ldots, X_{(m-1)d+1}) \tag{1.7.4}$$
and calculate the functions K and ε_m.

2. Show that for θ fixed, the state $(Y_m)_{m \geq 0}$ is asymptotically stationary.
(Hint: begin with $d = 2$; proceed with the help of a Taylor series expansion about θ_{m-1} to determine the residual term ε_m).

1.7.2 An Adaptive Control Example

The principle of minimum-variance adaptive control may be summarised in a simple case as follows. Let us consider the system

$$y_n = \sum_{i=1}^{p} a_i^0 y_{n-i} + u_{n-1} + \nu_n \tag{1.7.5}$$

where u_n is the control, y_n the observed output and (ν_n) an unknown perturbation, which is assumed to be a zero-mean Gaussian white noise.

The control objective is to effect a minimum-variance regulation (regulation means that we seek to maintain the system output at zero), that is to say to minimise $E(y_n)^2$ at each instant.

If
$$\theta_*^T = (a_1^0, \ldots, a_p^0)$$
is known, the solution is simple; we take as control rule
$$u_{n-1} = -\sum_{i=1}^{p} a_i^0 y_{n-i} \tag{1.7.6}$$

which always ensures the minimal variance, since $y_n = \nu_n$.

If now the parameter θ_* is unknown, the minimum variance regulator is formed by combining the estimation of θ_* by a least squares algorithm with the control law (1.7.6) applied with the estimate θ_{n-1}, to give the algorithm

$$\begin{aligned}
\theta_n &= \theta_{n-1} + \frac{1}{n}\Gamma_n^{-1}\Phi_n e_n \\
\Gamma_n &= \Gamma_{n-1} + \frac{1}{n}(\Phi_n \Phi_n^T - \Gamma_{n-1}) \\
\Phi_n^T &= (y_{n-1}, \ldots, y_{n-p}) \\
u_{n-1} &= -\Phi_n^T \theta_{n-1} \\
e_n &= y_n - \Phi_n^T \theta_{n-1} - u_{n-1} = y_n
\end{aligned} \tag{1.7.7}$$

1. Express algorithm (1.7.7) in the standard form and determine the function H, the state vector X_n and the additive term ε_n (follow closely the least squares transversal equaliser example).
2. If θ is fixed, does this in all cases guarantee that $X_n(\theta)$ is asymptotically stationary? (Determine the recursion satisfied by y_n for the control law $u_{n-1} = -\Phi_n^T \theta$.)
3. In your opinion, does the method of analysis proposed in this book provide relevant information about adaptive control?

1.7.3 Lattice Algorithms

The aim is to identify the unknown parameter $\theta_*^T = (a_1^0, \ldots, a_p^0)$ of an AR process of the form

$$y_n = \sum_{i=1}^{p} a_i^0 y_{n-i} + \nu_n \qquad (1.7.8)$$

where (ν_n) is a zero-mean, stationary white noise. Here we replace the expression (1.7.8) to describe y_n by another so-called **lattice** form; we shall see the advantages of this later.

Part 1. *Lattice filters.*

We associate with the recursion (1.7.8) the polynomial

$$A_p(z^{-1}) = 1 - \sum_{i=1}^{p} a_i z^{-i}$$

where z^{-1} is the delay operator.

1. Show that, given a sequence of so-called "reflection" coefficients $(k(i))_{1 \le i \le p}$, the recursive formulae

$$\begin{aligned}
A_i^+(z^{-1}) &= A_{i-1}^+(z^{-1}) - k(i) z^{-1} A_{i-1}^-(z^{-1}) \\
A_i^-(z^{-1}) &= z^{-1} A_{i-1}^-(z^{-1}) - k(i) A_{i-1}^+(z^{-1}) \\
A_0^\pm(z^{-1}) &= 1
\end{aligned} \qquad (1.7.9)$$

define a sequence of polynomials of the form

$$\begin{aligned}
A_i^+(z^{-1}) &= 1 - (\sum_{j=1}^{i-1} a_{ij} z^{-j}) - k(i) z^{-i} \\
&= z^{-i} A_i^-(z) \qquad (1.7.10)
\end{aligned}$$

2. Conversely, given a polynomial

$$A(z^{-1}) = 1 - \sum_{j=1}^{p} a_j z^{-j}$$

1.7 Exercises

show that it is possible (except in a special case to be defined) to calculate the sequence of reflection coefficients $(k(i))_{1 \leq i \leq p}$ such that formulae (1.7.9) give

$$A_p^+(z^{-1}) = A(z^{-1})$$

3. Deduce that it is possible to obtain the white noise ν_n in (1.7.8) from y_n using the formulae

$$\begin{aligned} e_n^+(i) &= e_n^+(i-1) - k(i)e_{n-1}^-(i-1) \\ e_n^-(i) &= e_{n-1}^-(i-1) - k(i)e_n^+(i-1) \\ e_n^\pm(0) &= y_n \\ \nu_n &= e_n^+(p) \end{aligned} \quad (1.7.11)$$

where $(k(i))_{1 \leq i \leq p}$ are the reflection coefficients associated with the polynomial $A(z^{-1})$ defined by the parameter θ_*. The rest of the problem concerns the identification of the reflection coefficients.

Part 2. *Lattice representation of an AR process.*

Show that, in formulae (1.7.11), the reflection coefficients are given by the following equivalent formulae where 'cor' denotes the correlation coefficient:

$$\begin{aligned} (i) \quad & k(i) = \mathrm{cor}(e_n^+(i-1), e_{n-1}^-(i-1)) \\ (ii) \quad & = \mathrm{cor}(y_n, e_{n-1}^-(i-1)) \\ (iii) \quad & = \mathrm{cor}(e_n^+(i-1), y_{n-i}) \end{aligned} \quad (1.7.12)$$

(Hint: define $e^\pm(i)$ to be the forward and backward prediction errors of order i:

$$\begin{aligned} e_n^+(i) &:= y_n - E(y_n | y_{n-1}, \ldots, y_{n-i}) \\ e_n^-(i) &:= y_{n-i} - E(y_{n-i} | y_n, \ldots, y_{n-i+1}) \end{aligned}$$

and show that (1.7.11) may be obtained using the reflection coefficients given by (1.7.12). Here $E(.|.)$ denotes the least squares estimate.)

Part 3. *Adaptive lattice algorithms.*

1. *Lattice gradient algorithms*
   (a) Show that

$$\begin{aligned} E(e^+(i)^2) &= E(e^-(i)^2) := \sigma_i^2 \\ \sigma_i^2 &= (1 - k(i)^2)\sigma_{i-1}^2 \end{aligned} \quad (1.7.13)$$

   (b) Using formula (1.7.12-i), the formulae (1.7.11) may be viewed as regression formulae where $k(i)$ is the parameter to be determined; estimate $k(i)$ using a least squares stochastic gradient method. This gives the algorithm

$$\begin{aligned}
e_n^+(i) &= e_n^+(i-1) - k_{n-1}(i)e_{n-1}^-(i-1) \\
e_n^-(i) &= e_{n-1}^-(i-1) - k_{n-1}(i)e_n^+(i-1) \\
k_n(i) &= k_{n-1}(i) + \gamma_n e_{n-1}^-(i-1)e_n^+(i) \\
e_n^\pm(0) &= y_n \qquad e_n^+(p) = e_n
\end{aligned} \qquad (1.7.14)$$

where the signal e_n is the usual prediction error used in least squares algorithms. Set

$$K_n^T = (k_n(1), \ldots, k_n(p))$$

Show that formulae (1.7.14) give rise to a sequence K_n which is determined by an adaptive algorithm, and identify the various terms: the function H, the state vector X_n, and the residual term ε_n. Show that for fixed K, the state vector $X_n(K)$ is asymptotically stationary.

2. *Burg type algorithm.*
   This uses formulae (1.7.12-i) and (1.7.13) directly to construct an estimator for $k(i)$:

$$\begin{aligned}
e_n^+(i) &= e_n^+(i-1) - k_{n-1}(i)e_{n-1}^-(i-1) \\
e_n^-(i) &= e_{n-1}^-(i-1) - k_{n-1}(i)e_n^+(i-1) \\
k_n(i) &= \frac{c_n(i)}{\sigma_n^2(i)} \\
c_n(i) &= c_{n-1}(i) + \gamma_n[e_n^+(i-1)e_{n-1}^-(i-1) - c_{n-1}(i)] \\
\sigma_n^2(i) &= \sigma_{n-1}^2(i) + \frac{\gamma_n}{2}[e_n^+(i-1)^2 + e_{n-1}^-(i-1)^2 - 2\sigma_{n-1}^2(i)]
\end{aligned}$$
$$(1.7.15)$$

Express this algorithm in the adaptive algorithm form; find the parameter on which the algorithm operates and determine the function H, the state X_n and the residual term ε_n of the standard form.

1.8 Comments on the Literature

General Comments.

The idea of determining a class of methods of estimation or identification, in the shape of *stochastic approximations* or *adaptive algorithms*, goes back to the 50s. The first article on this subject is (Robbins and Monro 1951) which concerns Robbins–Monro processes, the Soviet school led by Ya.Z.Tsypkin

1.8   Comments on the Literature   39

is to introduce the study of stochastic approximations as a corps doctrine (Tsypkin 1971). In parallel, in statistics, we see the development of research into *time series*, culminating in (Box and Jenkins 1970). In addition, the work of control scientists is beginning to create an interest in the *identification of dynamical systems* (Eykhoff 1974). Bringing all these ideas together in 1977, Ljung made an important step forward, by deriving the form of algorithm with conditionally linear dynamics which is particularly well suited to the identification of linear dynamical systems (Ljung 1977a,b). The first general form of adaptive algorithm which is suited to signal processing is given in (Benveniste, Goursat and Ruget 1980b): this takes simultaneous account of the dependencies and the discontinuities of the random vector field. The form of the algorithm used in this book is a generalisation of (Métivier and Priouret 1984); in parallel, the Markov form is also found in the excellent article (Kushner and Shwartz 1984), where the conditions given allow the treatment of discontinuities. Finally, the reference works on adaptive algorithms are at the moment (Ljung and Soderström 1983) and (Kushner 1984) the latter being a more theoretical work. One might also consult various articles on control science and systems in the Pergamon encyclopaedia.

References Used in this chapter.

In *Section 1.2* the description of the equaliser is drawn from (Benveniste, Bonnet, Goursat, Macchi and Ruget 1978); see also (Benveniste, Goursat and Ruget 1980a) and (Benveniste and Goursat 1984). For the phase-locked loop, a classical work is (Gardner 1979); see also (Falconer 1976) for a description which is closer to our own, which is borrowed from (Benveniste, Vandamme and Joindot 1979) and (Benveniste 1981).

Sections 1.3, 1.4, 1.5 rely on the form of algorithm described in (Métivier and Priouret 1984), from which the description in terms of the Markov representation is directly derived.

Exercises. Exercise 1 is original; the adaptive control example comes from (Aström, Borisson, Ljung and Wittenmark 1977); there are numerous articles on lattice algorithms, Exercise 3 depends upon the formulation in (Benveniste and Chauré 1981) and (Benveniste 1982a,b,c).

Chapter 2
Convergence: the ODE Method

2.1 Introduction

This chapter has a double purpose. Starting from a few informally stated theorems, backed up by appropriate heuristics, we shall present, firstly a guide for an initial coarse analysis (convergence) of the adaptive algorithms, and secondly, a guide to the essentials of algorithm design. The transient phase of the algorithm will be studied briefly.

The algorithms to be investigated are of the form introduced in Assumptions (A) of Chapter 1:

$$\theta_n = \theta_{n-1} + \gamma_n H(\theta_{n-1}, X_n) + \gamma_n^2 \varepsilon_n(\theta_{n-1}, X_n) \qquad (2.1.1)$$

Here the essential point is that the state vector X_n has a dynamic Markov representation controlled by θ, and so

$$P(\xi_n \in G \mid \xi_{n-1}, \xi_{n-2}, \ldots; \theta_{n-1}, \theta_{n-2}, \ldots) = \int_G \pi_{\theta_{n-1}}(\xi_{n-1}, dx)$$
$$X_n = f(\xi_n) \qquad (2.1.2)$$

where π_θ is the transition probability of a θ-dependent Markov chain ξ_n, f is a function, and E_θ is calculated with respect to the distribution under which (ξ_n) is a Markov chain with transition probability π_θ.

Recall also that for any fixed θ in the domain of operation of the algorithm (2.1.1), the Markov sequence (ξ_n) is asymptotically stationary with limiting distribution $\mu_\theta(d\xi)$.

The ODE associated with the algorithm is now introduced as follows: set

$$h(\theta) = \int H(\theta, f(\xi)) \mu_\theta(d\xi) = \lim_{n \to \infty} E_\theta(H(\theta, X_n)) \qquad (2.1.3)$$

then the ODE is given by:

(i) $$\dot{\theta} = h(\theta) \qquad \theta(0) = z$$
$$(2.1.4)$$

where $z = \theta_0$ is the initial value of the parameter θ. We shall denote the solution of the ODE by

(ii) $\qquad\qquad \theta(t) \quad$ or $\quad \theta(z,t), \quad t \geq 0$

2.2 Mathematical Tools: Informal Introduction

In order that the results may be well understood, we start with a heuristic examination of algorithm (2.1.1).

2.2.1 Convergence Heuristics

For clarity, we first discuss the case in which the gain γ_n is a small constant, denoted by γ. We have the following approximations, whose validity will be examined next, for fixed n and $N > 0$

$$\begin{aligned}
\theta_{n+N} &= \theta_n + \gamma \sum_{i=0}^{N-1} (H(\theta_{n+i}, X_{n+i+1}) + \gamma \varepsilon_{n+i}(\theta_{n+i}, X_{n+i+1})) \\
&\approx \theta_n + \gamma \sum_{i=0}^{N-1} H(\theta_{n+i}, X_{n+i+1}) \\
&\approx \theta_n + \gamma \sum_{i=0}^{N-1} H(\theta_n, X_{n+i+1}) \\
&= \theta_n + (N\gamma) \cdot \frac{1}{N} \sum_{i=0}^{N-1} H(\theta_n, X_{n+i+1}) \\
&\approx \theta_n + N\gamma h(\theta_n)
\end{aligned} \qquad (2.2.1)$$

Approximation 1.

Here we assume that γ is sufficiently small that the complementary term may be neglected; this approximation does not cause too great a problem.

Approximation 2.

Here we assume

1. that the function $\theta \to H(\theta, X)$ is regular;
2. that θ_{n+i} has been little modified in the N previous steps; this requires that N is not too large, and/or that γ is sufficiently small.

In a moment, we shall describe the problems presented by this algorithm when $H(\theta, X)$ is discontinuous (as is the case in several examples in Chapter 1). Discontinuities must be "quite rare" for Approximation 2 to be generally valid.

Approximation 3.

It is here, and only here, that the probabilities enter in. Since the parameter θ now has the fixed value θ_n, the state X_{n+i} is approximately stationary, if we wait long enough. Approximation 3 is then nothing other than the law of large numbers; assuming N sufficiently large.

In conclusion, it will be noted that Approximations 2 (N not too large) and 3 (N large) are apparently conflicting; this conflict is resolved by choosing γ to be "very small". Thus, summarising (2.2.1), at the two extremities, we have

$$\theta_{n+N} \approx \theta_n + N\gamma h(\theta_n) \qquad (2.2.2)$$

But then $\theta_0, \theta_N, \theta_{2N}, \ldots$ is the discretisation of the ODE (2.1.4) with step size $N\gamma$; similarly, $(\theta_n)_{n\geq 0}$ is now expressed in a form which is very close to the standard discretisation of the ODE (2.1.4), with step size γ:

$$\theta_{n+1} = \theta_n + \gamma h(\theta_n) + \text{perturbation} \qquad (2.2.3)$$

It follows that we have more or less the following

$$\theta_n = \theta(t_n) \qquad \text{with} \qquad t_n = n\gamma \qquad (2.2.4)$$

where $\theta(t)$ is defined in (2.1.4-ii), which formula links the continuous time of the ODE to the discrete time of the algorithm.

If a small but variable step size γ_n is used instead of the fixed step size γ, the same reasoning may be followed (using a less classical form of the law of large numbers to obtain Approximation 3), and then (2.2.4) will be replaced by

$$\theta_n = \theta(t_n) \qquad \text{with} \qquad t_n = \sum_{i=1}^{n} \gamma_i \qquad (2.2.5)$$

The inclusion of this background material will allow us to present the results which follow in such a way that their limitations may be better understood.

2.2.2 Result for a Finite Horizon

This type of result concerns the behaviour of the algorithm at instants n with

$$t_n \leq T \qquad (2.2.6)$$

where $T < \infty$ is fixed and t_n is defined in (2.2.5).

Few assumptions are needed.

Assumption (A.1). *There exists a fixed cylinder of diameter $\eta > 0$ containing the trajectory $(\theta(t))_{0 \leq t \leq T}$ of the ODE, in which Assumptions (A) of Chapter 1 are satisfied.*

2.2 Mathematical Tools: Informal Introduction

The first result now follows:

Theorem 1. Result for a finite horizon. Assumption (A.1). *Let*

$$\gamma := \max_{n:t_n \leq T} \gamma_n \qquad (2.2.7)$$

Then, for fixed $\varepsilon > 0$ and γ sufficiently small, we have

$$P\{\max_{n:t_n \leq T} \|\theta_n - \theta(t_n)\| > \varepsilon\} \leq C(\gamma, T) \qquad (2.2.8)$$

where, for fixed $T < \infty$, $C(\gamma, T)$ tends to zero as γ tends to 0.

In other words, with probability increasingly close to 1 as γ decreases, θ_n follows the trajectory $\theta(t_n)$ of the ODE, with uncertainty never greater than some arbitrary, a priori fixed $\varepsilon > 0$; thus we have **uniform convergence in probability**.

NOTE. The corresponding results are found in Part II of the book in Theorem 10 of Chapter 1, with more or less restrictive assumptions. The constants $C(\gamma, T)$ are given explicitly in that section. Similar results for weaker assumptions are found in Chapters 2 and 3 of Part II.

Because of its very generality, Theorem 1 is the fundamental theorem in the study of convergence. In fact, it allows an engineer to obtain directly the following very weak, yet general result. We shall use the following notation interchangeably

$$n := n_t \Leftrightarrow t = t_n \qquad (2.2.9)$$

Corollary 2. A weak convergence result. Assumption (A.1). *Suppose that the ODE has an attractor θ_*, i.e.*

$$\theta_* = \lim_{t \to \infty} \theta(t) \qquad (2.2.10)$$

For fixed $\varepsilon > 0$, choose T so that $\theta(T) - \theta_* \leq \varepsilon$. Theorem 1 then implies that

$$P\{\|\theta_{n_T}\| > 2\varepsilon\} \leq C(\gamma, T) \qquad (2.2.11)$$

Corollary 2 thus gives the convergence of a sort (very weak) of θ_n towards θ_*. This is, as we shall see, the best result that an engineer user may hope for.

2.2.3 Results for an Infinite Horizon

The phrase "infinite horizon" implies that we are interested in the behaviour of the algorithm as n tends to infinity. All the results which we shall state assume that the ODE is asymptotically stable; more precisely:

Assumption (A.2). *The ODE (2.1.4) has an attractor θ_*, whose domain of attraction we shall denote by D_*.*

In general θ_* reduces to a single point (but this restriction is not necessary, there may also be a limiting cycle or other objects), which is then simply a stable equilibrium point. The domain of attraction D_* is then the set of initial conditions z such that $\theta(z,t)$ converges to θ_* (the interior of D_* is non-empty).

We shall present two types of results: firstly for algorithms with "constant gain", and secondly for algorithms with "decreasing gain".

2.2.3.1 Algorithms with Constant Gain

We simply assume that

$$\gamma := \max_n \gamma_n < \infty, \qquad \sum_n \gamma_n = +\infty \qquad (2.2.12)$$

Assumption (A.2) must then be additionally strengthened as follows (see Chapter 4 of Part II for a formal statement of this assumption).

Assumption (A.2a). *The ODE is globally stable, with a unique stable equilibrium point θ_*; moreover $h(\theta)$ tends to infinity at at most a polynomial rate, and the growth of the random term $H(\theta, X_n) - h(\theta)$ towards infinity is controlled. Also Assumptions (A) of Chapter 1 are satisfied throughout the space.*

We then have the following theorem (cf. Theorem 15 of Chapter 4 of Part II for a formal result which implies Theorem 3; see also (Derevitskii and Fradkov 1974)):

Theorem 3. *Infinite horizon. Algorithms with constant gain. Assumptions (A.2a). For $\gamma > 0$ sufficiently small, for all $\varepsilon > 0$, there exists a constant $C(\gamma)$ such that*

$$\limsup_{n \to \infty} P\{\|\theta_n - \theta_*\| > \varepsilon\} \leq C(\gamma) \qquad (2.2.13)$$

where $C(\gamma)$ tends to zero as γ tends to zero.

This amounts to a strengthening of Corollary 2; which is significant, both mathematically, and from the user's point of view. The second condition of (2.1.12) ensures that $t_n \to \infty$, whence that $\theta(t_n) \to \theta_*$, which is quite minor.

We shall now examine why this theorem is the best that could be hoped for in the infinite horizon, constant gain case.

In the first place, it is well known that it is only possible to control the discretisation error at infinity in a differential equation of the type given in (2.2.3) when the differential equation is stable. This is a fortiori the case in this instance, where algorithm (2.1.1) may be considered as a perturbed discretisation of the ODE; moreover, the constants $C(\gamma, T)$ of Theorem 1 increase exponentially with T for γ fixed. Assumption (A.2) is therefore necessary when the horizon is infinite.

2.2 Mathematical Tools: Informal Introduction

A further difference between Theorems 1 and 3 is the position of sup; inside $P\{\ldots\}$ for Theorem 1, and outside it for Theorem 3. Unlike Theorem 1, Theorem 3 does not imply that the ODE–algorithm difference is uniform along the whole trajectory. Thus if $H(\theta, X_n)$ is observed for an infinitely long time, for θ close to θ_*, there will always exist arbitrary large values of n with $H(\theta, X_n)$ remote from the mean value $h(\theta)$. But then, at such instants, θ_n may be thrown far away from θ_*. Nonetheless, this does not prevent the probability that such an event occurs at time n, where n is fixed in advance, from being small, this explains the exact meaning of Theorem 3.

2.2.3.2 Algorithms with Decreasing Gain

Here the gains satisfy

$$\sum_n \gamma_n^\alpha < \infty \text{ for some } \alpha > 1, \quad \sum_n \gamma_n = +\infty \quad (2.2.14)$$

The second condition always guarantees that, $t_n \to \infty$, whilst the first condition will allow us to remove the residual fluctuation of the parameter θ asymptotically. Here the assumptions are as follows:

Assumption (A.2b). *Assumptions (A) of Chapter 1 are satisfied in the domain of attraction D_* introduced in (A.2)*

Theorem 4 now follows (cf. Theorem 14 of Chapter 1 of Part II, and Chapter 3 of Part II for weakened assumptions):

Theorem 4. *Infinite horizon. Algorithm with decreasing gain. Assumption (A.2b). Suppose that algorithm (2.1.1) is initialised with $\theta_0 = z \in Q$, where Q is a compact subset of D_* and $\xi_0 = \xi$ (cf. (2.1.2)). Then*
(i) we have

$$P\{\lim_{n\to\infty} \theta_n = \theta_*\} \geq 1 - C(\alpha, Q, |\xi|) \sum_n \gamma_n^\alpha \quad (2.2.15)$$

(ii) for any $\varepsilon > 0$ we have

$$P\{\max_n \|\theta_n - \theta(z, t_n)\| > \varepsilon\} < C(\alpha, Q, |\xi|) \sum_n \gamma_n^\alpha \quad (2.2.16)$$

Here $C(\alpha, Q, |\xi|)$ denotes a constant which depends on α, on the compact subset Q, and on the norm of the initial condition ξ: this constant increases with Q and ξ. Theorem 4 is very powerful, since it gives control over the error (either with respect to θ_* for (i), or with respect to the ODE for (ii)), which is uniform along the trajectory of the algorithm.

Various games can be played with Theorem 4, here are a few examples:

Corollary 5. Same assumptions as Theorem 4. *Suppose that at a given time $N > 0$, we again have $\theta_N \in Q$. Then*
(i) we have
$$P\{\lim_{n\to\infty} \theta_n = \theta_*\} \geq 1 - C(\alpha, Q, |\xi_N|) \sum_{n \geq N} \gamma_n^\alpha \qquad (2.2.17)$$
(ii) for fixed $\varepsilon > 0$
$$P\{\max_{n \geq N} \|\theta_n - \theta(t_n - t_N)\| > \varepsilon\} \leq C(\alpha, Q, |\xi_N|) \sum_{n \geq N} \gamma_n^\alpha \qquad (2.2.18)$$

In other words, if at time N, $\theta_N \in Q$ (in particular if the algorithm has not escaped from the domain of attraction D_*), then, the probability of convergence to θ_* is now controlled by the tail of the series $\sum \gamma_n^\alpha$ which (we recall) converges by virtue of (2.2.14). Corollary 5 is obtained from Theorem 4 by a simple application of the strong Markov property (neglecting the extra term $\varepsilon(\theta_{n-1}, X_n)$ in (2.1.1)) to the Markov pair (θ_n, ξ_n). This is equivalent to assuming that the algorithm is restarted at time N, with initial point θ_N.

The following corollary gives the classical result of Ljung (Ljung 1977a,b): this comes by repeated application of Corollary 5.

Corollary 6. Assumptions of Theorem 4. *For the set of trajectories (θ_n, ξ_n) which intersect a compact subset $Q \times \{|\xi| \leq M\}$ infinitely often, for some constant M fixed in advance,*
(i) we have
$$\theta_n \to \theta_* \quad a.s \qquad (2.2.19)$$
(ii) for any fixed $\varepsilon > 0$ we have
$$P\{\limsup_{n \to \infty} \|\theta_n - \theta(z, t_n)\| > \varepsilon\} = 0 \qquad (2.2.20)$$
(iii) if (X_n) is independent of θ, then (2.2.19, 2.2.20) are satisfied by the set of trajectories (θ_n) which intersect Q infinitely often.

This is a rather silly theorem which says that "if nothing goes wrong, all will be well". This theorem does not address the verification of the **boundedness condition**
$$P\{\theta_n \in Q \text{ infinitely often}\} = 1 \qquad (2.2.21)$$
Unfortunately, there is no general result which gives conditions under which (2.2.21) is satisfied; at present there are only specific results for each algorithm. (Ljung and Soderström 1983) suggested introducing a projection mechanism into the algorithm (2.1.1) itself, in order to guarantee (2.2.21). This is something not provided by most of the algorithms in (fully satisfactory!) practical use. Such a solution, justified only by considerations of mathematical inadequacy, is thus not satisfactory.

2.2 Mathematical Tools: Informal Introduction

The reader may wish to refer to (Eweda and Macchi 1984a,b) for an example of an algorithm for which (2.2.21) may be verified.

NOTE. Do not confuse the boundedness condition (2.2.21) with the **risk of explosion** mentioned in Subsection 1.4.1, which implies that there is a stable-state domain D_s, outside which the Markov chain π_θ representing the state (ξ_n) is no longer recurrent, but transient (and in general explosive). The risk of explosion must always be fought in practice (except in certain very special cases (Solo 1979), and in certain adaptive control algorithms (Goodwin, Ramadge and Caines 1980), where it is known that the algorithm will leave the explosive zone of its own accord). On the other hand, the absence of the risk of explosion does not imply the boundedness condition (2.2.21), a condition which the user will not be worried about in the absence of risk of explosion.

Finally, we point out the following negative result which is also given in (Ljung and Soderström 1983).

Theorem 7. Negative result for algorithms with decreasing gain. Assuming (2.2.14). *If there exists a point θ_* such that*

$$P\{\theta_n \to \theta_*\} > 0$$

then θ_* is necessarily a stable equilibrium point of the ODE.

In other words, an algorithm with decreasing gain may only converge (if it converges) to a stable equilibrium point of the ODE.

2.2.4 Summary of the Mathematical Tools

A summary of all the theorems stated might be given as follows: **after the transient phase, the behaviour of algorithm (2.1.1) is represented to a first approximation by that of the ODE (2.1.4).**

Here, "to a first approximation" has a double meaning. In the first instance, it means that "rare events", having a small probability, are ignored (these will be briefly studied in Exercise 13 of this chapter). Also, secondly, it means that the error is over-estimated in an essentially coarse way, and that the issues associated with the rate of convergence (which, as we shall see, play a major role in the tracking of slow variations via adaptive algorithms) are not examined.

The second conclusion is that, for our part, we prefer to emphasise results for algorithms with "constant gain", since such algorithms are the only ones which have the ability to track non-stationary parameters, and are thus the only ones used in practice.

2.3 Guide to the Analysis of Adaptive Algorithms

We first describe the guide itself, then we give several examples of its use. Since it is primarily concerned with convergence theorems, this guide will not provide detailed information (rate of convergence, rare events) about the algorithm.

2.3.1 The Guide

The guide is described in 3 stages.

Stage 1. *Express the algorithm in the general form and verify that the theory is applicable.*

The first step is the exercise undertaken by us in Chapter 1, and to be undertaken by you when you solve the problems at the end of this chapter; it amounts to the identification of the objects in expression (2.1.1)—the function H, the residue ε_n (if necessary), the gain γ_n, the state (X_n) and the augmented state (ξ_n).

We try to establish whether the gain is decreasing or constant; in the latter case, we must be certain that the gain may be considered to be small. An examination of the heuristics described in 2.2.1 shows that "small" means that

$$\|H(\theta_n, X_{n+1}) - H(\theta_{n-1}, X_{n+1})\| \ll \|H(\theta_n, X_{n+1})\| \qquad (2.3.1)$$

where θ_n is obtained from θ_{n-1} by one step of the algorithm; the sign $\ll$ means that, based on lengthy practical experience, the ratio of the left-hand side to the right-hand side is of the order of at most 10^{-3} (except in certain specific applications), barring exceptional events (such as crossing a discontinuity).

In order to verify that the theory is applicable, the main issues are:

1. To identify (if they exist) values of the parameter θ for which the state (ξ_n) is explosive. If such values exist, to examine if they are attainable by θ_n; if so, then the rest of the guide will describe the behaviour of the algorithm, provided it has not exploded (this condition may be observed a posteriori); if not, then all is well.

2. To check that the state (ξ_n) has a unique stationary asymptotic behaviour. Broadly speaking, this amounts to checking that for large N, the variables ξ_n and ξ_{n+N} are approximately independent (so called "mixing" properties; other slightly weaker properties will be used in Part II of the book).

Stage 2. *Calculation of the mean vector field $h(\theta)$.*

There is nothing to say, other than that this calculation determines the ODE.

2.3 Guide to the Analysis of Adaptive Algorithms

Stage 3. *Study of the ODE.*

This is a classical analysis of a differential equation. Since we are essentially interested in the asymptotic behaviour of the algorithm, the analysis is qualitative and centres mainly around a **study of attractors and their domains of attraction**. An examination of the trajectories of the ODE will then allow us to predict the behaviour of the algorithm in accordance with Theorems 1 to 7 of this chapter.

One thing to be investigated whenever possible when studying the ODE is the potential J (if it exists) from which the vector field $h(\theta)$ is derived.

$$h(\theta) = -\nabla J(\theta) = -\frac{d}{d\theta}J(\theta) \qquad (2.3.2)$$

The qualitative investigation then becomes a **consideration of the local minima of J, and of their domains of attraction**. Of course, the existence of such a potential is not necessarily generally guaranteed (except, clearly, when θ is a scalar!).

2.3.2 Examples

These were introduced in Chapter 1, and the reader may wish to refer to the descriptions in that chapter.

2.3.2.1 The Equaliser in the Learning Phase

Here, we shall examine the two versions, corresponding to formulae (1.2.13) and (1.2.15) (respectively).

The Transversal Equaliser (Formulae (1.2.13)).

Stage 1. *Expression of the algorithm in the general form.*

The algorithm is in the form (2.1.1), with no perturbation term, and with

$$\begin{aligned} X_n^T &= (Y_n^T, a_n) \\ H(\theta; X) &= H(\theta; Y, a) = -Y(Y^T\theta - a) \end{aligned} \qquad (2.3.3)$$

H is known to be regular. On the other hand, if (Y_n) has a Markov representation, then so does (X_n), as we saw in Proposition 1 of Chapter 1. If we assume that the channel is unknown, but has a rational, stable transfer function, then (y_n) will be an ARMA signal, and so, the same will be true of (Y_n). Note that the conditional distribution for the state (X_n) does not depend on θ; moreover, since an ARMA signal is asymptotically stationary, it follows that Assumptions (A) are satisfied for all θ. Thus the behaviour of the algorithm is summed up by its ODE, the precise form of the approximation obtained in this way depends upon the nature (decreasing or constant) of the gains chosen.

Stage 2. *Calculation of the ODE.*

This comes from calculating the mean field

$$h(\theta) = E(H(\theta, X_n)) = -E(Y_n(Y_n^T\theta - a_n)) \qquad (2.3.4)$$

where E refers to the asymptotic law under which (X_n) is stationary.

Stage 3. *Analysis of the ODE.*

An easy calculation now gives

$$\begin{aligned} h(\theta) &= \nabla J(\theta) = \frac{d}{d\theta}J(\theta) \\ J(\theta) &= E(e_n^2(\theta)) \\ e_n(\theta) &= a_n - c_n(\theta) = a_n - Y_n^T\theta \end{aligned} \qquad (2.3.5)$$

In other word, the trajectories of the ODE are the lines of steepest descent of the functional J, which measures the mean square error between the true message and the output of the equaliser. If

$$\theta_* := \arg\min_\theta J(\theta) \qquad (2.3.6)$$

then the ODE is globally stable with a unique attractor θ_*, since J is a quadratic (whence convex) functional.

Conclusion.

θ_n converges to θ_* as per one of the senses described in the 7 theorems. Note that in this case, the results are for an infinite horizon, for algorithms with constant gain. The same approach may be applied, without a great deal of modification, to the recursive equaliser corresponding to formulae (1.2.14).

The Transversal Equaliser with the Least Squares Algorithm, Formulae (1.2.15).

Stage 1. *Expression of the algorithm in the general form.*

This was carried out in Paragraph 1.3.1.2, the results being given in formulae (1.3.2, 1.3.3).

There is nothing new to say about the state X_n. Since (y_n) is asymptotically stationary, the perturbation term in (1.3.2-i) will not pose a problem if

$$R_* := E(Y_n Y_n^T) > 0 \qquad (2.3.7)$$

We assume this to be the case. The theory once again applies.

Stage 2. *Derivation of the ODE.*

If

$$\Theta^T = (\theta^T, \mathrm{col}(R)^T) \qquad (2.3.8)$$

2.3 Guide to the Analysis of Adaptive Algorithms

then from (1.3.2) we have
$$h^T(\Theta) = (h_\theta^T(\Theta), h_R^T(R)) \tag{2.3.9}$$
where

(i) $\quad h_\theta(\Theta) = -R^{-1}E(Y_n(Y_n^T\theta - a_n))$
(ii) $\quad h_R(R) = -\text{col}(R - E(Y_nY_n^T)) \tag{2.3.10}$

In formulae (2.3.10), E again refers to the asymptotic distribution under which X_n is stationary. If we rewrite the second component of Θ in matrix form, we now obtain the ODE

(i) $\quad \dot{\theta} = -R^{-1}(R_*\theta - \Lambda_*)$
(ii) $\quad \dot{R} = -(R - R_*)$
(iii) $\quad R_* = E(Y_nY_n^T), \quad \Lambda_* = E(Y_na_n) \tag{2.3.11}$

Stage 3. *Analysis of the ODE, an example of a Newtonian stochastic method.*
We now introduce the functional
$$\boldsymbol{J}(\Theta) = J(\theta) + K(R) = E(e_n^2(\theta)) + \|\text{col}(R - R_*)\|^2 \tag{2.3.12}$$
where $e_n(\theta)$ is defined in (2.3.5). Thus we have only added the correction term $K(R)$ to the functional $J(\theta)$ which we used before. Note that $\boldsymbol{J}$ is still quadratic, and that
$$\Theta_* := \arg\min_\Theta \boldsymbol{J}(\Theta) = (\theta_*, R_*) \tag{2.3.13}$$
where θ_* was given in (2.3.6). Now we shall minimise $\boldsymbol{J}$ by a so-called "quasi-Newtonian" method: to obtain the field of lines of descent of $\boldsymbol{J}$, the usual gradient is multiplied by a "value approximating" the inverse of the Hessian (second derivative) at the point in question. In our case, for the gradient we have

(i) $\quad \dfrac{\partial}{\partial \theta}\boldsymbol{J}(\Theta) = \dfrac{d}{d\theta}J(\theta) = E(Y_n(Y_n^T\theta - a_n))$
(ii) $\quad \dfrac{\partial}{\partial(\text{col}(R))}\boldsymbol{J}(\Theta) = \text{col}(R - R_*) \tag{2.3.14}$

and for the Hessian

(i) $\quad \dfrac{\partial^2}{\partial \theta^2}\boldsymbol{J}(\Theta) = R_*$
(ii) $\quad \dfrac{\partial^2}{\partial\theta\partial(\text{col}(R))}\boldsymbol{J}(\Theta) = 0$
(iii) $\quad \dfrac{\partial^2}{\partial(\text{col}(R))^2}\boldsymbol{J}(\Theta) = I \tag{2.3.15}$

The vector field corresponding to an application of Newton's method (exact Hessian) with R in matrix form, is then

(i) $$\dot{\theta} = -R_*^{-1}(R_*\theta - \Lambda_*)$$
(ii) $$\dot{R} = -(R - R_*) \qquad (2.3.16)$$

Note that (2.3.16) is the same as (2.3.11) with R_* replacing R in (2.3.16-i). Thus two important points may be made:

- in the first instance, since R is positive symmetric, (2.3.16-i) always defines a line of descent of the potential J; on the other hand, for t sufficiently large, (2.3.16-ii) gives

$$R(t) \approx R_* \qquad (2.3.17)$$

and so (2.3.16) approximates to Newton's method. This explains our use of the term quasi-Newtonian;

- the importance of such a modification is evident from (2.3.16-i), which may be rewritten in the form

$$\dot{\theta} = -(\theta - R_*^{-1}\Lambda_*)$$

where the new expression shows clearly that, unlike (2.3.5), (2.3.16) goes straight to the target. Figure 7 (below) illustrates, for the component θ only, the difference in the behaviour of the two ODEs.

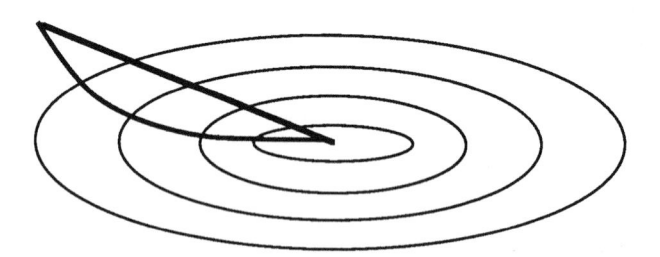

Figure 7. Gradient method and Newton's method.

2.3 Guide to the Analysis of Adaptive Algorithms

Conclusion.

Once again, θ_n converges to θ_*. The functional minimised was the same one as in the previous example: however, it was minimised by a Newtonian stochastic method which we shall mention again later.

2.3.2.2 The Self-Adaptive Equaliser

This corresponds to formulae (1.2.13, 1.2.16). We shall only examine here the transversal version corresponding to the learning phase formulae (1.2.13).

Self-Adaptive Equaliser, Formulae (1.2.13) and (1.2.16).

Stage 1.

The algorithm is of the form (2.1.1), without complementary term, with

$$\begin{aligned} X_n &= Y_n \\ H(\theta, Y) &= Y[Y^T\theta - Q(Y^T\theta)] \end{aligned} \quad (2.3.18)$$

where Q is the quantiser mentioned in (1.2.16). There is nothing new to say, except that the function $H(.,Y)$ is discontinuous at all points θ such that $Q(Y^T\theta)$ is discontinuous. It is harder to prove that these discontinuities satisfy the required theoretical conditions. This case is analysed in detail in Part II, Chapter 2. The conclusion is that, if in (1.2.5) we have $\nu_n \neq 0$, then the discontinuity of H is acceptable; in particular, the mean field h will be smooth. Thus we may apply the theory.

Stage 2. *Calculation of the ODE.*

We have

$$h(\theta) = E[Y_n(Y_n^T\theta - Q(Y_n^T\theta))] \quad (2.3.19)$$

where the symbol E still refers to the stationary distribution.

Stage 3. *Analysis of the differential equation.*

Figure 8 (below) shows the graph of Q for the case in which the message (a_n) is uniformly distributed over $\{\pm 1, \pm 3\}$. The extrapolation to the general case $\{\pm 1, \pm 3, \ldots, \pm 2K+1\}$ is immediate.

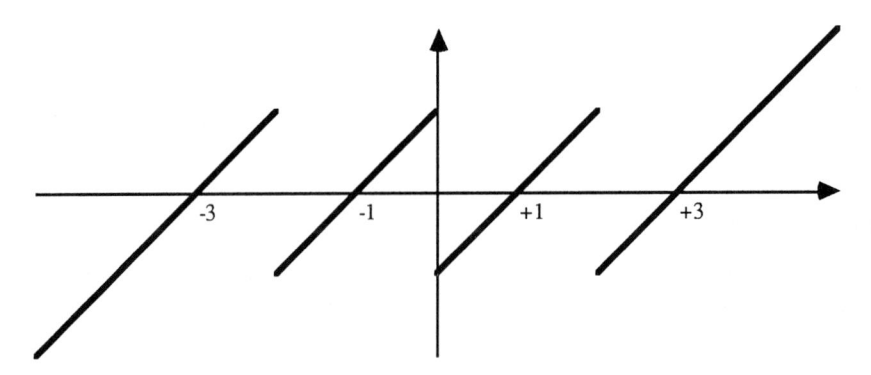

Figure 8. Graph of Q for $a_n \in \{\pm 1, \pm 3\}$.

In Exercise 4, the reader is guided through a proof of the following formulae

$$h(\theta) = -\nabla J(\theta)$$
$$J(\theta) = E[\hat{e}_n^2(\theta)]$$
$$\hat{e}_n(\theta) = c_n(\theta) - \hat{a}_n(\theta) \qquad (2.3.20)$$

where we recall that $\hat{a}_n = Q(c_n)$. In other words, the vector field $h(\theta)$ is again the derivative of a potential, the so-called **pseudo mean square error** (pseudo because a_n is replaced by the reconstructed message $\hat{a}_n$). The problem now becomes an investigation of the potential to be minimised, J: all the local minima must be identified, together with their domains of attraction. Unfortunately, at the time of writing, there is no complete study of this potential; (Verdu 1984) even gives a negative result, by exhibiting parasitic minima in certain cases.

Conclusion.

The study of the ODE appears extremely difficult, there is, so-to-speak, no complete analysis of this algorithm. The ODE however is still the best tool for tackling the problem. The decision-feedback equaliser corresponding to formulae (1.2.17) has an even more difficult ODE, since, in this case, the mean vector field $h(\theta)$ is no longer the gradient of a potential. In the next chapter, we shall see a simpler example of a recursive equaliser; the reader might also refer to Exercise 6.

2.4 Guide to Adaptive Algorithm Design

This guide does not claim to be a universal bible: we shall simply describe two procedures which have been proven by lengthy practical use. We would strongly advise users to adhere to one or other of these methods. Firstly, we shall describe the two methods, then we shall illustrate their use with several examples. The various exercises will allow users to try their own hands.

This guide does not approach the problem from basics; it assumes that the roles of the observed signal, and of the tunable parameter θ are given. The definition of these respective roles is usually called the **modelling task**, this is often difficult. The modelling task is thoroughly examined in Ljung and Soderström's book. We have preferred to omit that here, since our interest is not restricted to linear models; modelling considerations should be neither too general, nor too abstract.

2.4.1 First Guide: the ODE Method

Broadly speaking, this consists in reversing the stages of the method of analysis; in what follows P_θ denotes the distribution under which X_n is stationary, with Markov representation controlled by θ.

Stage 1. *Choice of functional to be minimised of the form $E_\theta[j(\theta, X_n)]$, or choice of equilibrium condition of the form $E_\theta[H(\theta, X_n)] = 0$.*

There is little to say about this stage: it is entirely the responsibility of the user, and requires a good knowledge of the application. The two objects to be chosen are the state vector X_n, formed from the observations, and the function j or H. We stress that there is a constraint on the expression $E(.)$: the symbol E may only appear on the outside of the formula. No other expression may be used as a starting point for an adaptive algorithm. In Stage 2, we shall consider only the case in which the functional is to be minimised: this is the most commonly used method.

Stage 2. *Choice of a method to minimise $J(\theta) = E_\theta[j(\theta, X_n)]$.*

Two types of methods are most widely used in practice; gradient methods and Newtonian methods. These correspond to the following differential equations:

Gradient method.
$$\begin{aligned}\dot{\theta} &= h_G(\theta), \\ h_G(\theta) &= -\nabla J(\theta)\end{aligned} \quad (2.4.1)$$

Newtonian method.
$$\begin{aligned}\dot{\theta} &= h_N(\theta), \\ h_N(\theta) &= -\left(\frac{\partial^2}{\partial \theta^2} J(\theta)\right)^{-1} \cdot \nabla J(\theta)\end{aligned} \quad (2.4.2)$$

Newtonian methods are more expensive, but generally lead more directly to the result. At this point, we have a differential equation which behaves in the required way (converges to the desired points).

If we were to seek to satisfy an equilibrium condition $E_\theta[H(\theta, X_n)] = 0$, the differential equation might be

$$\dot{\theta} = h(\theta)$$
$$h(\theta) = E_\theta[H(\theta, X_n)] \qquad (2.4.3)$$

In this case, it is necessary to verify that the desired points of convergence θ_* are indeed stable equilibrium points of (2.4.3).

Stage 3. *Derivation of an initial algorithm.*

If we take (2.4.1) or (2.4.3) as our starting point, then, the procedure is simple: the algorithm is :

$$\theta_n = \theta_{n-1} + \gamma_n H(\theta_{n-1}, X_n) \qquad (2.4.4)$$

where (γ_n) is a sequence of suitably chosen gains, and X_n is constructed from θ_{n-1}. If we start from (2.4.1), (2.4.4) is known as a **stochastic gradient method**. Note that $H(\theta_{n-1}, X_n)$, is the simplest estimator of the true gradient of $J(\theta)$ imaginable; more elaborate estimators could be used, but, as we shall see in our investigation of multistep algorithms, such techniques are rare in practice. In (2.4.2), the gradient part is treated as before: a simple estimator is chosen. If the Hessian has to be inverted (which is difficult), a more elaborate estimator should be used, as required for any particular case.

Stage 4. *Development and subsequent analysis of the definitive algorithm.*

In certain cases (very frequently in signal processing), the algorithm obtained after Stage 3 is judged to be too complex. This algorithm is then simplified or readjusted according to heuristic considerations. The final algorithm is then tested in simulations, it may equally well be returned for a further application of the given method of analysis.

Conclusions and Comments.

Of course, some stages may be short-circuited; we shall see examples of this. However we note that 2 step design methods (Stages 1 to 3, then Stage 4) are currently very frequently used. Beginning with a functional provides a solid, and generally clearly defined basis: moreover, with experience, one can guess judicious simplifications or modifications. The end result is often an algorithm which it would have been difficult to design directly.

2.4.2 Second Guide: the Use of Criteria Defined on Each Trajectory (Likelihood and Other Methods)

The procedure here differs from that previously followed.

2.4 Guide to Adaptive Algorithm Design

Stage 1. *Choice of criterion.*

The starting point here is a given finite sample of n observations. The engineer must then

1. choose the sequence of state vectors $(X_k(\theta))_{1 \leq k \leq n}$ to represent these observations;

2. choose a criterion. This could be either

$$\min_\theta J_n(\theta)$$
$$J_n(\theta) = \sum_{k=1}^{n} j_k[\theta, X_k(\theta)] \qquad (2.4.5)$$

(cf. the likelihood methods introduced in the example of Auto Regressive and ARMA identification in Subsection 3.3.3)
or

$$\begin{aligned} h_n(\theta) &= 0 \\ h_n(\theta) &= \sum_{k=1}^{n} H_k(\theta, X_k(\theta)) \end{aligned} \qquad (2.4.6)$$

for which the prototype will be the method of instrumental variables used in Exercise 7.

Stage 2. *Derivation of the adaptive algorithm.*

For fixed n, Stage 1 provides a value θ_n. Next, θ_n must be expressed as a function of θ_{n-1}, in the form (2.1.1). There is no general recipe for this; this operation is at times feasible, at times feasible modulo some modifications to the formulae, and at times infeasible; see the examples.

Stage 3. *Analysis of the algorithm.*

Taking into account the previous remarks, this stage is necessary, except when

1. the behaviour of θ_n given by Stage 1 is already theoretically known (this is the case for likelihood methods),

2. Stage 2 was carried out without modifying the results of Stage 1.

In fact, these conditions are seldom satisfied.

Comments.

We decided to include this type of approach because of the importance of likelihood methods, which are of this type. The so-called "prediction error" methods, favoured by control scientists, and described in detail in Ljung and Soderström's book, are derived from likelihood methods, and are also of this type. We shall see that it is quite common for two different routes to lead to the same algorithm.

2.4.3 Examples

Again, we shall concentrate on the applications described in Chapter 1. In addition, we give several academic examples, to complete the picture.

2.4.3.1 The Phase-Locked Loop: Design by the ODE Method

See Paragraph 1.2.1.1 for a description of this application. We have chosen this example to illustrate design methods, since algorithms (1.2.25, 1.2.26) do not appear intuitive; an understanding of the design processes on the other hand will help throw some light on this.

Stage 1. *Choice of criterion.*

The rule of the game is that the criterion should involve only the received signal (y_n), and not the emitted message a_n: thus (in the terminology used for the equaliser), there is no learning phase, and the operation is immediately self-adaptive. We introduce the following two functionals.:

(i) $$J_c(\phi) = E|\, y_n^4 e^{-i4\phi} + 1\,|^2$$
(ii) $$J_r(\phi) = E|\, y_n e^{-i\phi} - \hat{a}_n(\phi)\,|^2$$
$$\hat{a}_n(\phi) = \text{sgn}[\text{Re}(y_n e^{-i\phi})] + i\,\text{sgn}[\text{Im}(y_n e^{-i\phi})] \qquad (2.4.7)$$

These two functionals are shown in Fig. 9, below. Note that here $\hat{a}_n^4(\phi) = a_n^4 = -1$.

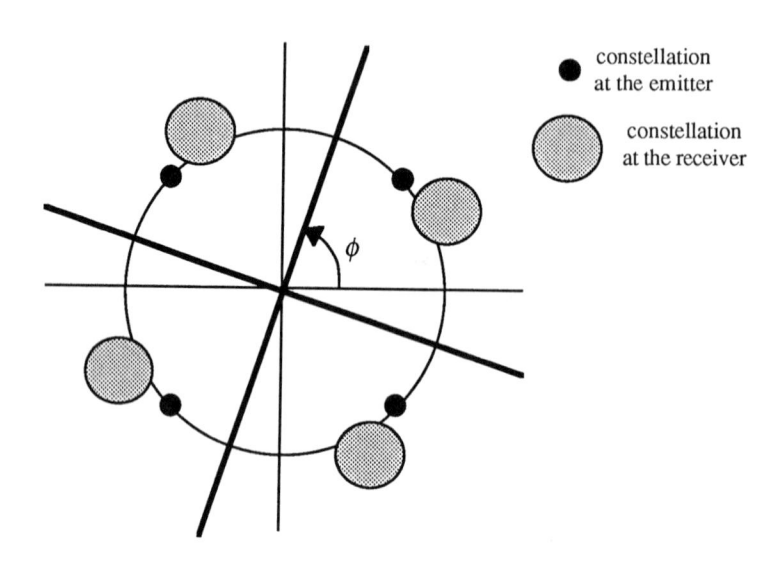

Figure 9. Illustration of the functionals J_c and J_r.

2.4 Guide to Adaptive Algorithm Design

The introduction of the fourth power in J_c removes the indeterminacy modulo $\pi/2$. These two functionals are used to tune the adjustable phase ϕ so that the black arrows have a minimal length, in the mean squares sense: this minimises the risk that the corrected points $y_n e^{-i\phi}$ cross the decision thresholds. These two criteria clearly make good sense. Figure 10 (below) illustrates the nature of these functionals (cf. Exercise 5).

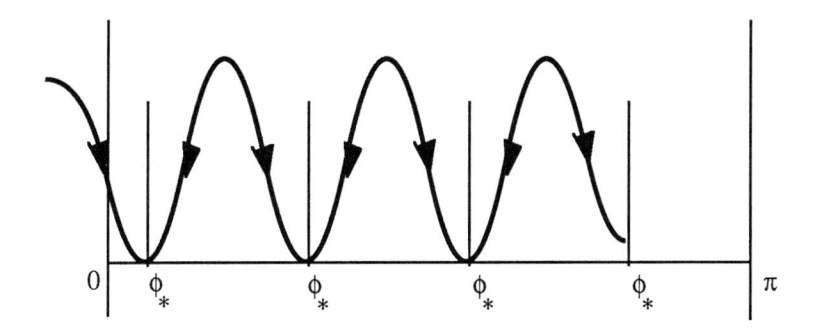

Figure 10. Graph of $J_c(\phi)$ or $J_r(\phi)$.

The functionals are periodic of period $\pi/2$, and in each period, there is a unique local minimum, which we denote by $\phi_* + k.\pi/2$. Each minimum represents a good choice for ϕ; it is not necessary to know the "true" phase value of the channel, since the message is encoded using the phase **transitions** of a_n, it is sufficient to determine the phase modulo $\pi/2$ (cf. Subsection 1.2.2).

Stage 2. *Choice of minimisation method.*

We shall use a simple gradient method. The gradients are given by

$$(i) \qquad \frac{d}{d\phi} J_c(\phi) = 8.E[\mathrm{Im}(y_n^4 e^{-i4\phi})]$$

$$(ii) \qquad \frac{d}{d\phi} J_r(\phi) = -2E[\mathrm{Im}(y_n e^{-i\phi} \overline{\tilde{a}}_n(\phi))] \qquad (2.4.8)$$

where $\overline{z}$ denotes the complex conjugate of z.

The corresponding ODEs are then

$$\dot\phi = -\nabla J_{c,r}(\phi) \qquad (2.4.9)$$

Stage 3. *Derivation of the algorithm.*

Formally: the standard discretisation scheme of (2.4.9) gives

$$\phi_n = \phi_{n-1} - \gamma_n \nabla J_{c,r}(\phi_{n-1}) \qquad (2.4.10)$$

If in (2.4.10), we replace the gradient by the current observation (a very coarse estimate), we obtain the two algorithms previously given in formulae (1.2.25) and (1.2.26), namely

(i) $$\phi_n = \phi_{n-1} - \gamma_n \operatorname{Im}(y_n^4 e^{-i4\phi_{n-1}}) \qquad (2.4.11)$$

for the Costas loop, and

(ii) $$\phi_n = \phi_{n-1} + \gamma_n \operatorname{Im}(y_n e^{-i\phi_n} \overline{\tilde{a}}_n(\phi_{n-1}))$$

for the decision-feedback loop. Here, a constant gain γ is commonly used.

Stage 4. *Analysis.*

There is nothing to add, we have simply derived the stochastic gradient methods of minimising J_c and J_r, and the algorithms we have described satisfy Assumptions (A) of Chapter 1 for all ϕ. The only possible comment is that since J is periodic, and thus has an infinite number of distinct domains of attraction, **only the results for a finite horizon are applicable in the case of constant gain**. Moreover, if we use an algorithm with decreasing gain, Corollary 6 does not allow us to predict the domains of attraction for θ_n. The long term behaviour of the algorithm is thus entirely open to study.

Stage 5 (optional).

The following simplified versions of (2.4.11) are sometimes used

$$\phi_n = \phi_{n-1} + \gamma \operatorname{sgn} \varepsilon_n(\phi_{n-1}) \qquad (2.4.12)$$

where ε_n is derived from (2.4.11-i or ii). The reader is invited to examine the algorithm thus obtained.

2.4.3.2 Some Versions of the Blind Equaliser: Design by the ODE Method

See Subsection 1.2.1 for a description of this problem.

Stage 1. *Choice of a functional.*

This must not be the functional (2.3.20) associated with the self-adaptive equaliser. We do not propose to cover this problem in full, and would refer the reader to (Benveniste, Goursat and Ruget [BGR] 1980a). We shall simply summarise the results of that paper. We recall here, that the equalisation process is as described in Fig. 5; we use the notation from that figure. The results of [BGR] do not apply directly to the present situation, since the distribution of a_n is by assumption different from that here, and since the noise ν_n is neglected.

We shall say that the variables a_n are **sub-Gaussian**, if they are uniformly distributed over the interval $[-d, +d]$, or if they have a probability density of the form

$$e^{-g(x)} \qquad (2.4.13)$$

where g is an even function which is differentiable, except possibly at the origin, and $g(x)$ and $g'(x)/x$ are strictly increasing in $\mathbb{R}_+$ (for example

2.4 Guide to Adaptive Algorithm Design

$g(x) = |x|^\beta, \beta > 2$); if $g'(x)/x$ is strictly decreasing in $\mathbb{R}_+$ (for example $g(x) = |x|^\beta, \beta < 2$) and the remaining properties are as above, then we shall say that the variables a_n are **super-Gaussian**.

We shall make use of the {channel–equaliser} transfer function given by

$$T(z^{-1}) = \theta(z^{-1}) \cdot S(z^{-1}) \qquad (2.4.14)$$

The aim is to tune θ so as to invert S; this may be expressed in the form

$$T_*(z^{-1}) := \theta_*(z^{-1}) \cdot S(z^{-1}) = \pm z^{-N} \qquad (2.4.15)$$

for some unknown delay N. Note that in the absence of any reference, only the **global** message $(a_n)_{n \in \mathbb{Z}}$ can be reconstituted; there is no statistical distinction between $(c_{n+N}(\theta))_{n \in \mathbb{Z}}$ and $(c_n(\theta))_{n \in \mathbb{Z}}$, nor even between $(c_n(\theta))_{n \in \mathbb{Z}}$ and $(-c_n(\theta))_{n \in \mathbb{Z}}$, since $(a_n)_{n \in \mathbb{Z}}$ and $(-a_n)_{n \in \mathbb{Z}}$ have the same joint distribution. Condition (2.4.15) is thus the best one could hope for.

The following two functionals are of interest for our purpose:

$$J_S(\theta) = \sigma \cdot E[c_n(\theta) - \alpha \operatorname{sgn} c_n(\theta)]^2 \qquad (2.4.16)$$

where $\sigma = \pm 1$ and $\alpha \in \mathbb{R}$ will be defined later, and

$$J_\nu(\theta) = E[g(c_n(\theta))] \qquad (2.4.17)$$

when the (a_n) have a probability density of the form (2.4.13).

A proof of the following theorem is given in [BGR]:

Theorem [BGR].
(i) *If the (a_n) are* **sub-Gaussian**, *then if*

$$\alpha = \frac{E a_n^2}{E |a_n|}, \qquad \sigma = \pm 1 \qquad (2.4.18)$$

all the local minima of the functional J_S satisfy (2.4.15).
(ii) *If now, we restrict the two functionals J_S and J_ν to θ such that*

$$\|T\|^2 := \frac{1}{2\pi} \int_0^{2\pi} \|T(e^{i\omega})\|^2 d\omega = 1 \qquad (2.4.19)$$

(i.e. the global channel has energy 1), then J_ν satisfies (2.4.15) in both the sub-Gaussian and the super-Gaussian cases, and J_S satisfies (2.4.15) with $\alpha \geq 0$ and $\sigma = +1$ in the sub-Gaussian case and with $\alpha \geq 0$ and $\sigma = -1$ in the super-Gaussian case.

Point (i) of Theorem [BGR] leads to the equaliser, whilst point (ii) gives a method of deconvolution in non-standard cases, as used by (Goursat 1984) in geophysics. Note also that (2.4.17) is strongly related to the likelihood, which, by an observation due to L.Ljung, is of the form

$$\sum_{k=1}^n g(c_k(\theta)) \qquad (2.4.20)$$

for a sample of size n. In any case, since a minimum of (2.4.20) cannot be explicitly determined, we shall, even in this case, stick to our original design method.

Stage 2. *Gradient method of minimisation.*
One open problem remains, the actual form of the equaliser (all zeroes or transversal, poles–zeroes or recursive) is not yet fully determined; in fact, formulae (2.4.16) and (2.4.17) do not require it to be. We juggle with two things simultaneously

- with the parameterisation of the equaliser (θ will have various meanings);
- with the energy constraint (2.4.19). In this case, several possible gradients will be obtained. Furthermore, we shall concentrate our attention on J_S, as defined in (2.4.16), since there is nothing to be gained by studying J_ν.

Transversal Equaliser, Criteria (2.4.16, 2.4.18).

The filter is now synthesised according to the classical formula of (1.2.7), namely $c_n(\theta) = Y_n^T \cdot \theta$. Thus, by taking derivatives inside the expectation, we obtain

$$
\begin{align}
(i) \quad & h(\theta) := -\nabla J_S(\theta) = E(Y_n \varepsilon_n(\theta)) \\
(ii) \quad & \varepsilon_n(\theta) = \alpha \, \mathrm{sgn}(c_n(\theta)) - c_n(\theta) \\
(iii) \quad & c_n(\theta) = Y_n^T \theta
\end{align}
\tag{2.4.21}
$$

Recursive Equaliser, Criteria (2.4.16, 2.4.18).

The filter is now a poles–zeroes filter of the form

$$
\begin{align}
(i) \quad & c_n(\theta) = \Phi_n(\theta)^T \cdot \theta \\
(ii) \quad & \Phi_n^T(\theta) := (y_{n+N}, \ldots, y_{n-N}; c_{n-1}(\theta), \ldots, c_{n-P}(\theta)) \\
(iii) \quad & \theta^T := (\theta'(-N), \ldots, \theta'(+N); \theta''(1), \ldots, \theta''(P))
\end{align}
\tag{2.4.22}
$$

In order to calculate the gradient, we need to calculate $\frac{\partial}{\partial \theta} c_n(\theta)$. We leave the reader to check that if we differentiate (2.4.22-i) term-by-term, we have

$$
\begin{align}
(i) \quad & \frac{\partial}{\partial \theta} c_n(\theta) = \frac{1}{K(z^{-1})} \Phi_n(\theta) \\
(ii) \quad & K(z^{-1}) := 1 - \sum \theta''(j) z^{-j}
\end{align}
\tag{2.4.23}
$$

where the $\theta''(j)$ are the coefficients of the "pole" part of the filter θ, defined in (2.4.22-iii). In (2.4.23-i), the filter $K^{-1}(z^{-1})$ is applied to each component of Φ_n. The gradient is given by

$$
h(\theta) := -\nabla J_S(\theta) = E\left(\frac{1}{K(z^{-1})} \Phi_n(\theta) \cdot \varepsilon_n(\theta)\right)
\tag{2.4.24}
$$

2.4 Guide to Adaptive Algorithm Design

where $\varepsilon_n(\theta)$ is as in (2.4.21-ii) and the filter K corresponds to the poles of the equaliser θ.

Transversal Equaliser: Criterion with Constraint Corresponding to (2.4.16, 2.4.19). This is a more interesting case which leads us to an algorithm with equality constraints, in which the parameter θ_n evolves, not in a Euclidean space, but in a variety.

Before we calculate the gradient of the restriction of J_S to those θ which satisfy (2.4.19), we translate the implicit condition (2.4.19) into an explicit condition on θ. This is obtained by decomposing the equaliser as a cascade of two successive filters, as shown in Fig. 11.

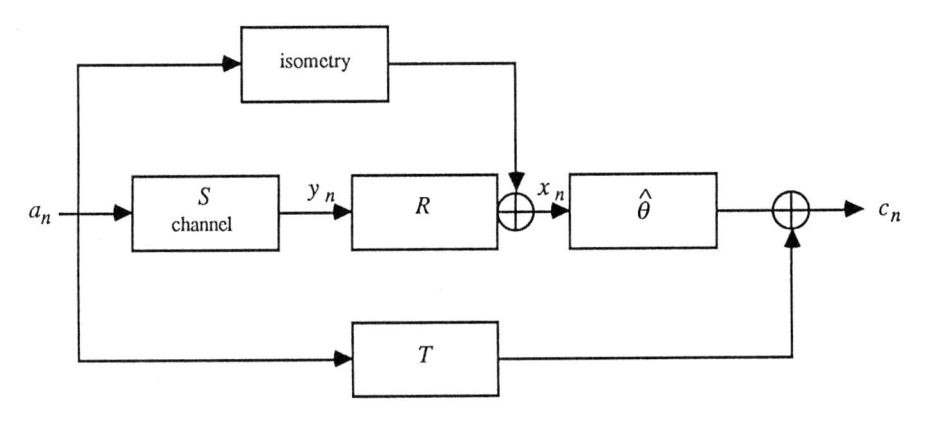

Figure 11. Equaliser decomposition.

The new equaliser is actually decomposed into the form

$$c_n(\hat{\theta}) = \hat{\theta}(z^{-1}) \cdot R(z^{-1})y_n = \hat{\theta}(z^{-1}) \cdot x_n \qquad (2.4.25)$$

where only $\hat{\theta}$ is adjustable, and R remains fixed. The fixed filter $R(z^{-1})$ is chosen so as to ensure that

$$| R(e^{i\omega}) \cdot S(e^{i\omega}) | = 1 \quad \forall \omega \qquad (2.4.26)$$

In other words, $R \cdot S$ is an **all-pass filter**. The realisation of R is a classical problem in signal processing; one easy method is to construct

$$\hat{x}_n = y_n - \hat{y}_{n|n-1}$$

where $\hat{y}_{n|n-1}$ is the optimal linear prediction of y_n, given $y_{n-1}, y_{n-2}, \ldots$; $\hat{x}_n$ is thus a white noise (i.e. a sequence of uncorrelated, but not necessarily independent, variables) and it remains to normalise $\hat{x}_n$ so that the signal x_n has the same variance as a_n. The remaining equaliser, denoted by θ in Fig. 11, must invert the phase

$$\arg\left(S(e^{i\omega}) \cdot R(e^{i\omega})\right)$$

of the all-pass filter $R \cdot S$, this explains the nomenclature in Fig. 11. Thus we may reformulate the constraint (2.4.19) in terms of the new $\hat{\theta}$, since, by virtue of (2.4.26)

$$\|\hat{\theta}\|^2 = \|T\|^2 \qquad (2.4.27)$$

We now denote the restriction of J_S to equalisers of energy 1 by

$$V(\hat{\theta}) := J_S(\hat{\theta})_{\|\hat{\theta}\|=1} \qquad (2.4.28)$$

In order to obtain the gradient V, (2.4.21) is modified in the following way:

$$h_c(\hat{\theta}) = -\nabla V(\hat{\theta}) = -(h(\hat{\theta}) - \hat{\theta} \cdot (\hat{\theta}^T h(\hat{\theta}))) \qquad (2.4.29)$$

where $\|\hat{\theta}\| = 1$ and h is defined in (2.4.21); $h_c(\hat{\theta})$ is just the projection of $h(\hat{\theta})$ onto the sphere $\|\hat{\theta}\| = 1$. We recall that the equaliser $\hat{\theta}$ is synthesised to have the transversal form

$$c_n(\hat{\theta}) = \sum_{i=-N}^{+N} \hat{\theta}(i) x_{n-i} \qquad (2.4.30)$$

and thus (self-explanatory notation), we have

$$\|\hat{\theta}\|^2 = \sum_i \hat{\theta}(i)^2 \qquad (2.4.31)$$

Stage 3. *Derivation of a stochastic gradient algorithm; an example of an algorithm with constraints.*

We have several candidate ODEs, and we apply a procedure consisting of

1. the discretisation of the ODEs;
2. the replacement of the mean vector field by the simplest estimator of this field, namely its current value.

Blind Equaliser Without Constraints, Transversal Form.

The ODE is formed from the vector field $h(\theta)$ given in (2.4.21). The stochastic gradient algorithm is thus

$$\begin{aligned} \theta_n &= \theta_{n-1} + \gamma Y_n \varepsilon_n(\theta_{n-1}) \\ \varepsilon_n(\theta) &= \alpha \operatorname{sgn} c_n(\theta) - c_n(\theta) \\ c_n(\theta) &= Y_n^T \theta \end{aligned} \qquad (2.4.32)$$

where α is given in (2.4.18).

Blind Equaliser Without Constraints, Recursive Form.

Here, the ODE corresponds to the mean field given in (2.4.24). Since the state vector $\Phi_n(\theta)$ depends on θ, we use the following procedure to obtain the algorithm from the ODE: observations strictly prior to the present time n are

2.4 Guide to Adaptive Algorithm Design

used to calculate the various signals of interest $(c_{n-1}, c_{n-2}, \ldots, c_{n-P})$, these same signals are calculated at time n using formulae (2.4.22) with $\theta = \theta_n$, and the calculated signal values prior to time n. We obtain the following algorithm:

(i) $\qquad \theta_n = \theta_{n-1} + \gamma \tilde{\Phi}_n \varepsilon_n$

(ii) $\qquad \varepsilon_n = \alpha \operatorname{sgn} c_n - c_n$

(iii) $\qquad c_n = \Phi_n^T \theta_{n-1}$

$$\tilde{c}_n = \sum_{j=1}^{P} \theta''_{n-1}(j) \tilde{c}_{n-j} + c_n$$

(iv) $\qquad \tilde{y}_{n+N} = \sum_{j=1}^{P} \theta''_{n+N-j} + \tilde{y}_{n+N}$

$$\tilde{\Phi}_n^T = (\tilde{y}_{n+N}, \ldots, \tilde{y}_{n-N}; \tilde{c}_{n-1}, \ldots, \tilde{c}_{n-P}) \qquad (2.4.33)$$

Let us now examine algorithm (2.4.33) more closely. It is of the form (2.1.1) with the state X_n given by

$$X_n = \begin{pmatrix} \Phi_n \\ \tilde{\Phi}_n \end{pmatrix} \qquad (2.4.34)$$

The theory needed to derive the semi-Markov form of X_n is in every respect analogous to our investigation of the recursive equaliser in Paragraph 1.3.2.1. Provided that the filter $K(z^{-1})$ associated with θ in (2.4.33-ii) has a **stable inverse**, the vector $X_n(\theta)$ obtained from formulae (2.4.22) and (2.4.33-iv) is asymptotically stationary, for fixed θ. This stationary distribution is used in formula (2.4.24) to define the mean field, and also in formulae (2.4.10).

In conclusion, algorithm (2.4.33) satisfies Assumptions (A), moreover

1. its ODE satisfies the assumptions, since it is derived from the potential J_S,

2. there is a risk of explosion in (2.4.33) if θ leaves the domain of stability of the filter $K^{-1}(z^{-1})$.

Blind Equaliser with Constraints, Transversal Form: a Stochastic Gradient with Constraints.

Here, the ODE is given by (2.4.29) and its trajectories lie on the unit sphere $\|\hat{\theta}\| = 1$. In order to apply the usual procedure, we should, in the first instance, be aware that a differential equation on a variety cannot be discretised simply by a brute force application of Taylor's formula: omission of higher order terms may cause the discretised trajectory to leave the variety. Figure 12 (below) illustrates our discretisation procedure (the chosen discretisation step in this figure is $\gamma = 1$).

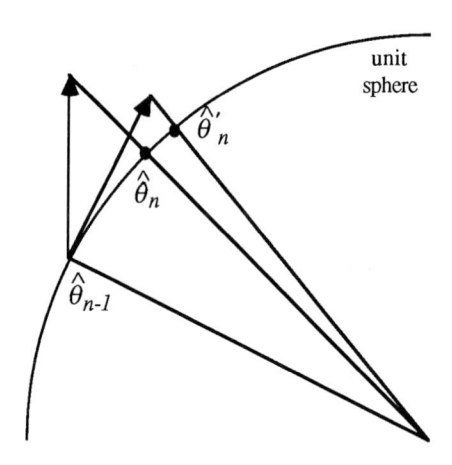

Figure 12. Discretisation scheme for the ODE $\dot{\theta} = h_c(\hat{\theta})$.

The normal procedure would give

$$\hat{\theta}'_n = \text{proj}(\hat{\theta}_{n-1} + \gamma h_c(\hat{\theta}_{n-1})) \qquad (2.4.35)$$

where proj denotes the projection onto the unit sphere; but the formula

$$\hat{\theta}_n = \text{proj}(\hat{\theta}_{n-1} + \gamma h(\hat{\theta}_{n-1})) \qquad (2.4.36)$$

enables us to obtain an alternative discretisation more cheaply using the (easily checked) inequalities

$$\begin{aligned}
\|\hat{\theta}_n - \gamma h_c(\hat{\theta}_{n-1})\| &\leq \|\hat{\theta}_n - \hat{\theta}'_n\| + \|\hat{\theta}'_n - \gamma h_c(\hat{\theta}_{n-1})\| \\
&\leq C_1 \cdot \gamma^2 + C_2 \cdot \gamma^2
\end{aligned} \qquad (2.4.37)$$

Thus we construct the adaptive algorithm from (2.4.32). For a small, constant gain γ, the usual procedure now gives

$$\begin{aligned}
\hat{\theta}_n &= \lambda_n(\hat{\theta}_{n-1} + \gamma Y_n \varepsilon_n(\hat{\theta}_{n-1})) \\
\varepsilon_n(\hat{\theta}) &= \sigma \cdot (\alpha \operatorname{sgn} c_n(\hat{\theta}) - c_n(\hat{\theta})) \\
c_n(\hat{\theta}) &= X_n^T \cdot \theta \\
\lambda_n^{-1} &= \|\hat{\theta}_{n-1} + \gamma Y_n \varepsilon_n(\hat{\theta}_{n-1})\|
\end{aligned} \qquad (2.4.38)$$

where σ and α are chosen as indicated in point (ii) of Theorem [BGR], c_n is as in (2.4.30) and the scalar λ_n forces the projection onto the unit sphere. This algorithm is clearly of the form (2.1.1) **with a non-zero complementary term** in γ^2. Here, we have a second example of an algorithm where the complementary term is useful. Algorithm (2.4.38) is applied by (Goursat 1984), in its super-Gaussian form, to problems of blind deconvolution in seismic reflection.

Stage 4. *Simplification of the recursive version.*

The algorithms using a transversal formulation of the equaliser have been deemed to be satisfactory (if necessary, they may be simplified by using only the signs of the components of the vector field). We shall now concentrate our attention on the recursive formulation (2.4.33). This formulation is much more complex as a result of the calculation involved and of the extra memory required to obtain $\tilde{\Phi}_n$. The simplified version omits the filter $K^{-1}(z^{-1})$:

$$\theta_n = \theta_{n-1} + \gamma \Phi_n \varepsilon_n \qquad (2.4.39)$$

and everything else is as in (2.4.33-ii,iii). Note that the algorithm has a structure similar to that of the self-adaptive recursive equaliser given in formula (1.2.17). As always with signal feedback in a recursive filter, there is a risk of explosion. The ODE associated with (2.4.39) is complicated to study; only local stability results exist, see Exercise 6 for a similar, simpler problem.

Conclusion to this Example.

In this case, this procedure has been shown to be the only successful way of constructing an algorithm: the procedure was effectively used to derive an algorithm whose effect was not easy to guess. It is also worth pointing out that uniform distributions on the finite sets of the form $\{\pm 1, \ldots, \pm 2K+1\}$, as used for a_n in data transmission, do not come under the heading of sub-Gaussian distributions: they are only very crudely equivalent to the uniform distribution on the corresponding interval. Thus this investigation, which we have effectively used as a base from which to develop the blind equaliser, does not provide an analysis (in the mathematical sense of the term) of the data transmission algorithm.

Finally, we illustrate this example with several diagrams taken from the article [BGR].

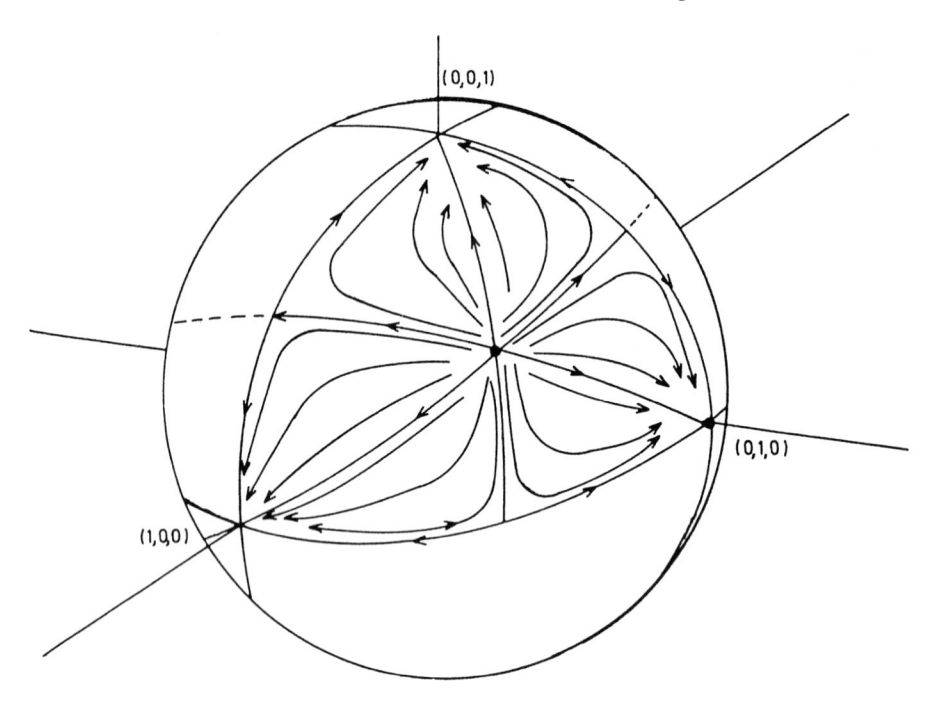

Figure 13. Trajectories of the ODE of the algorithm with constraints (3 components).

2.4 Guide to Adaptive Algorithm Design 69

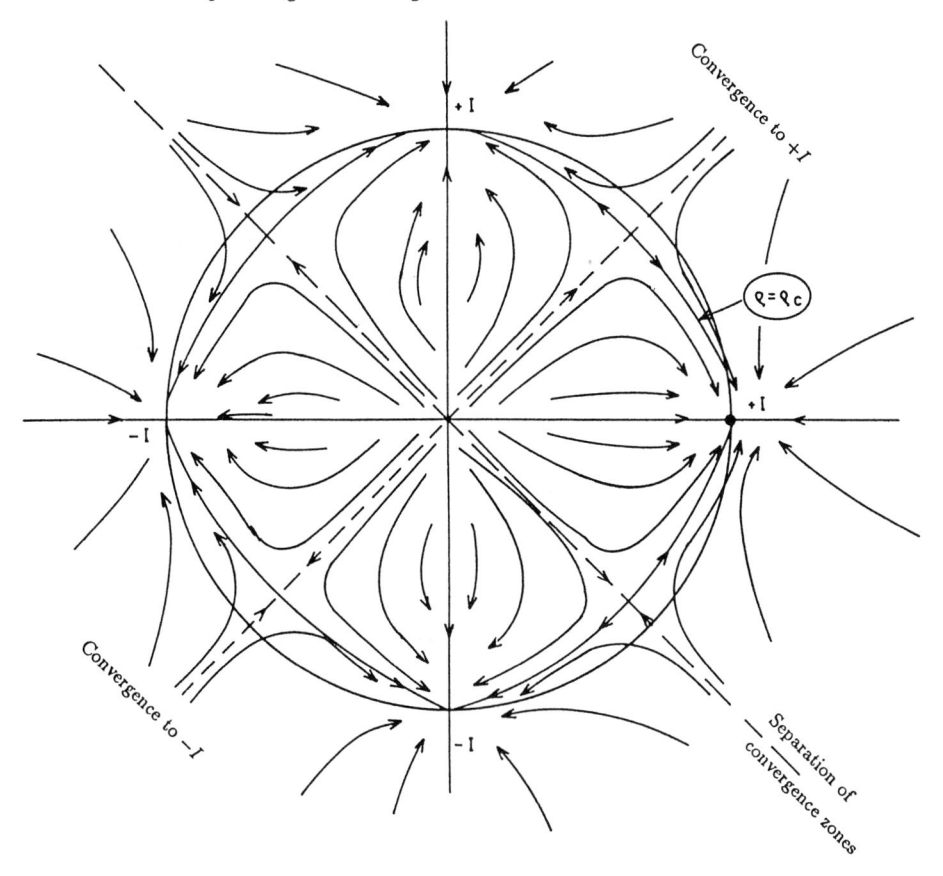

Figure 14. Trajectories of the ODE of the algorithm without constraints (2 components).

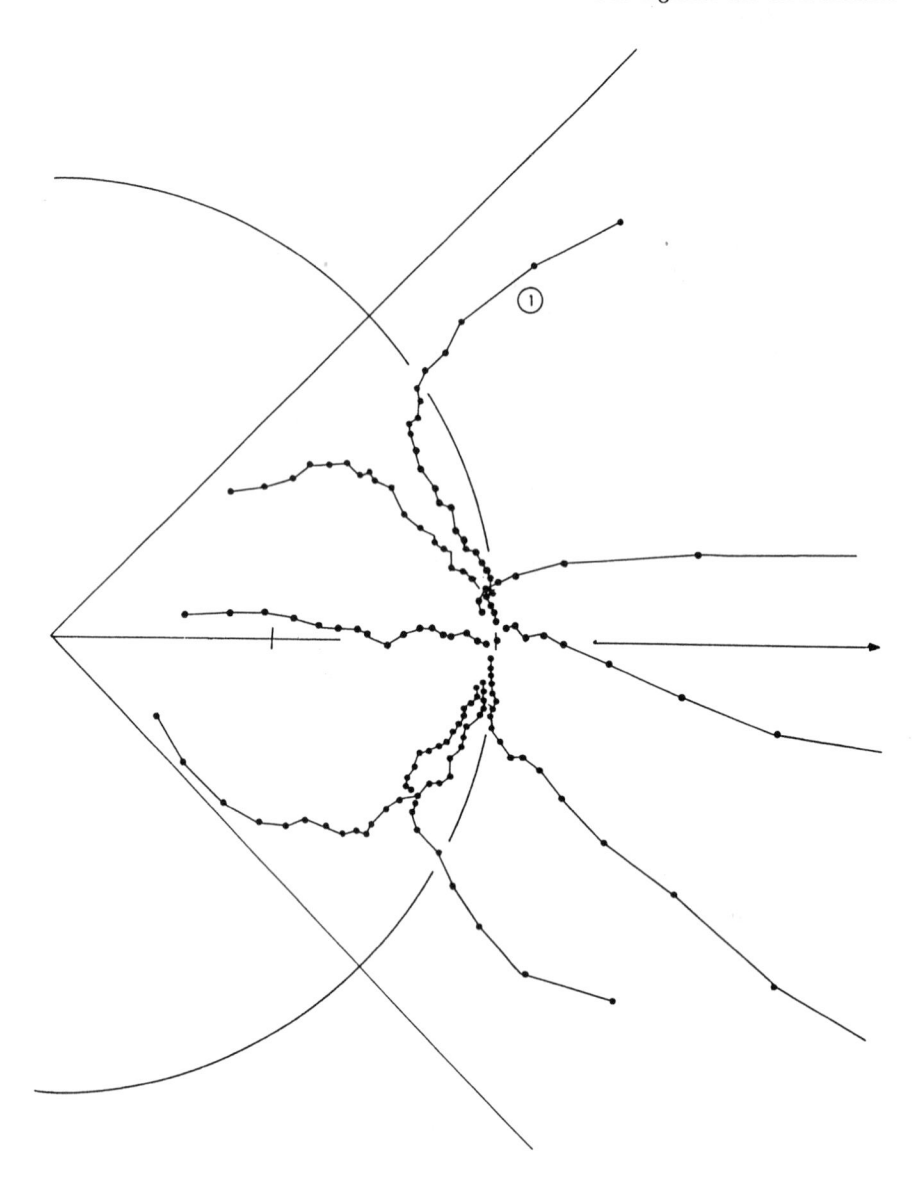

Figure 15. Trajectories of the algorithm without constraints.

Figures 14 and 15 should be compared.

2.4 Guide to Adaptive Algorithm Design

2.4.3.3 Identification by an AR or ARMA Model: Likelihood Method

It is desired to express an observed signal (y_n) in the form

$$y_n = \Phi_n^T \theta_* + \nu_n \qquad (2.4.40)$$

where (ν_n) is a white noise with stationary, Gaussian distribution, and where

(i)
$$\Phi_n^T = (y_{n-1}, \ldots, y_{n-p})$$
$$\theta_*^T = (a_*(1), \ldots, a_*(p))$$

in the case of an Auto-Regressive (AR) model, and

(ii)
$$\Phi_n^T = (y_{n-1}, \ldots, y_{n-p}; \nu_{n-1}, \ldots, \nu_{n-q})$$
$$\theta_*^T = (a_*(1), \ldots, a_*(p); b_*(1), \ldots, b_*(q)) \qquad (2.4.41)$$

in the case of an ARMA (Auto Regressive Moving Average) model. We shall illustrate the second method.

Stage 1. *Choice of a functional.*

Suppose we are give a sample $y_1, \ldots, y_n$. If the AR and ARMA models are parameterised by the 2-tuple (θ, σ), where θ is the collection of AR or ARMA parameters, and σ^2 is the variance of the excitation noise, then minus the log-likelihood of the sample $y_1, \ldots, y_n$ is given by

$$-\log L\{y_1, \ldots, y_n \mid (\theta, \sigma)\} = \frac{1}{\sigma^2} \sum_1^n e_k(\theta)^2 + n \log \sigma\sqrt{2\pi} \qquad (2.4.42)$$

where $e_k(\theta)$ is the prediction error, given in the AR case by

(i)
$$e_k(\theta) = y_k - \sum_{i=1}^p a_i y_{k-i}$$
$$\theta^T = (a_1, \ldots, a_p)$$

and in the ARMA case by

(ii)
$$e_k(\theta) = y_k - \sum_{i=1}^p a_i y_{k-i} - \sum_{j=1}^q b_j e_{k-j}(\theta)$$
$$\theta^T = (a_1, \ldots, a_p; b_1, \ldots, b_q) \qquad (2.4.43)$$

Formulae (2.4.43) assume that initial conditions on (y_k) and (e_k) are given (or chosen). The method of maximum likelihood estimation of θ leads to the minimisation of the **least squares** functional

$$J_n(\theta) := \sum_{k=1}^n e_k(\theta)^2 \qquad (2.4.44)$$

If the (y_n) cannot be assumed to be Gaussian, it is still possible to attempt to minimise (2.4.44); however, in this case, the least squares method is no longer the same as the maximum likelihood method.

Stage 2. *Derivation of the adaptive algorithm.*

We distinguish between the AR and the ARMA cases, which are different in complexity.

AR case.

We seek to minimise (2.4.44) for the model (2.4.43-i). Since the functional J_n is quadratic, the result θ_n is easily calculated.

$$\theta_n = \left(\sum_{k=1}^{n} \Phi_k \Phi_k^T\right)^{-1} \left(\sum_{k=1}^{n} \Phi_k y_k\right) \qquad (2.4.45)$$

As we shall see, θ_n is readily expressed as a function of θ_{n-1}. If we set

$$R_n := \frac{1}{n} \sum_{k=1}^{n} \Phi_k \Phi_k^T \qquad (2.4.46)$$

then we have

$$\begin{aligned} nR_n\theta_n &= (n-1)R_{n-1}\theta_{n-1} + \Phi_n y_n \\ &= (nR_n - \Phi_n\Phi_n^T)\theta_{n-1} + \Phi_n y_n \\ &= nR_n\theta_{n-1} + \Phi_n(y_n - \Phi_n^T\theta_{n-1}) \end{aligned}$$

Finally, we obtain the **Recursive Least Squares** algorithm (RLS):

(i) $\qquad \theta_n = \theta_{n-1} + \frac{1}{n} R_n^{-1} \Phi_n (y_n - \Phi_n^T \theta_{n-1})$

(ii) $\qquad R_n = R_{n-1} + \frac{1}{n}(\Phi_n\Phi_n^T - R_{n-1}) \qquad (2.4.47)$

Usually (2.4.47-ii) is replaced by an equation in R_n^{-1}, thereby dispensing with the need for matrix inversion in (2.4.47-ii): this is the well-known Riccati equation, and there is a wealth of literature on this topic, see (Ljung and Soderström 1983).

ARMA case.

For given (y_n), and fixed θ, the prediction error

$$\begin{aligned} e_n(\theta) &= y_n - \left(\sum_{i=1}^{p} a(i)y_{n-i} + \sum_{j=1}^{q} b(j)e_{n-j}(\theta)\right) \\ \theta^T &= (a(1),\ldots,a(p); b(1),\ldots,b(q)) \end{aligned} \qquad (2.4.48)$$

is defined using a recursion involving θ; $e_n(\theta)$ is no longer linear in θ, and so the least squares criterion is no longer quadratic. The arguments which follow are taken from Ljung and Soderström's book.

2.4 Guide to Adaptive Algorithm Design

For n sufficiently large, θ_n will be close to θ_{n-1}, and we may derive θ_n from a local examination of the criterion $J_n(\theta)$ in a neighbourhood of θ_{n-1}. A second order Taylor expansion (using the fact that $o(x)/x$ tends to zero as x tends to zero) gives

$$J_n(\theta) = J_n(\theta_{n-1}) + [\nabla J_n(\theta_{n-1})]^T \cdot (\theta - \theta_{n-1})$$
$$+ \frac{1}{2}(\theta - \theta_{n-1})^T [\text{Hess } J_n(\theta_{n-1})](\theta - \theta_{n-1}) + o\|\theta - \theta_{n-1}\|^2 \quad (2.4.49)$$

where Hess denotes the Hessian (matrix of second derivatives). Minimisation of both components of (2.4.49) with respect to θ gives

$$\theta_n = \theta_{n-1} - (\text{Hess } J_n(\theta_{n-1}))^{-1} \cdot \nabla J_n(\theta_{n-1}) + o\|\theta_n - \theta_{n-1}\| \quad (2.4.50)$$

The algorithm is then given by neglecting the final term in (2.4.50), and making an approximate estimate of the gradient and the Hessian. Note that (2.4.50) is similar to a step of a quasi-Newtonian method.

Before computing the gradient and the Hessian, we must first calculate $\frac{\partial}{\partial \theta}\Phi_k(\theta)$, where $\Phi_k(\theta)$ includes the previous signals y_{k-i} and prediction errors e_{k-j}. We have already carried out this calculation in (2.4.23, 2.4.24), and we restate these results below

$$\Phi_k^T(\theta) := (y_{k-1}, \ldots, y_{k-p}; e_{k-1}(\theta), \ldots, e_{k-q}(\theta))$$
$$\tilde{\Phi}_k(\theta) := \frac{\partial}{\partial \theta}\Phi_k(\theta)$$
$$\tilde{\Phi}_k(\theta) = \sum_{j=1}^{q} b(j)\tilde{\Phi}_{k-j}(\theta) + \Phi_k(\theta) \quad (2.4.51)$$

where the $b(j)$ are the MA coefficients of θ.

Then we have

$$\nabla J_n(\theta) = \sum_{k=1}^{n} \tilde{\Phi}_k(\theta)[\Phi_k^T(\theta) \cdot \theta - y_k] \quad (2.4.52)$$

and for the Hessian, the more complicated formula

$$\text{Hess } J_n(\theta) = \sum_{k=1}^{n} \tilde{\Phi}_k(\theta)\tilde{\Phi}_k^T(\theta) + \sum_{k=1}^{n} \left(\frac{\partial}{\partial \theta}\tilde{\Phi}_k(\theta)\right)(\Phi_k^T(\theta) \cdot \theta - y_k) \quad (2.4.53)$$

Once again, $\frac{\partial}{\partial \theta}\tilde{\Phi}_k(\theta)$ is given by the action of linear filters on the observations $y_{k-1}, y_{k-2}, \ldots$ prior to time k. But then for $\theta = \theta_n$, the prediction error

$$e_k = y_k - \Phi_k^T \cdot \theta_n \quad (2.4.54)$$

calculated using the sample $y_1, \ldots, y_n$ is orthogonal to the data strictly prior to time k; and so the second term of (2.4.53) is approximately zero for $\theta = \theta_n$, whence

$$\text{Hess } J_n(\theta_n) \approx \sum_{k-1}^{n} \tilde{\Phi}_k(\theta_n) \cdot \tilde{\Phi}_k^T(\theta_n) \quad (2.4.55)$$

To determine the algorithm itself from (2.4.50, 2.4.51, 2.4.52, 2.4.54), we cheat a little by calculating the gradient and the Hessian just as in the previously described procedure for the blind recursive equaliser. Finally, we obtain the **Recursive Maximum Likelihood (RML)** algorithm :

(i) $$\theta_n = \theta_{n-1} + \frac{1}{n} R_n^{-1} \tilde{\Phi}_n e_n$$

(ii) $$R_n = R_{n-1} + \frac{1}{n}(\tilde{\Phi}_n \tilde{\Phi}_n^T - R_{n-1})$$

(iii) $$\tilde{\Phi}_n = (\tilde{y}_{n-1}, \ldots, \tilde{y}_{n-p}; \tilde{e}_{n-1}, \ldots, \tilde{e}_{n-q})$$

(iv) $$\tilde{y}_n = \sum_{j=1}^{q} b_n(j) \tilde{y}_{n-j} + y_n$$

(v) $$\tilde{e}_n = \sum_{j=1}^{q} b_n(j) \tilde{e}_{n-j} + e_n$$

(vi) $$e_n = \sum_i a_{n-1}(i) y_{n-i} + \sum_j b_{n-1}(j) e_{n-j}$$

(vii) $$\theta_n^T = (a_n(1), \ldots, a_n(p); b_n(1), \ldots, b_n(q)) \qquad (2.4.56)$$

If we eliminate the filter on the state vector, as before, we obtain the **Extended Least Squares (ELS)** algorithm.

$$\theta_n = \theta_{n-1} + \frac{1}{n} R_n^{-1} \Phi_n e_n$$
$$R_n = R_{n-1} + \frac{1}{n}(\Phi_n \Phi_n^T - R_{n-1}) \qquad (2.4.57)$$

where the undefined objects are described in formulae (2.4.56).

Again, we note that these algorithms are implemented somewhat differently, using the inverse R_n^{-1}, with a Riccati equation or other fast algorithms.

Stage 3. *Analysis of the algorithms.*

We have already met algorithms of this form, for example in Paragraph 1.3.1.2. These algorithms may be written in the form (2.1.1), relative to the extended vector

$$\Theta_n = \begin{pmatrix} \theta_n \\ \text{col } R_n \end{pmatrix} \qquad (2.4.58)$$

with the explicit inclusion of the higher order complementary term; gains are decreasing. The least squares algorithm satisfies all the assumptions and carries no risk of explosion. The RML and ELS algorithms also satisfy the assumptions, but may explode as a result of the recursion (2.4.56-vi).

As in Paragraph 1.3.1.2 (least squares equaliser algorithm), RLS and RML are **stochastic Newtonian methods**, which means that their ODE minimises the least squares functional

$$J(\theta) = E(e_n^2(\theta)) \qquad (2.4.59)$$

in a quasi-Newtonian fashion. Note that whilst RLS is actually a maximum likelihood method (whence the classical statistical results (Box and Jenkins 1970) apply), the same is not true of RML; and therefore the latter must be analysed in more detail.

As far as ELS is concerned, its ODE is not the derivative of a potential. This ODE was studied in depth by Ljung (Ljung 1977b) (cf. Exercise 6). A direct full study of ELS, which investigated the risk of explosion, using stochastic stability methods specific to this algorithm, is given in (Solo 1979).

2.4.4 Conclusion and Summary of Subsection 2.4.3

We have given two guides to adaptive algorithm design, and illustrated the use of these guides via examples. These guides take a "top-down" approach; beginning with a synthetic object (a criterion), they follow a well-defined mechanism to end with an algorithm. The initial synthetic object is in general, much easier to guess and conceive than the end product. Engineers may use this first algorithm as a basis for simplifications, or well tested variations, as required; we have seen several examples of this (omission of a filter, replacement of the components of the vector field by their signs,...).

The exercises illustrate this point. Some of the algorithms given in the exercises are original, or not yet fully understood up to today (in particular, Exercises 11 and 12 on adaptive quantisation).

Finally, note that we have deliberately not covered Bayesian methods for determining algorithms; such methods are actually completely dependent on the modelling methodology, and we do not wish to touch upon this. However, we shall meet Bayesian methods (leading to a Kalman filter) incidentally, in the next chapter.

2.5 The Transient Regime

Suppose that the algorithm is intended to converge to a given value θ_*. In practice, the transient regime is taken to mean the behaviour of the algorithm away from θ_*.

We shall simply make a number of observations which are part of the folklore of adaptive processing, at the same time, we stress the primary importance of simulation.

Constant gain.

Since γ is necessarily small, the algorithm follows the ODE from the start. If we anticipate the convergence of the algorithm to a value θ_*, the transient regime (which in practice corresponds to the behaviour of the algorithm away from θ_*) is completely described by the ODE. In fact, there is no transient phase, in the mathematical sense.

Variable gain.

This presupposes nothing about the chosen asymptotic behaviour of the gain ($\gamma_n \to 0$ or not). If it is desired that $\theta_n \to \theta_*$, then it may be best to use relatively large gains in the transient phase. The trajectory of the ODE (which converges to θ_*) will then be subject to larger fluctuations, but more rapidly convergent. Specific details of this in specific case of linear systems are given by Ljung and Söderström.

2.6 Conclusion

We have given an informal description of the mathematical tools relevant to the study of convergence, which is the first stage in the study of adaptive algorithms. These tools provide the user with (generally satisfactory) qualitative information about the system. We have also taken good care to point out those things which the theorems do not tell us: thus the user is forewarned that certain types of behaviour predicted by the ODE may not be realised by the algorithm.

Then we used these tools to simplify the analysis and the algorithm design phases as far as possible, though these phases may still be complicated. However, in our experience, the algorithm design procedures described, facilitate the task in many different cases.

When you have worked through some of the exercises at the end of this chapter, you will be ready to move on to the next chapter, which considers algorithm optimisation and the tracking of non-stationary systems.

2.7 Exercises

2.7.1 Algorithms with Cyclic Update

This exercise follows on from Exercise 1.7.1, to which the reader should refer for notation.

Determine the ODE of the algorithm with cyclic update (1.7.2), and compare it with that of the original algorithm. What conclusions can be drawn?

2.7.2 An Adaptive Control Algorithm

Sequel to Exercise 1.7.2

1. Prove that the ODE of algorithm (1.7.7) is of the form

$$\dot{\theta} = \Gamma^{-1} h(\theta)$$
$$\dot{\Gamma} = R(\theta) - \Gamma$$

   and calculate $h(\theta)$ and $R(\theta)$. Can you define the ODE for arbitrary points (θ, Γ)?

2.7 Exercises

2. Show that this ODE has a unique equilibrium point $(\theta_\infty, \Gamma_\infty)$, and calculate this 2-tuple as a function of the true parameter θ_* which characterises the system.

3. We shall study the stability of the equilibrium point $(\theta_\infty, \Gamma_\infty)$ as follows:
   (i) show that we need only study the derivative $h_\theta(\theta_\infty)$;
   (ii) show that the signal $\partial y_n / \partial a_i$ (where we recall that $\theta^T = (a_1, \ldots, a_p)$) satisfies a recursion to be calculated;
   (iii) deduce that $h_\theta(\theta_\infty) = -I$.

4. What does the above analysis tell us about the original adaptive control problem? Do you think that this analysis will reassure an engineer presented with algorithm (1.7.7)?

2.7.3 Lattice Algorithms

Sequel to Exercise 1.7.3.

2.7.3.1 Lattice Gradient Algorithm (1.7.14)

(i) Show that the ODE is a differential system of the form

$$\begin{aligned}
\dot{k}(1) &= h_1(k(1)) \\
\dot{k}(2) &= h_2(k(2), k(1)) \\
&\cdots \\
\dot{k}(p) &= h_p(k(p), \ldots, k(1))
\end{aligned} \quad (2.7.1)$$

and calculate the functions h_i.

(ii) If K_* denotes the set of reflection coefficients associated with the true AR model (formulae (1.7.12)), show that K_* is the unique equilibrium point of ODE (2.7.1), and that this equilibrium point is stable.

2.7.3.2 Burg's Algorithm (1.7.15)

Imitate the above analysis for this algorithm.

2.7.4 Analysis of the Functional Associated with the Self-Adaptive Equaliser

See Subsection 2.3.2 of this chapter for notation.
Obtain formula (2.3.20) when the variable Y_n has a continuous probability density function $f(Y)$ where Y is a generic point of $\mathbb{R}^{2N+1}$.
Hint. Denote the Lebesgue measure of $\mathbb{R}^{2N+1}$ by dY, and prove that :

$$J(\theta) = \int_{Y^T\theta < -2} (Y^T\theta + 3)^2 f(Y) dY$$
$$+ \int_{-2 \leq Y^T\theta < 0} (Y^T\theta + 1)^2 f(Y) dY$$
$$+ \int_{0 \leq Y^T\theta < 2} (Y^T\theta - 1)^2 f(Y) dY$$
$$+ \int_{2 \leq Y^T\theta} (Y^T\theta - 3)^2 f(Y) dY$$

then calculate the derivative with respect to θ of the second term of this equation; don't forget that the limits of the integrals also involve θ.

2.7.5 Analysis of the Potential of Phase-Locked Loops

Notation of Paragraphs 2.4.3.1 and 1.2.2.
1. Prove formulae (2.4.8).
2. Consider the functionals J_c and J_r:
   (i) investigate their periodicity;
   (ii) let ϕ_{*c} and ϕ_{*r} denote arbitrary global minima of J_c and J_r, respectively. Show that

$$\phi_{*c} = \frac{1}{4} \arg\left((\sum_k s_k^4) e^{i4\phi_*} \right) \mod \frac{\pi}{2}$$

and that if we make the approximation $\hat{a}_n \approx a_n$ (weak error rate), then

$$\phi_{*r} \approx \phi_*$$

(ϕ_* is the true channel phase shift, cf. formula (1.2.22)). Calculate ϕ_{*c}, assuming a weak distortion $\sum_{k \neq 0} |s_k| \ll s_0$.

2.7.6 Analysis of the ODE of the Extended Least Squares ARMA Identification Algorithm

See Paragraph 2.4.3.3 for the notation.
This ODE is given by

$$\dot{\theta} = R^{-1} h(\theta)$$
$$\dot{R} = \Sigma(\theta) - R$$

where

$$h(\theta) = E(\Phi_n(\theta) e_n(\theta))$$
$$\Sigma(\theta) = E(\Phi_n(\theta) \Phi_n^T(\theta))$$

Suppose that (y_n) is an ARMA process of orders (p, q) (i.e. p and q are the minimal orders of a representation of (y_n)).

2.7 Exercises

1. Let $(\theta_\infty, R_\infty)$ be an equilibrium point of the ODE. Show that in all cases $\theta_\infty = \theta_*$.

2. Let $B_*(z^{-1}) := 1 - \sum_{j=1}^{q} b_*(j) z^{-j}$, and

$$\tilde{\Phi}_n(\theta) := \frac{1}{B_*(z^{-1})} \Phi_n(\theta)$$
$$\Sigma(\theta) := E(\Phi_n(\theta) \Phi_n^T(\theta))$$
$$\tilde{\Sigma}(\theta) := E(\Phi_n(\theta) \tilde{\Phi}_n^T(\theta))$$

Derive the formulae

$$B_*(z^{-1})(e_n(\theta) - \nu_n) = \Phi_n^T(\theta)(\theta - \theta_*)$$
$$h(\theta) = \tilde{\Sigma}(\theta)(\theta - \theta_*)$$

3. Define the function

$$V(\theta, R) := (\theta - \theta_*)^T R (\theta - \theta_*)$$

Show that

$$\frac{d}{dt} V(\theta(t), R(t)) = -(\theta(t) - \theta_*)^T$$
$$[\tilde{\Sigma}(\theta(t)) + \tilde{\Sigma}^T(\theta(t)) - \Sigma(\theta(t)) + R(t)](\theta(t) - \theta_*)$$

where $(\theta(t), R(t))$ is a trajectory of the ODE.

4. Let L be a vector in $\mathbb{R}^{p+q}$, and set $u_n(\theta) := L^T \Phi_n(\theta)$, $\tilde{u}_n(\theta) := L^T \tilde{\Phi}_n(\theta)$. Show that

$$L^T (\tilde{\Sigma}(\theta) + \tilde{\Sigma}^T(\theta) - \Sigma(\theta)) L = 2 \int_{\pi}^{-\pi} S_{u_\theta}(\omega) \operatorname{Re}\left(\frac{1}{B_*(e^{i\omega})} - \frac{1}{2}\right) d\omega$$

where $S_{u_\theta}(\omega)$ is the spectral density of the signal $u_n(\theta)$.

5. Give an argument using Lyapunov functions to deduce from Questions 3 and 4 that (θ_*, R_*) is the unique **global** attractor for the ODE, when $B_*^{-1}(z^{-1}) - \frac{1}{2}$ is **positive-real**:

$$\operatorname{Re}\left(B_*^{-1}(e^{i\omega}) - \frac{1}{2}\right) > 0 \; \forall \omega$$

2.7.7 Instrumental Variables Method

Here, the aim is to identify the (vector) parameter θ_* of the system

$$y_n = \Phi_n^T \theta_* + w_n$$

where the stationary process w_n is not observed, and is **not decorrelated** from the regression vector Φ_n. For a sample $(y_1, \Phi_1; \ldots; y_n, \Phi_n)$, the instrumental variables (IV) estimator is then given by the following formula:

$$\theta_n := \left(\frac{1}{n}\sum_{k=1}^{n} Z_k \Phi_k^T\right)^{-1} \left(\frac{1}{n}\sum_{k=1}^{n} Z_k y_k\right) \qquad (2.7.2)$$

where (Z_k) is a sequence of vectors of the same dimension as Φ_k, constructed by the user from the known observations: Z_k is called the **instrument**.

1. Express algorithm (2.7.2) in the form of an adaptive algorithm (by the method used for the least squares algorithm).
2. Calculate the ODE of this algorithm.
3. Show that if the instrument (Z_n) is chosen so that

$$R := E(Z_n \Phi_n^T) \text{ is invertible}$$
$$E(Z_n w_n) = 0$$

then θ_* is the unique equilibrium point of the ODE, and this equilibrium point is stable.

2.7.8 Adaptive Noise Suppression

In many cases, a required broad-band signal is perturbed by a narrow-band signal. We shall consider the extreme case in which:

- the required signal is a white noise $\{\nu_t\}$ $t \in \mathbb{Z}$, and
- the perturbation is a sinusoid :

$$s_t = C \cos(\omega_0 t + \Phi); \quad t \in \mathbb{Z}$$

where ω_0 is the normalised frequency (sampling period=1), and Φ is a uniform phase angle.

We wish to suppress the perturbation $\{s_t\}$ in the observed signal $y_t = s_t + \nu_t$, to leave the required signal $\{\nu_t\}$.

Part 1.

If the frequency ω_0 of the sinusoid is known, the problem is simple, we have only to identify the phase and the amplitude, and then subtract the reconstructed sinusoid. This can be implemented as shown in the following diagram, where ω_1 and ω_2 are the parameters to be tuned.

2.7 Exercises

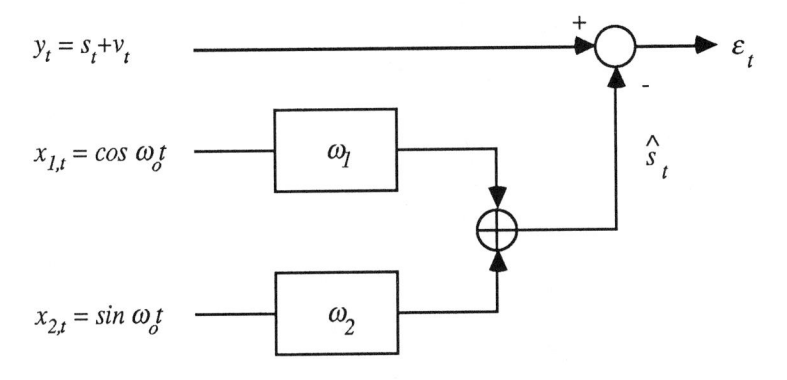

Figure 16

a. Explain why minimisation of $E(\varepsilon_t^2)$ with respect to w_1 and w_2 gives the desired solution.

b. Calculate the gradient of ε_t^2 with respect to w_1 and w_2.

c. Give a recursive stochastic gradient algorithm, with decreasing scalar gain, which minimises $E\{\varepsilon_t^2\}$.

d. Calculate the equilibrium point(s) of the algorithm, by determining the values w_1^* and w_2^* for which the mean gradient is zero.

e. Check your results using the identity: $\cos(a+b) = \ldots$.

Part 2.

In general, the frequency ω_0 is unknown, and we can approximate the previous solution using a scheme of the form:

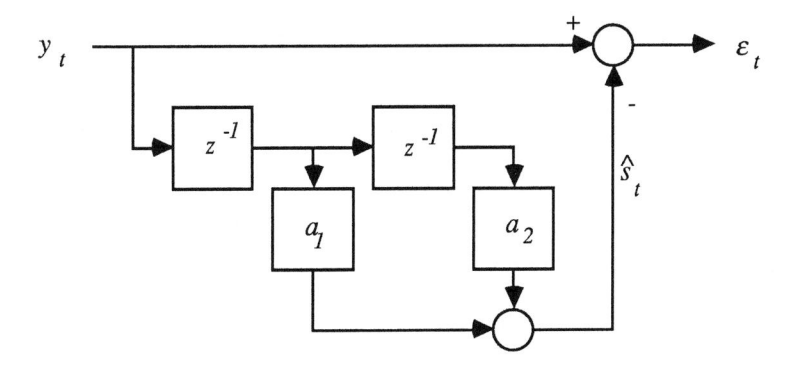

Figure 17

where $z^{-1} s_t = s_{t-1}$ and a_1, a_2 are the parameters to be determined.

a. Explain why, if the required signal is absent ($\nu_t \equiv 0$), we are in a situation analogous to the previous one (Part 1). Give the optimal values of a_1 and a_2 in this case.

b. If ν_t is not identically zero, there is a degradation of performance. Express the gradient of ε_t^2 in terms of a_1 and a_2, and give a recursive algorithm of the previous type (cf. Part 1c) to determine these parameters.

c. Calculate the equilibrium point of the algorithm.

d. Do we necessarily improve the (asymptotic) performance of the scheme by increasing the number of parameters to be tuned?:

$$s_t = a_1 y_{t-1} + a_2 y_{t-2} + a_3 y_{t-3} + \ldots$$

Part 3.

In Part 2, we used an AR(2) model to estimate/predict the perturbation

$$\hat{s}_t = z^{-1}(a_1 + a_2 z^{-1}) y_t = z^{-1} A(z^{-1}) y_t$$

We could equally well use the ARMA models:

$$s_t = z^{-1} \frac{A(z^{-1})}{C(z^{-1})} y_t$$

and interpret the proposed solutions, as a filtering problem, in the following way. Set:

$$s_t = G(z^{-1}) y_t$$
$$\varepsilon_t = (1 - G(z^{-1})) y_t = H(z^{-1}) y_t$$

Ideally, we would like a filter $G(z^{-1})$ which only passed the sinusoid, or analogously, a filter $H(z^{-1})$ which only removed the perturbation/sinusoid. For filters of comparable order, an ARMA filter $G(z^{-1})$ thus seems preferable to an AR filter.

Largely for reasons connected with the rate of convergence of identification algorithms, it is best to identify as few parameters as possible, and so, instead of using an ARMA(2,2) filter for $G(z^{-1})$, we consider only structures of the following form:

$$G(z^{-1}) = -z^{-1} \frac{C_2 + C_1 z^{-1} + z^{-2}}{1 + C_1 z^{-1} + C_2 z^{-2}}$$

2.7 Exercises

a. Calculate the zeroes of $H(z^{-1}) = 1 - G(z^{-1})$. Determine their modules.

b. Find the condition on C_1 and C_2 for which $H(z^{-1})$ eliminates the frequency ω_0 to the optimal extent.

c. Find the condition on C_1 and C_2 for $H(z^{-1})$ to be stable.

d. Express the signal recursions in the form

$$\varepsilon_t = y_t + y_{t-3} - \Phi_t^T \theta$$

with

$$\theta = \begin{bmatrix} C_1 \\ C_2 \end{bmatrix},$$

and define the components of Φ_t.

e. Use this representation to calculate the gradient of ε_t^2 with respect to θ (care!). Set:

$$\Psi_t = -\left[\frac{d\varepsilon_t}{d\theta}\right]^T$$

and define a means of generating Ψ_t as a function of Φ_t and of the estimates θ_{t-1} of θ at time $t-1$.

f. Give an algorithm to determine θ.

2.7.9 Adaptive Echo Cancellation

Two different techniques are used in telephone transmission systems:

- 2-line circuits for local networks, both directions using the same channel;
- 4-line circuits for long-distance calls, using a different 2-line circuit for each direction.

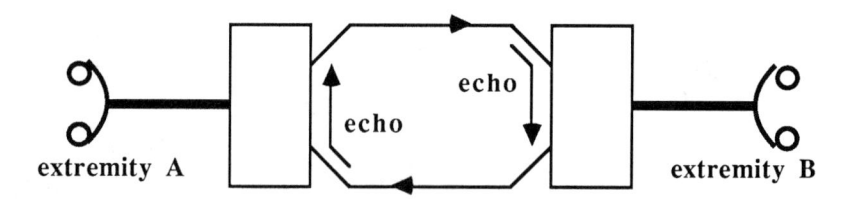

Figure 18. Echo phenomenon.

The connection between 2-line circuits and 4-line circuits is via a differential transducer, which produces an echo (as shown in Fig. 18).

An echo-canceller may be installed at Extremity A in order to prevent the formation of the echo y_n, even when the extremity B is emitting; thus the unpleasant audible echo at the extremity A may be suppressed. We now describe the system as perceived at Extremity A, with no echo-cancellation.

The signal received at A is (cf. Figure 19)

$$x_n = b_n + y_n = b_n + \sum_{k \geq 0} s_k a_{n-k}$$

where b_n is the only signal required. The echo canceller is thus an adaptive filter installed at A, which works as follows (cf. Figure 20):

$$\hat{b}_n = -\sum_{k=0}^{p} \theta(k) a_{n-k} + x_n$$

Suppose that the messages (a_n) and (b_n) are stationary, zero-mean, **decorrelated** signals.

1. Consider the functional

$$J(\theta) := E\left(x_n - \sum_{k=0}^{p} \theta(k) a_{n-k}\right)^2$$

Investigate J (convexity, localisation of minimum or minima).

2. Give a stochastic gradient algorithm to minimise J; you have an adaptive echo-canceller.

2.7 Exercises

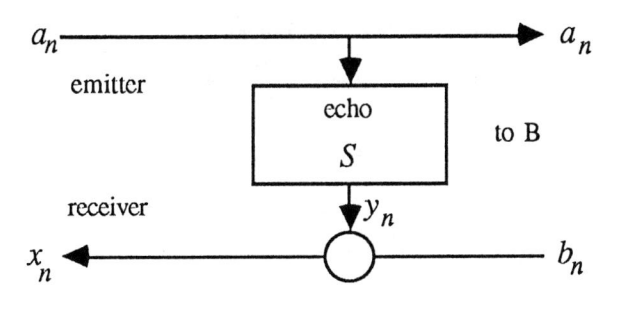

[Figure 19: diagram showing emitter a_n → to B (a_n), echo S, receiver x_n ← y_n ← b_n]

Figure 19

2.7.10 Joint Equalisation and Phase Recovery

The notation is as in Equation (1.2.22). The aim is simultaneous channel equalisation and phase recovery. We shall consider three cases; the learning phase, the self-adaptive phase, and blind equalisation.

The received signal has the following form

$$y_n = \left(\sum_{k=-\infty}^{+\infty} s_k a_{n-k} + \nu_n \right) e^{i\phi_*}$$

The receiver has the following structure

$$\begin{aligned} c_n(\theta, \phi) &= Y_n^T \theta \cdot e^{-i\phi} \\ Y_n^T &:= (y_{n+N}, \ldots, y_n, \ldots, y_{n-N}) \\ \theta^T &:= (\theta(-N), \ldots, \theta(0), \ldots, \theta(+N)) \end{aligned}$$

where y_n and θ are complex, and

$$\hat{a}_n := \text{sgn}\,[\text{Re}\,c_n] + i\,\text{sgn}\,[\text{Im}\,c_n] \qquad (2.7.3)$$

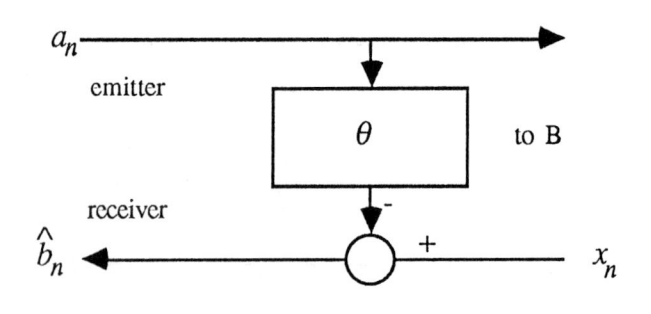

Figure 20

Part 1. *Learning phase.*

The emitted signals (a_n) are known at the receiver, and are used to tune θ and ϕ. The optimal values (θ_*, ϕ_*) are obtained by minimising the functional

$$J_1(\theta, \phi) := E|\, c_n(\theta, \phi) - a_n\,|^2$$

1. Is the optimal pair (θ_*, ϕ_*) unique?
2. Show that a stochastic gradient method of minimising J_1 leads to an algorithm of the form

$$\begin{aligned} \theta_n &= \theta_{n-1} + \gamma \overline{Y}_n e^{i\phi_{n-1}} e_n(1) \\ \phi_n &= \phi_{n-1} + \gamma \mathrm{Im}(c_n \overline{e}_n(1)) \end{aligned} \quad (2.7.4)$$

(where $\overline{z}$ is the conjugate of z), and calculate the error signal $e_n(1)$.

Part 2. *Self-adaptive equaliser.*

Proceed as usual: in the error signal $e_n(1)$, replace the true message a_n by the estimate $\hat{a}_n$ defined in (2.7.3); thus obtain the error signal $e_n(2)$. Show that the algorithm thus derived from (2.7.4) is a stochastic gradient method of minimising the functional

$$J_2(\theta, \phi) = E|\, e_n - \hat{a}_n\,|^2$$

2.7 Exercises

Part 3. *Blind equaliser.*

Here, we are interested in Quadrature Amplitude Modulation (QAM), the only difference now is that the complex message a_n satisfies

$$\operatorname{Re} a_n \in \{\pm 1, \pm 3\}$$
$$\operatorname{Im} a_n \in \{\pm 1, \pm 3\}$$

where each of the messages $\operatorname{Re} a_n$ and $\operatorname{Im} a_n$ constitutes a sequence of independent variables uniformly distributed over the set $\{\pm 1, \pm 3\}$, and the two messages $\operatorname{Re} a_n$ and $\operatorname{Im} a_n$ are independent. We will try to extend the blind equaliser, as defined for amplitude modulation, to this case.

1. Show that the right generalisation, in the complex case, of the functional (2.4.16) is

$$J_3(\theta, \phi) := E|\, c_n - \alpha \operatorname{sgn} c_n\,|^2$$
$$\alpha = 2.5$$
$$\operatorname{sgn} c_n := \operatorname{sgn}[\operatorname{Re} c_n] + i \operatorname{sgn}[\operatorname{Im} c_n]$$

   (Hint. Reduce to the real case by applying Theorem [BGR] of Chapter 2 to the sequence (b_n) given by $b_{2n} = \operatorname{Re} a_n$, $b_{2n+1} = \operatorname{Im} a_n$).

2. Deduce that the blind equaliser with phase recovery is again of the form (2.7.4), with error signal $e_n(3)$ to be determined.

Subsidiary question.

Taking into account the answer to the Question 1 of Part 1, what is, in your opinion, the advantage of joint equalisation and phase recovery?

2.7.11 Adaptive Quantisation

Given a stationary sequence of real valued variables $(x_n)_{n \in \mathbb{Z}}$ from an unknown distribution $f(x)dx$, it is desired to determine an optimal quantiser of size p (p reconstruction points) for this signal.

More precisely, a quantiser Q consists of

- p points $\theta(1), \ldots, \theta(p) \in \mathbb{R}$, called the "signal reconstruction points", and
- a function $Q : \mathbb{R} \to \{\theta(1), \ldots, \theta(p)\}$, i.e. a partition of $\mathbb{R}$.

If there is no ambiguity, we simply denote

$$\hat{x}_n = Q(x_n) \tag{2.7.5}$$

Associated with the quantiser is its **distortion**, defined by

$$J(Q) = E(x_n - \hat{x}_n)^2 \tag{2.7.6}$$

We seek to minimise this distortion.

1. Show that, for fixed $\theta^T := (\theta(1),\ldots,\theta(p))$, the best quantiser Q corresponds to the function defined by choosing the nearest neighbour:
$$Q(x) = \arg\min_{\theta(i)} (x - \theta(i))^2 \qquad (2.7.7)$$
In what follows, we shall suppose that for given θ, Q is as defined by (2.7.7); in this case, the distortion J may be viewed as function of θ.

2. Show that, if we suppose that the x_n have a continuous density function $f(x)$, then J is differentiable, and its derivative is given by
$$\frac{\partial}{\partial \theta_i} J(\theta) = -E(x_n - \theta(i) \mid \hat{x}_n = \theta(i)) \cdot P(\hat{x}_n = \theta(i))$$
Does the equation $\frac{\partial}{\partial \theta} J(\theta) = 0$ have a unique solution?

3. Deduce that the following algorithm is a stochastic gradient method of minimising J:
$$\text{if} \qquad \hat{x}_n(\theta_{n-1}) = \theta_{n-1}(i)$$
then
$$\begin{aligned}\theta_n(j) &= \theta_{n-1} \text{ for } j \neq i \\ \theta_n(i) &= \theta_{n-1}(i) + \gamma_n(x_n - \theta_{n-1}(i))\end{aligned} \qquad (2.7.8)$$
($\hat{x}_n(\theta)$ is of course the reconstructed value when the quantiser is given by θ). The algorithm described in (2.7.8) is an **adaptive quantiser**.

4. For fixed θ, set
$$\begin{aligned}a_i &= P(x_n = \theta(i)) \text{ for } i = 1,\ldots,p \\ b_i &= \frac{\theta(i+1) - \theta(i)}{2} \cdot f\left(\frac{\theta(i+1) + \theta(i)}{2}\right) \text{ for } i = 1,\ldots,p-1.\end{aligned}$$
Show that the Hessian of J is a triple-diagonal matrix given by
$$\text{Hess } J(\theta) = \begin{pmatrix} a_1 - b_1 & -b_1 & & & \\ -b_1 & a_2 - (b_1 + b_2) & \ddots & & \\ & \ddots & \ddots & \ddots & \\ & & \ddots & a_{p-1} - (b_{p-2} + b_{p-1}) & -b_{p-1} \\ & & & -b_{p-1} & a_p - b_{p-1} \end{pmatrix}$$

5. Investigate the behaviour of the algorithm when $f(x)dx$ is nearly uniform on the finite set $\{-1, 0, +1\}$, when (unfortunately), we chose $p = 2$.

2.7.12 Vector Quantisation: Lloyd's Adaptive Algorithm

Here, the problem is to extend the previous example to the case of vector variables. Thus we have a stationary sequence of variables $(X_n)_{n \in \mathbb{Z}}$ which take values in $\mathbb{R}^d$, and are distributed according to the law $f(X)dX$ ($dX = dx_1, \ldots, dx_d$) and $X = (x_1, \ldots, x_d)$), for some continuous function f.

A vector quantiser Q then consists of :

- p points $\theta(1), \ldots, \theta(p) \in \mathbb{R}^d$, and
- a function $Q : \mathbb{R}^d \to \{\theta(1), \ldots, \theta(p)\}$.

Here again, if there is no ambiguity, we denote

$$\hat{X}_n := Q(X_n)$$

It is again desired to minimise the distortion associated with the quantiser Q:

$$J(Q) := E\|\hat{X}_n - X_n\|^2 \qquad (2.7.9)$$

1. Show that given $\theta^T = (\theta^T(1), \ldots, \theta^T(p))$, the optimal choice of Q corresponds to the nearest neighbour function given by

$$Q(X) = \arg\min_i \|X - \theta(i)\|^2 \qquad (2.7.10)$$

In what follows, Q will be chosen to be a function of θ, in accordance with (2.7.10). The distortion is then also a function of θ, which we denote by $J(\theta)$.

2. Show that we again have

$$\frac{\partial}{\partial \theta(i)} J(\theta) = -E(X_n - \theta(i) \mid \hat{X}_n = \theta(i)) \cdot P(\hat{X}_n = \theta(i)) \qquad (2.7.11)$$

3. Deduce that the following is an adaptive algorithm to minimise J: if

$$\hat{X}_n(\theta_{n-1}) = \theta_{n-1}(i)$$

then

$$\begin{aligned}\theta_n(j) &= \theta_{n-1}(j) \quad \text{for } j \neq i \\ \theta_n(i) &= \theta_{n-1}(i) + \gamma_n(X_n - \theta_{n-1}(i))\end{aligned}$$

This algorithm is an adaptive version of Lloyd's algorithm (also known as [LBG]), see (Gray 1984).

2.7.13 Large Deviations in ALOHA: a Counterexample to the Predictions of the ODE

ALOHA was the first protocol for satellite data transmission which used a probabilistic rule to allocate the transmission bearer to users. The principle of ALOHA is as follows. A satellite transmission channel is to be shared amongst a very large number M of users. At each discrete time $n, n+1, \ldots$, each user may wish to send a packet of data. If two users emit at the same time, the two packets are lost, and re-emission is necessary; the users in question are said to be *blocked*.

The originality of ALOHA lies in the nature of the re-emission protocol, the objective of which was to avoid a collapse of the system due to an excess of simultaneous requests to emit. Here is the protocol. *When a user is blocked at time n_0, he tosses a coin, which is weighted with probability $p = P\{heads\} \ll 1/2$, and he re-emits the packet when he throws a head.* The tosses of different users are considered to be independent.

To complete the description of the system, we also need a model of the new emissions. Such a model is again given by tosses of a weighted coin with bias $q \ll p$, where the users are again considered to be independent.

1. Denote the number of blocked users at time n by θ_n. Show that θ_n satisfies a recursion of the form

$$\theta_n = \theta_{n-1} + H(\theta_{n-1}, X_n) \qquad (2.7.12)$$

where X_n denotes the outcome of the set of all the tosses (both emissions and re-emissions) at time n. Prove that θ_n is a Markov chain.

2. Calculate the distribution of the random variable $H(\theta, X_n)$, for fixed θ.

3. We are interested in the case in which there is a large number of users. In this case, we introduce the following asymptotic framework

$$M \to \infty, \qquad Mq \to \alpha, \qquad \alpha > 0 \text{ fixed} \qquad (2.7.13)$$

Prove that in this case we have

$$\begin{aligned}
P[H(\theta, X_n) < -1] &= 0 \\
P[H(\theta, X_n) = -1] &= \theta p (1-p)^{\theta-1} e^{-\alpha} \\
P[H(\theta, X_n) = 0] &= (1-p)^\theta \alpha e^{-\alpha} + [1 - \theta(1-p)^{\theta-1}] e^{-\alpha} \\
P[H(\theta, X_n) = 1] &= [1 - (1-p)^\theta] \alpha e^{-\alpha} \\
P[H(\theta, X_n) = k] &= \frac{\alpha^k}{k!} e^{-\alpha} \text{ for } k > 1
\end{aligned} \qquad (2.7.14)$$

4. Prove that for α and p sufficiently small, the convergence theorems of this chapter may be assumed to apply, i.e. a gain of 1 is sufficiently small that algorithm (2.7.12) is well approximated by its ODE. Consider the ODE, and prove that it has stable equilibrium points

   - θ_*, with domain of attraction $[0, \theta_c)$ where $\theta_c \gg \theta_*$;
   - $+\infty$, with domain of attraction $(\theta_c, +\infty)$.

5. Show that the Markov chain θ_n is *transient*. Proceed as follows. Suppose that the Markov chain has an invariant probability $\pi = (p_k)_{k \geq 0}$.

   a. Prove that this assumption is equivalent to the following conditions

   $$(i) \quad e^\alpha p_k = p_k + \alpha p_{k-1} + \sum_{i=2}^{k} \frac{\alpha^i}{i!} p_{k-i}$$
   $$+ (k+1)p(1-p)^k p_{k+1}$$
   $$- kp(1-p)^{k-1} p_k$$
   $$- \alpha(1-p)^k p_k$$
   $$- \alpha(1-p)^{k-1} p_{k-1}$$
   $$(ii) \quad \sum p_k = 1, \quad p_k \geq 0 \qquad (2.7.15)$$

   b. In (2.7.15-i), sum the equalities for 1 to K, and deduce that

   $$(K+1)p(1-p)^{K+1} e^{-\alpha} p_{K+1} \geq p_K(1-e^{-\alpha}) \qquad (2.7.16)$$

   and thus that (2.7.16) and (2.7.15-ii) are contradictory.

6. What conclusions can you draw from a comparison of the answers to Questions 4 and 5?

2.7.14 Neural Networks and Gradient Back-Propagation

The aim of this exercise is to construct a family of functions, with parameter θ (or sometimes Θ) in $\mathbb{R}^d$, which will allow us to represent any function

$$\Phi : \{0,1\}^N \to \{0,1\}^M \qquad (2.7.17)$$

from an N-cube to an M-cube. In what follows, such a function will be said to be *symbolic*. Any function of the form

$$g_\theta : \{0,1\}^N \to \{0,1\}$$
$$g_\theta(X) = \mathbf{1}_{\{\theta^T X > \frac{1}{2}\}}, \quad \theta \in \mathbb{R}^N \qquad (2.7.18)$$

where $\mathbf{1}_\Pi$ takes the value 1 if the predicate Π is satisfied, and 0 otherwise, is said to be a *neuron*. The components of the vector θ are called the *synaptic weights*. A neuron may be represented by a directed graph (its "block diagram"), with input nodes the components of the input X, and output node $y = g_\theta(X)$, with edges joining each input node to this output node. Thus a *neural network* is a finite, cycle-free, directed graph, obtained by assembling the neurons two-by-two, identifying the output of one with the inputs to the other, as shown in the diagram below (analogies with the brain strongly implied).

This problem will examine how neural networks may be used to represent symbolic functions, and how such a network may be "taught".

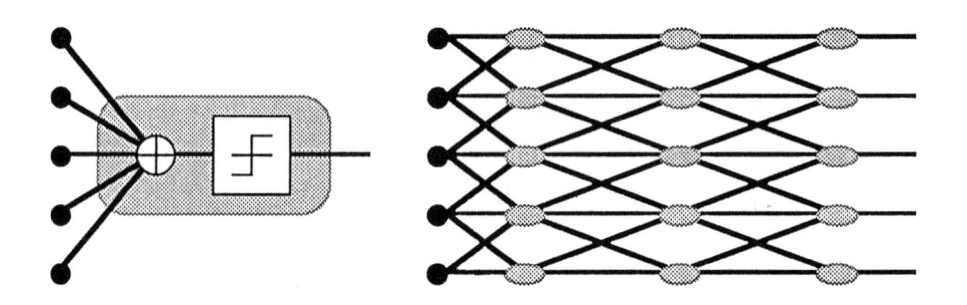

Figure 21. A network of neurons.

Preliminaries.

1. Show that any symbolic function from the N-cube to the M-cube may be represented by a symbolic function Φ from the $(N+1)$-cube to the M-cube, which satisfies the additional property $\Phi(0,\ldots,0) = (0,\ldots,0)$. In what follows, we shall only consider functions of this type: from now on, these will be called *symbolic functions*.

2. Suppose $M = 1$. Does the family of *neurons* (functions of the form (2.7.18)) allow us to represent all symbolic functions from the N-cube to $\{0,1\}$? (Hint. Consider the case N=2).

3. Show that any vertex X of the N-cube may be separated from the other vertices of the N-cube by a hyperplane with equation

$$\theta^T Y = \frac{1}{2}$$

In the sequel, θ_X will denote the parameter θ of such a hyperplane.

4. Set $M = 2^N$. Let

$$G_\Theta : \{0,1\}^N \to \{0,1\}^M$$

2.7 Exercises

denote the function whose coordinates are the neurons g_{θ_X}, where X runs over the vertices of the N-cube. Show that any symbolic function Φ of the N-cube to $\{0,1\}$ may be represented in the form of a 2-layer neural network

$$X \xrightarrow{G_\Theta} \mathbb{Z} \xrightarrow{g_\theta} y = \Phi(X) \qquad (2.7.19)$$

where G_Θ was defined above, and where g_θ is a neuron from the M-cube into $\{0,1\}$, to be defined.

Preliminary conclusions.

In what follows, any function from the N-cube to the M-cube whose coordinates are neurons will be called a *layer of neurons*. The main point to be made is that in order to represent arbitrary symbolic functions, a neural network requires (at least) two layers. We shall study networks with an arbitrary number (P) of layers, and we shall discuss techniques for determining the synaptic weights.

In what follows, we shall assume that we are given a sequence of independent, identically distributed random variables (X_n), on the N-cube, with distribution μ. We suppose also that we observe the pair (X_n, Y_n), where Y_n is the image of X_n under the (unknown) symbolic function Φ which we desire to learn.

Part 1. *Use of a single layer* $(P = 1)$.

We suppose here that the symbolic function to be learnt may be represented by 1 neuron (this is not in general the case, as we saw above). In this case, the output Y_n is a scalar.

1. Give a stochastic gradient algorithm to minimise the functional

$$J(\theta) = E[Y_n - \theta^T X_n]^2$$

   and investigate this algorithm. If the algorithm converges to an equilibrium point θ_*, is the corresponding neuron g_{θ_*} equal to Φ?

2. Next we consider the following algorithm for the same purpose as in 1.

$$\theta_n = \theta_{n-1} + \gamma_n X_n e_n(\theta_{n-1}) 1_{\{g_{\theta_{n-1}}(X_n) \neq Y_n\}}$$
$$e_n(\theta) = Y_n - X_n^T \theta$$

   Correction is made only when the input/output correspondence is incorrect. Investigate this algorithm, and associate it with the functional

$$J(\theta) = E[u(\theta; X_n, Y_n)]$$
$$u(\theta; X, Y) = \{[Y + X^T\theta - 2YX^T\theta - \frac{1}{2}]_+ + \frac{1}{2}\}^2$$

   where $z_+ = \max\{z, 0\}$.

Part 2. *Arbitrary P, use of sigmoid functions.*

In order to avoid singularities of the gradient field, we shall now regularise the sign which appears in the definition of neurons. Thus, any function of the form

$$\hat{g}_\theta(X) = s[X^T\theta]$$
$$s[x] = \frac{1}{1+\exp[-\frac{1}{\epsilon}(x-\frac{1}{2})]}$$

such that $\mathbf{1}_{\{x>\frac{1}{2}\}}$ is regularised by $s[x]$ for $\epsilon > 0$ small, will be called a *lazy neuron* (such a neuron reacts slowly). Note that

$$s'[x] = \frac{1}{\epsilon} s[x](1-s[x])$$

We also introduce the notion of a *lazy layer of neurons* $\hat{G}_\Theta$, which is a layer of neurons as previously defined, in which all the neurons are lazy.

1. Here we use a 2-layer network of lazy neurons, as given by (2.7.19). Give a stochastic gradient algorithm to minimise

$$J[\Theta; \theta] = E(Y_n - \hat{g}_\theta(Z_n))^2$$
$$Z_n = \hat{G}_\Theta(X_n)$$

   and investigate this algorithm.

2. Generalise this method to case of an arbitrary number of (lazy) layers. Explain the term *gradient back-propagation* which is given to such an algorithm. Proceed as follows. Denote the successive layers of neurons by

$$X = Z^{(0)}, Z^{(1)}, \ldots, Z^{(k)}, \ldots, Z^{(K)} = y$$

   The essential feature in the calculation of the gradient is the quantity

$$\frac{\partial Z_n^{(K)}}{\partial w}$$

   where w denotes one of the synaptic weights of the network, for the layer k_*. In order to calculate this quantity, we let Z_n^* denote the output of the neuron which involves the given weight w. Regarding the network as a directed graph, we let

$$\lambda = Z^*, Z^{(k_*+1)}_{\lambda(k_*+1)}, \ldots, Z^{(l)}_{\lambda(l)}, \ldots, y$$

   denote a path from Z^* to y; Λ^* will denote the set of all such paths. Prove that (dropping the time index)

$$\frac{\partial Z^{(K)}}{\partial w} = \sum_{\lambda \in \Lambda^*} \left[\prod_{l=k_*+1}^{K} \frac{\partial Z^{(l)}_{\lambda(l)}}{\partial Z^{(l-1)}_{\lambda(l-1)}} \right] \frac{\partial Z^*}{\partial w} \qquad (2.7.20)$$

2.7 Exercises

Derive the formulae

$$\frac{\partial \hat{g}_\theta(X)}{\partial \theta} = \frac{1}{\epsilon} X \hat{g}_\theta(X)[1 - \hat{g}_\theta(X)]$$

$$\frac{\partial \hat{g}_\theta(X)}{\partial X} = \frac{1}{\epsilon} \theta \hat{g}_\theta(X)[1 - \hat{g}_\theta(X)]$$

Lastly, use a dynamic programming argument to calculate (2.7.20) for layers of decreasing index.

Part 3. *Arbitrary P, randomised neurons.*

In Part 2, we used lazy neurons to avoid singularities in the mean field; but we know that there are cases in which the singularities due to quantisers are smoothed thanks to the presence of regular probability distributions for the random fields. We shall investigate this possibility. By a *randomised neuron*, we mean a random variable with values in $\{0, 1\}$ which has the form

$$\tilde{g}_\theta(X) = \begin{cases} 1 \text{ with probability } \hat{g}_\theta(X) \\ 0 \text{ with probability } 1 - \hat{g}_\theta(X) \end{cases}$$

where $\hat{g}_\theta$ is the lazy neuron introduced above. Show that, although this algorithm is different from that in Part 2, it has the same ODE.

2.7.15 Markov Fields on Graphs, and Simulated Annealing Algorithms

Part 1. *Markov fields on graphs.*

We begin by setting the scene (this is a little long).

The graph.
In this exercise, we assume we have a *finite (undirected) graph* $\mathcal{G} = \{\mathcal{T}, \mathcal{E}\}$, where $\mathcal{T}$ is the set of vertices or *sites* of the graph, and $\mathcal{E}$ is the set of edges. Sites will be denoted generically by the letter t. Two sites t and t' are said to be *neighbours* if they are connected by an edge. The set of neighbours of a site t will be denoted by N_t. The subset $\mathcal{C}$ of $\mathcal{T}$ is said to be a *clique*, if every pair of sites of $\mathcal{T}$ are neighbours.

Configurations.
Every site t is labelled by a symbol ω_t from a finite alphabet S. The set ω of ω_t for all sites t is said to be a *configuration*. For $A \subset \mathcal{T}$, we denote the restriction of ω to sites in A by ω_A. Lastly, we set

$$\omega_t^A = \omega_t \text{ if } t \in A, \ o \text{ otherwise,}$$

where o is a distinguished element of the alphabet S.

Examples.

The most common examples are the finite subsets of the graphs $\mathbb{Z}^d$, in which two sites are neighbours if the sum of the pairwise differences of their coordinates is ≤ 1 (2^d neighbourhood). The most popular case is $d = 2$, the alphabet S being the set of grey levels; a configuration is then an *image*.

However, we might also consider the case of a *circuit-free, directed graph*, in which the set of labels of S lying along a given path is interpreted as a *word*; this corresponds to the case of speech recognition.

Random field on $\{\mathcal{G}, S\}$.

We denote the set of configurations by Ω. A *random field* on $\{\mathcal{G}, S\}$ is simply a probability measure P on Ω. The field is said to be *Markov* if the condition

$$P(\omega_t | \omega_{T-\{t\}}) = P(\omega_t | \omega_{N_t})$$

holds, i.e. the conditional probability of ω_t given the restriction of the configuration to the other sites, depends only on the restriction to the neighbours of t.

Potential, energy, local potential.

A family of functions $(V_A(\omega))$, such that V_A depends only on ω_A, where A runs over the set of subsets of T, is said to be a *potential*. With each potential, we associate an *energy function*

$$U(\omega) = -\sum_A V_A(\omega)$$

A potential (V_A) is said to be *local* if $V_A = 0$ whenever A is not a clique.

Gibbs measure associated with a potential.

We associate with each potential (V_A) a probability

$$P(\omega) = \frac{1}{Z} e^{-U(\omega)}$$
$$Z = \sum_\omega e^{-U(\omega)}$$

Here Z is the *partition function* of the field, and P is the *Gibbs measure* associated with the potential.

1. Show that if the potential is local, the Gibbs measure is that of a Markov field.

2. Derive the combinatorial formula

$$\sum_{A \supset B} (-1)^{|A-B|} = 1 \text{ if } B = T, \quad 0 \text{ otherwise}$$

where A and B are subsets of T, and $|\ldots|$ denotes the cardinality of the given set.

2.7 Exercises

3. Given a random field P, show that the *inversion formula*

$$V_A(\omega) = \sum_{B \subset A} (-1)^{|A-B|} \log P(\omega^B)$$

defines a potential associated with P.

4. Suppose now that the field P is Markov. Derive the formula

$$V_A(\omega) = \sum_{B \subset A} (-1)^{|A-B|} [\log P(\omega_a^B \mid \omega_{N_a}^B) + \log P(\omega_{T-\{a\}}^B)]$$

where a is a fixed point in A. Deduce that the potential of this field is local, and that it is given by the formula

$$V_{\mathcal{C}}(\omega) = \sum_{B \subset \mathcal{C}} (-1)^{|\mathcal{C}-B|} \log P(\omega_a^B \mid \omega_{N_a}^B), \quad a \in \mathcal{C}$$

where $\mathcal{C}$ is a clique. Thus we have shown that a potential is local if and only if the field which it defines is Markov. Note that the potential defined in the above formula is normalised by the constraint $V_A(\varrho) = 0$ where ϱ denotes the constant configuration equal to o everywhere.

Part 2. *Analysis of a particular Markov chain: the Gibbs sampler.*

We consider the following Markov chain defined on a finite state space Ω. We are given an *energy function* U on Ω, and an *irreducible, symmetric* matrix of transition probabilities Q, with zero entries on its diagonal (i.e. $Q(\omega,\omega) = 0$). We suppose also that we are given a parameter $T > 0$ called the *temperature*. The Markov chain (ω_n) is then given by the following algorithm

- given ω_{n-1},
- select a random w from the distribution $Q(\omega_{n-1}, w)$,
- select an exponential random variable Z_n, independent of the previous random variables, and set

$$\omega_n = w \text{ if } U(w) - U(\omega_{n-1}) \leq T Z_n \text{ otherwise } \omega_n = \omega_{n-1}$$

1. Calculate the transition probability $P(\omega, \omega')$ of this Markov chain.
2. Show that it has a unique invariant probability given by

$$\pi(\omega) = \frac{1}{Z} e^{-\frac{U(\omega)}{T}}$$
$$Z = \sum_{\omega} e^{-\frac{U(\omega)}{T}}$$

Determine the limit of this invariant probability as $T \to 0$.

Part 3. *Identification of Markov fields on graphs.*

Here, we are given a graph $\mathcal{G} = \{\mathcal{T}, \mathcal{E}\}$, and a Markov field $P^{\theta*}$ on this graph, with an associated local potential $V_C^{\theta*}$. This field is assumed to be a regular function of $\theta \in \mathbb{R}^d$, in the sense that the function $\theta \to V_C^\theta(\omega)$ is of class C^1 for any configuration ω. We observe a "perturbed configuration" w, which has the conditional distribution

$$P(w \mid \omega) = \exp[-\sum_t h(w_t, \omega_t)] \tag{2.7.21}$$

The aim in this part is to define and study an algorithm to determine the unknown parameter θ_* which is taken to represent the "true" model for the underlying Markov field.

Example.

We refer back to out earlier discussion of images, and assume now that ideal images correspond to a model of the form

$$V_C^\theta = \sum_{i=1}^d \theta(i) V_C^i \tag{2.7.22}$$

for some given family of potentials V_C^i and parameter $\theta \in \mathbb{R}^d$. A perturbed version w of an image associated with this model is observed, and it is desired to determine the corresponding "true" value of θ_* which will enable us to "restore the image", i.e. to reconstruct an estimate of the unperturbed image $\hat{\omega}$.

Notation.

We set

$$U_\theta(\omega) = -\sum_{C:clique} V_C^\theta(\omega)$$
$$W_\theta(\omega) = \sum_{t \in \mathcal{T}} h(w_t, \omega_t) - \sum_{C:clique} V_C^\theta(\omega)$$

and we denote the Gibbs measure of the Markov field defined by the energy function W_θ (as described above) by P_W^θ. Lastly, we denote the normalisation constant appearing in P_W^θ by Z_W^θ.

1. Show that $P_W^\theta(\omega)$ is the conditional probability of the configuration ω, given the observation w.

2. We wish to determine θ_* using maximum likelihood methods. Show that for the exponential model (2.7.22) we must solve the equation

$$Z_W^\theta - \sum_{C:clique} V_C(w) = 0 \tag{2.7.23}$$

2.7 Exercises

where V_C denotes the vector with coordinates V_C^i, and that in the general case we must solve the equation

$$E_W^\theta[\frac{\partial}{\partial \theta}W_\theta] - \frac{\partial}{\partial \theta}W_\theta(w) = 0 \qquad (2.7.24)$$

Of course the difficulties are associated with the first term of the left-hand side of this equation, the calculation of which requires exhaustive exploration of the configurations (a practically impossible task).

3. Following Younes, we shall use a stochastic gradient method to estimate the constants, with the help of a Gibbs sampler. First of all we let $P_\theta(\omega, \omega')$ denote the transition probability of the Gibbs sampler, associated (as indicated in Question 1 of Part 2) with the energy function $W_\theta(\omega)$, for temperature $T = 1$, and for $Q(\omega, \omega')$ with law

$$Q(\omega, \omega') = \pi(\omega_t, \omega'_t)\mu(t)1_{\{\omega_s = \omega'_s, s \neq t\}}$$

where μ is a probability distribution on the set of sites $\mathcal{T}$, and π is a symmetric, irreducible matrix of transition probabilities with zero diagonal on the alphabet S.

The following algorithms are proposed for the exponential case (corresponding to (2.7.23))

$$\theta_n = \theta_{n-1} - \gamma_n(\sum_{C:clique} V_C(\omega_n) - \sum_{C:clique} V_C(w))$$
$$P[\omega_n \mid \omega_{n-1}, \ldots; \theta_{n-1}, \ldots] = P_{\theta_{n-1}}(\omega_{n-1}, \omega_n)$$

and for the general case (corresponding to (2.7.24))

$$\theta_n = \theta_{n-1} - \gamma_n(\frac{\partial}{\partial \theta}W_{\theta_{n-1}}(\omega_n) - \frac{\partial}{\partial \theta}W_{\theta_{n-1}}(w))$$
$$P[\omega_n \mid \omega_{n-1}, \ldots; \theta_{n-1}, \ldots] = P_{\theta_{n-1}}(\omega_{n-1}, \omega_n)$$

Study these algorithms, and verify that they carry out the desired function, i.e. that they converge to the maximum likelihood estimator $\hat{\theta}$ of θ_*.

Part 4. *The Boltzmann machine case, estimation of the true parameter θ_*.*

In Part 3, we simply showed how to calculate the maximum likelihood estimator $\hat{\theta}$ of the unknown parameter, based upon the observation of a single noisy sample w of the random field. Here, in the Boltzmann machine case, which we shall see is a special case of the exponential model, we shall determine the "true" parameter from a sequence of independent, noise-free observations (w_n) of the field. Firstly we introduce the *Boltzmann machine* (or *spin glass*) model. This is a particular case of the model introduced in (2.7.22) and above.

Here, the alphabet S is the set of *spins*, represented by $\{-1, +1\}$. The energy U_θ is given by the following quadratic form in ω

$$U_\theta = \sum_t \theta_t \omega_t - \frac{1}{2} \sum_{s,t} \theta_{s,t} \omega_s \omega_t$$

where $\theta_{t,t} = 0$ and θ represents the collection of the θ_t and the $\theta_{s,t}$ for all sites s and t.

1. Suppose we have a noise-free observation $w = \omega_0$ of the field corresponding to the unknown value θ_* of the parameter which it is desired to determine. Use the algorithm of Part 3 to calculate the maximum likelihood estimator for this case (note that the conditional distribution of the observations (2.7.21) is here given by $P(w \mid \omega) = \delta_\omega(w)$ where δ is the Dirac measure).

2. Next suppose we have a sequence of independent, noise-free observations (w_n) of this field. Suggest an amendment to the previous algorithm, which will ensure that the estimator θ_n converges to the *true* value θ_*. This is a delicate question with a number of possible answers. Try two solutions.

   The first solution should make simultaneous use of 2 copies of the graph $\mathcal{G}$ associated with the field. The first copy will serve to receive the observations w_n, whilst the second copy should be used by the Gibbs samplers to determine the ω_n.

   The second solution should only use one copy of the graph $\mathcal{G}$. This copy should be used to receive alternate "bursts" of the w_n (during the so-called *learning* phases) and of the ω_n (during the so-called *hallucination* phases), according to a protocol to be described.

2.8 Comments on the Literature

General Comments.

In the 1950s, following the article (Robbins and Monro 1951), considerable advances in the study of stochastic approximations for independent vector fields were made by the Russian school; (Tsypkin 1971) is a basic reference work on this topic.

This work led to an initial series of general theorems on stochastic approximations for stationary, possibly dependent vector fields with parameter θ. (Khas'minskii 1966) gives a convergence result for algorithms with constant gain and smooth dependent vector fields with parameter θ (this result is given for continuous-time stochastic approximations, and is presented

2.8 Comments on the Literature

as a stochastic version of the "averaging principle" in mechanics (Bogolyubov and Metropol'skii 1961)). The same mixing assumptions are then weakened in (Borodin 1977); see (Billingsley 1968) for various information about mixing. In the same vein (Derevitskii and Fradkov 1974) give convergence results for algorithms of constant step size, for independent vector fields with parameter θ, possibly including discontinuities. This article is important, insofar as it gives results for both finite and infinite horizons. We must also refer to the important work of Kushner and Huang, which is fully described in (Kushner 1984). In parallel, (Benveniste, Goursat and Ruget 1980b) give the first example of a convergence result adapted for use in signal processing (i.e. simultaneously taking into account the dependencies of the vector field and the discontinuities due for example to the use of quantisers); see also (Eweda and Macchi 1983, 1984a,b) for a more complete analysis of the least squares algorithm with constant scalar gain.

L.Ljung was the first to formulate a general convergence result in a Markov type setting, more exactly, for the case of conditionally linear dynamics, with decreasing gains. The essential references are the original articles (Ljung 1977a,b) and (Ljung 1978), which describes extensive applications of the method, and lastly the book (Ljung and Soderström 1983); also, we should refer to the article (Davis and Vinter 1985). In generalising and systematising the Markov approach, (Métivier and Priouret 1984) gave general results for algorithms with decreasing gain; a similar formulation is also found in (Kushner 1984). These results only apply to smooth (and in general Lipschitz) vector fields. The first joint investigation of both the Markov approach and of the discontinuities was the excellent article (Kushner and Shwartz 1984), where an elegant weak convergence technique was used for algorithms with decreasing step size, and for algorithms with constant step size in a finite horizon. The method used in Part II of this book, for the same assumptions is different; it depends upon *Poisson's equation* for Markov chains. This method immediately exhibits a martingale with all its accompanying limit theorems. We note, in passing, the renewed interest in deterministic averaging methods using singular perturbation techniques. This approach is fully described in (Anderson et al. 1986), which also contains results similar to those of this chapter, and which introduces the possibility of defining the concept of the ODE (which the authors call "slow manifold") for values of the tunable parameter for which the state X_n is explosive.

We must also mention papers which give a direct treatment of difficult problems. For a full treatment of the boundedness condition, we refer to the papers (Eweda and Macchi 1983, 1984a,b, 1985) on the least squares algorithm. On the risk of explosion, we refer for example to (Solo 1979, 1981) (for convergence of the ELS) and to (Goodwin, Ramadge and Caines 1980) (for an adaptive control algorithm); these papers use random, Lyapunov function methods, adapted to the problems in hand.

References Used in this Chapter.

Sections 2.2, 2.3 and 2.4 are original, except for the examples. The phase-locked loop example is taken from (Benveniste, Vandamme and Joindot 1979), the blind equaliser example is taken from (Benveniste, Goursat and Ruget 1980a) and from (Benveniste and Goursat 1984), and the AR or ARMA identification algorithm is from (Ljung and Soderström 1983).

As for the exercises, Exercise 2.7.2 is from (Ljung and Soderström 1983), 2.7.3 follows (Benveniste and Chauré 1981), 2.7.5 relies heavily on (Benveniste, Vandamme and Joindot 1979), 2.7.6 is from (Ljung and Soderström 1983), and 2.7.8 was supplied to us by J.J.Fuchs. Exercise 2.7.12 is taken from (Delyon 1986). Exercise 2.7.13 was largely given to us by M.Cottrell and J.C.Fort, see (Cottrel, Fort and Malgouyres 1983), (Azencott 1980) and (Kushner 1984) for more information on large deviations. Exercise 2.7.14 concerns a subject in vogue: "neural networks"; here, we give an example of the so-called back-propagation algorithm, in order to show that in our opinion, many aspects of this theory are related to general results on stochastic approximations given in this book. The details of the algorithm are original; for more detailed background information, refer to the excellent collective work (Fogelman-Soulié, Robert and Tchuente 1987). Lastly, Exercise 2.7.15, which deals with learning problems for Gibbs fields on graphs, comes from an idea due to (Younes 1988a,b). The rapid description of Gibbs fields is taken directly from (Gelfand 1987), other useful references on this subject are (Hajek 1985), (Kinderman and Snell 1980) and (Prum 1986).

Chapter 3
Rate of Convergence

In this chapter, the rate of convergence of the algorithm to its ODE and/or to the desired value θ_* is described in more detail. The analysis is again asymptotic.

The first section provides an informal description of the requisite mathematical tools, namely various results about asymptotic normality. These results allow us to observe the behaviour of the difference $\theta_n - \theta(t_n)$ between the algorithm and its ODE as though through a magnifying glass. On this occasion, we shall see that the behaviour of algorithms with constant gain is qualitatively different from that of algorithms with decreasing gain.

Then we will be ready to tackle our first problem; the optimisation of algorithms with decreasing gain for the identification of a fixed parameter θ_*.

This chapter is thus a chapter of transition, designed to take us to the heart of the matter, which will be discussed in Chapter 4, namely the tracking of time-varying parameters.

Throughout this chapter, the notation and the objects are as before; for reference, see either the beginning of Chapter 2, or the final summary (statement of Assumptions (A)) at the end of Chapter 1.

3.1 Mathematical Tools: Informal Description

Since the corresponding results are very different in nature, we shall separate those relating to algorithms with constant gain from those concerning algorithms with decreasing gain. The particular importance of algorithms with constant gain, which are able to track non-stationary parameters must be emphasised.

3.1.1 Algorithms with Constant Gain

We now suppose that
$$\gamma_n = \gamma \qquad \forall\, n \qquad (3.1.1)$$
Later, we shall indicate when and how this assumption may be weakened as in the case of the convergence results given in the previous chapter. As usual, we study the asymptotic behaviour as $\gamma \to 0$.

3.1.1.1 Heuristics

For simplicity, we consider here the case in which there is no complementary term ε_n. Recall that, from (3.1.1), we have $t_n = n \cdot \gamma$. The notation is as in Chapter 2. We have

$$\begin{aligned}\theta_n - \theta(t_n) = \theta_{n-1} - \theta(t_{n-1}) &+ \gamma(H(\theta_{n-1}, X_n) - H[\theta(t_{n-1}), X_n]) \\ &+ \gamma(H[\theta(t_{n-1}), X_n] - h[\theta(t_{n-1})]) \\ &- (\theta(t_n) - \theta(t_{n-1}) - \gamma h[\theta(t_{n-1})])\end{aligned} \quad (3.1.2)$$

where $\theta(t)$ is the solution of the ODE with the initial condition $z = \theta_0$. We now set

$$\tilde{\theta}_n^\gamma = \gamma^{-\frac{1}{2}}[\theta_n - \theta(t_n)] \quad (3.1.3)$$

This is the difference between the algorithm and the ODE magnified by a factor of $\gamma^{-\frac{1}{2}}$. If we introduce $\tilde{\theta}_n^\gamma$ into (3.1.2), and sum from 1 to n, we obtain

$$\begin{aligned}\tilde{\theta}_n^\gamma &= \tilde{\theta}_0^\gamma \\ &+ \gamma^{\frac{1}{2}} \sum_{k=0}^{n-1}(H(\theta_k, X_{k+1}) - H[\theta(t_k), X_{k+1}]) \\ &+ \gamma^{\frac{1}{2}} \sum_{k=0}^{n-1}(H[\theta(t_k), X_{k+1}] - h[\theta(t_k)]) \\ &- \gamma^{-\frac{1}{2}}\left(\theta(t_n) - \theta(0) - \gamma\sum_{k=0}^{n-1} h[\theta(t_k)]\right) \\ &= 0 + [a] + [b] + [c]\end{aligned} \quad (3.1.4)$$

We shall study the three terms separately.

Term [c].

This term is of the form $\gamma^{-\frac{1}{2}} \cdot \varepsilon(\gamma)$, where $\varepsilon(\gamma)$ is the error arising from the discretisation of the ODE; since $\varepsilon(\gamma)$ is of the same order as γ, [c] tends to 0 with γ, and when we look at the terms [a] and [b], we shall see that we may neglect [c].

Term [a].

We use the first order approximation

$$H(\theta_k, X_{k+1}) - H[\theta(t_k), X_{k+1}] \approx H_\theta[\theta(t_k), X_{k+1}] \cdot (\theta_k - \theta(t_k)) \quad (3.1.5)$$

where

$$H_\theta(\theta, X) := \frac{\partial}{\partial \theta} H(\theta, X) \quad (3.1.6)$$

Approximation (3.1.5) presupposes that $H(., X)$ is sufficiently smooth; we shall see however, that once again the theorems tolerate discontinuities. Thus

3.1 Mathematical Tools: Informal Description

from (3.1.5), and using (3.1.3), we have

$$[a] \approx \gamma \sum_{k=0}^{n-1} H_\theta[\theta(t_k), X_{k+1}] \cdot \tilde{\theta}_k^\gamma \tag{3.1.7}$$

Using the same argument as for the approximations of formula (2.2.1) (convergence heuristics), we may replace

$$H_\theta(\theta(t_k), X_{k+1})$$

in (3.1.7) by

$$h_\theta(\theta(t_{k+1})) = E_{\theta(t_{k+1})}[H_\theta(\theta(t_{k+1}), X_{k+1})] \tag{3.1.8}$$

Once again, we have used the fact that $\theta(t)$ is slowly varying with respect to the random X_{k+1}, whence a law of large numbers may be applied. Suppose now that t is a fixed real number $t > 0$, and let

$$n_t = \text{ largest integer } \leq \frac{t}{\gamma} \tag{3.1.9}$$

Using the approximation (3.1.8), (3.1.7) may be rewritten in the form

$$[a] \approx \gamma \sum_{k=0}^{n_t} h_\theta[\theta(t_{k+1})] \cdot \tilde{\theta}_k^\gamma \approx \int_0^t h_\theta(\theta(s)) \cdot \tilde{\theta}_s^\gamma ds \tag{3.1.10}$$

where $\tilde{\theta}_t^\gamma$ is, for example, given by linear interpolation of $\tilde{\theta}_n^\gamma$ between the points $t_n = n \cdot \gamma$.

Term [b]

We set

$$S_t^\gamma := \sum_{k=1}^{n_t} (H[\theta(t_k), X_{k+1}] - h[\theta(t_k)]) \tag{3.1.11}$$

We choose δt to be sufficiently small that $\theta(t)$ may be considered to be fixed. From (3.1.11), we have

$$S_{t+\delta t}^\gamma - S_t^\gamma \approx \sum_{k=n_t+1}^{n_{t+\delta t}} (H[\theta(t), X_k] - h[\theta(t)]) \tag{3.1.12}$$

But now, a version of the central limit theorem for dependent variables yields that

$$R_{t,\delta t}^{-\frac{1}{2}} \cdot (S_{t+\delta t}^\gamma - S_t^\gamma) \xrightarrow[\gamma \to 0]{} N(0, I)$$

$$R_{t,\delta t} := \text{cov}\,(S_{t+\delta t}^\gamma - S_t^\gamma) \tag{3.1.13}$$

On the other hand, let

$$R(\theta) := \sum_{k=-\infty}^{+\infty} E_\theta([H(\theta, X_k) - h(\theta)][H(\theta, X_0) - h(\theta)]^T)$$

$$= \sum_{k=-\infty}^{+\infty} \text{cov}_\theta[H(\theta, X_k), H(\theta, X_0)] \tag{3.1.14}$$

where E_θ is the expectation with respect to the stationary asymptotic distribution of X_k; note that if the X_k are independent, (3.1.14) reduces to the covariance matrix of the random field $H(\theta, X_k)$. Then we have

$$R_{t,\delta t} \approx \frac{\delta t}{\gamma} R(\theta(t)) \tag{3.1.15}$$

If, further, we accept that for $s < t < s' < t'$, the variables $S_t^\gamma - S_s^\gamma$ and $S_{t'}^\gamma - S_{s'}^\gamma$ are approximately independent for small γ, then for fixed $t < t'$ formulae (3.1.11), (3.1.12), (3.1.13) and (3.1.15) give

$$\gamma^{\frac{1}{2}}(S_{t'}^\gamma - S_t^\gamma) \xrightarrow[\gamma \to 0]{} \int_t^{t'} R^{\frac{1}{2}}(\theta(s))dW_s \tag{3.1.16}$$

where (W_t) is a (vectorial) Wiener process (with covariance matrix the identity).

Final Heuristics.

Finally (3.1.4), (3.1.10) and (3.1.16) show that for small γ, $\tilde{\theta}_t^\gamma$ is an approximate solution of the stochastic linear differential equation

$$d\tilde{\theta}_t = h_\theta(\theta(t)) \cdot \tilde{\theta}_t dt + R^{\frac{1}{2}}(\theta(t)) \cdot dW_t \tag{3.1.17}$$

It will be noted that these heuristics depend upon far more results than simply the law of large numbers as used in Chapter 2. The reader will not therefore be surprised that the rigorous study of this "invariance principle" in Part II of the book uses sophisticated (but elegant) methods.

3.1.1.2 Results for a Finite Horizon

As in Chapter 2, we distinguish between the two cases of finite and infinite horizons.

Notation and Assumptions.

The notation is as in Chapter 2. We further define:

$$h_\theta := \frac{d}{d\theta} h$$

$$R(\theta) := \sum_{n=-\infty}^{+\infty} \text{cov}_\theta[H(\theta, X_n), H(\theta, X_0)] \tag{3.1.18}$$

where cov_θ denotes the covariance when (X_n) is asymptotically stationary; suitable assumptions about the Markov chain π_θ ensure the convergence of the series (3.1.18) (cf. Part II, Chapter 4). Lastly, we set

$$\tilde{\theta}_{t_n}^\gamma = \gamma^{-\frac{1}{2}}(\theta_n - \theta(t_n)) \tag{3.1.19}$$

3.1 Mathematical Tools: Informal Description

and let $(\tilde{\theta}_t^\gamma)$ denote the trajectory obtained by linear interpolation of (3.1.19). Then we have the following theorem, which we have already justified informally:

Theorem 1. *Constant gain, finite horizon T, Assumption (A.1) of Chapter 2. As γ tends to 0, the difference $(\tilde{\theta}_t^\gamma)_{0 \leq t \leq T}$ converges weakly towards the solution $(\tilde{\theta}_t)_{0 \leq t \leq T}$ of the stochastic linear differential equation*

$$d\tilde{\theta}_t = h_\theta[\theta(t)] \cdot \tilde{\theta}_t dt + R^{\frac{1}{2}}[\theta(t)] \cdot dW_t \tag{3.1.20}$$

where (W_t) is a standard vector Wiener process, and $R(\theta)$ is as defined in (3.1.18).

The corresponding theorem in Part II is Theorem 7 of Chapter 4.

3.1.1.3 Results for an Infinite Horizon

We assume that the ODE has a unique stable equilibrium point θ_* (Assumption (A.2a) of Chapter 2). Thus, for large t, we may replace $\theta(t)$ by θ_* in (3.1.20); and so for large t, we have the approximation

$$d\tilde{\theta}_t = h_\theta(\theta_*) \cdot \tilde{\theta}_t dt + R^{\frac{1}{2}}(\theta_*) \cdot dW_t \tag{3.1.21}$$

In other words, $\tilde{\theta}$ is the output of a filter with transfer function

$$[sI - h_\theta(\theta_*)]^{-1} R^{\frac{1}{2}}(\theta_*) \tag{3.1.22}$$

which is excited by white noise. The necessary and sufficient condition for (3.1.21) to define a stationary process is that *all the eigenvalues of the matrix $h_\theta(\theta_*)$ have strictly negative real parts.* We express this condition as follows

$$\operatorname{Re} \lambda(h_\theta(\theta_*)) < 0 \tag{3.1.23}$$

Under these conditions, the stationary process $(\tilde{\theta}_t)$ defined by (3.1.21) has a marginal distribution

$$\tilde{\theta}_t \sim N(0, P) \tag{3.1.24}$$

where $P > 0$ is the unique symmetric solution of the **Lyapunov equation**

$$h_\theta(\theta_*) \cdot P + P \cdot h_\theta^T(\theta_*) + R(\theta_*) = 0$$

Whence we have the following theorem, corresponding to Theorem 15 of Part II, Chapter 4:

Theorem 2. *Constant gain, infinite horizon, Assumption (A.2a) of Chapter 2, conditions (3.1.23). As $\gamma_n \downarrow 0$ and $t_n \uparrow \infty$, we have*

$$\lim_{n \to \infty} \tilde{\theta}_{t_n}^{\gamma_n} = N(0, P) \qquad (3.1.25)$$

where P is defined in (3.1.24), and "lim" refers to weak convergence.

Theorem 2 is not a simple consequence of Theorem 1. In fact, the weak convergence of the trajectories involves viewing the trajectories in an arbitrary but fixed and finite horizon: it gives no information as $t \to \infty$.

We shall use Theorems 1 and 2 as follows: when the ODE has a stable equilibrium point θ_*, then we use the following approximation

$$\gamma^{-\frac{1}{2}}(\theta_n - \theta(t_n)) \approx \tilde{\theta}_{t_n} \qquad (3.1.26)$$

where $(\tilde{\theta}_t)$ satisfies (3.1.21) for γ small and $t_n \geq 0$ (arbitrarily large if necessary).

Generalisation.

The gains (γ_n) do not have to be constant; see (Derevitskii and Fradkov 1974), where the correct normalisation is given in this case.

3.1.2 Algorithms with Decreasing Gain

We assume, as always in this case, the existence of a stable equilibrium point θ_* for the ODE. The gains are of the form

$$\gamma_n \approx \frac{1}{n} \qquad (3.1.27)$$

where the notation means that $n \cdot \gamma_n$ converges to a constant; note that this condition is more restrictive than condition (2.2.14), but the results are different if (3.1.27) is replaced by a different condition. Refer to Part II, Section 4.5.

3.1.2.1 Heuristics

Our main aim is to point out the differences from the results for algorithms with constant gain. Thus we simplify matters as far as possible by assuming that

1. θ is a scalar parameter;

2. the variables (X_n) are independent and identically distributed.

3.1 Mathematical Tools: Informal Description

These assumptions correspond to Robbins–Monro processes in the scalar case. For n sufficiently large we have:

$$\begin{aligned}\theta_n - \theta_* &= \theta_{n-1} - \theta_* + \frac{1}{n}(H(\theta_{n-1}, X_n) - h(\theta_{n-1})) + \frac{1}{n}(h(\theta_{n-1}) - h(\theta_*)) \\ &\approx \theta_{n-1} - \theta_* + \frac{1}{n} h_\theta(\theta_*) \cdot (\theta_{n-1} - \theta_*) + \frac{1}{n}(H(\theta_{n-1}, X_n) - h(\theta_{n-1}))\end{aligned}$$
(3.1.28)

If in (3.1.28) we set

$$\tilde{\theta}_n := n^{\frac{1}{2}}(\theta_n - \theta_*) \tag{3.1.29}$$
$$\beta := -h_\theta(\theta_*) \tag{3.1.30}$$

then by iteration we obtain

$$\begin{aligned}\tilde{\theta}_n &\approx n^{\frac{1}{2}} \prod_{k=1}^{n}\left(1 - \frac{\beta}{k}\right) \tilde{\theta}_0 + n^{\frac{1}{2}} \sum_{k=1}^{n} \left(\prod_{i=k}^{n}\left(1 - \frac{\beta}{i}\right)\right) \cdot \frac{1}{k}[H(\theta_{k-1}, X_k) - h(\theta_{k-1})] \\ &\approx \frac{1}{n^{\beta - 1/2}} \tilde{\theta}_0 + \frac{1}{n^{\beta - 1/2}} \sum_{k=1}^{n} k^{\beta-1}[H(\theta_{k-1}, X_k) - h(\theta_{k-1})]\end{aligned}$$
(3.1.31)

where we have used the approximation

$$\prod_{k=1}^{n}\left(1 - \frac{\beta}{k}\right) \approx \frac{1}{n^\beta} \quad \text{for } \beta > 0$$

Equations (3.1.31) clearly illustrate the necessity of the condition

$$h_\theta(\theta_*) = -\beta < -\frac{1}{2} \tag{3.1.32}$$

which is stronger than condition (3.1.23), this is needed to eliminate the influence of the non-zero initial condition. Moreover, if we denote the variance of the vector field at the equilibrium point θ_* by

$$R := \mathrm{var}_{\theta_*}(H(\theta_*, X_k)) \tag{3.1.33}$$

then the variance of the last summand in (3.1.31) is approximately

$$\frac{1}{n^{2\beta-1}}\sum_{k=1}^{n} k^{2(\beta-1)}(H(\theta_{k-1}, X_k) - h(\theta_{k-1}))^2 \approx \frac{R}{n^{2\beta-1}}\sum_{k=1}^{n} k^{2(\beta-1)} \xrightarrow[n \to \infty]{} \frac{R}{2\beta - 1}$$
(3.1.34)

But then, using (3.1.31), (3.1.32) and (3.1.34), together with an appropriate version of the central limit theorem, we finally obtain

$$n^{\frac{1}{2}}(\theta_n - \theta_*) \xrightarrow[n \to \infty]{} N\left(0, \frac{R}{2\beta - 1}\right)$$
$$\beta = -h_\theta(\theta_*) \tag{3.1.35}$$

3.1.2.2 Results for Algorithms with Decreasing Gain

The assumptions are those given in Chapter 2 for algorithms with decreasing gain, namely Assumption (A.2b). To these are added the existence of $R(\theta)$ (cf. (3.1.18)) for $\theta = \theta_*$, and the following essential stability condition

$$\text{Re}\,\lambda(h_\theta(\theta_*)) < -\frac{1}{2} \qquad (3.1.36)$$

which thus reinforces condition (3.1.23). The following theorem corresponds to Theorem 13 of Part II, Chapter 4.

Theorem 3. *Algorithm with gain $\frac{1}{n}$; Assumption (A.2b) of Chapter 2, conditions (3.1.36). We have*

$$n^{\frac{1}{2}}(\theta_n - \theta_*) \xrightarrow[\gamma \to 0]{} N(0, P) \qquad (3.1.37)$$

where P is the unique symmetric (positive thanks to (3.1.36)) solution of the Lyapunov equation

$$\left(\frac{I}{2} + h_\theta(\theta_*)\right) P + P\left(\frac{I}{2} + h_\theta(\theta_*)\right)^T + R(\theta_*) = 0 \qquad (3.1.38)$$

In Part II of the book, we describe the behaviour of the algorithm when $\gamma_n \approx n^{-\alpha}$ for $\alpha \neq 1$.

Conclusion of Section 3.1.

We shall make constant use of the mathematical tools described above; both in what remains of Chapter 3, and in the following two chapters. If the ODE is seen as providing the first term in the approximate description of the algorithm, the results on asymptotic normality given in this chapter may be considered as the next higher order term in this description.

3.2 Applications to the Design of Adaptive Algorithms with Decreasing Gain

3.2.1 Status of the Problem

Several stages in the design of adaptive algorithms using the ODE method were identified in Chapter 2. We now consider these same stages when the departure point is the minimisation of a functional. The stages are

1. choice of the functional J;
2. choice of the direction of descent to minimise J, and derivation of the corresponding ODE;
3. derivation of the algorithm.

3.2 Algorithms with Decreasing Gain

We omit Stage 4, which is of no interest here. The choices made in Stages 1 and 2 clearly have an effect on the algorithm derived in Stage 3. Here we shall study the optimisation problem of Stage 2, namely the **optimisation of the search direction** (we use this phrase rather than the expression "direction of descent" since the latter over-emphasises the minimisation of a functional).

By way of illustration, note that the only difference between the transversal equaliser algorithms (formulae (1.2.13)) and the least squares algorithms (formulae (1.2.15)) is that, as shown in our analysis in Paragraph 2.3.2.1, they follow different search directions. On the other hand, the phase-locked loops (cf. formulae (1.2.25) and (1.2.26) and the investigation in Paragraph 2.4.3.1) use different criteria from the outset.

In order to explicitly highlight the search direction as an aspect of adaptive algorithm design, we shall modify the general expression (1.1.1). Neglecting the complementary term in ε_n, we shall work with the formula

$$\theta_n = \theta_{n-1} + \frac{1}{n}\Gamma h(\theta_{n-1}, X_n) \qquad (3.2.1)$$

in which the only new point is the introduction of the gain matrix Γ to define the search direction.

In all that follows, the random vector field $H(\theta_{n-1}, X_n)$ is assumed given; on the other hand, the gain matrix Γ is to be chosen. We shall examine the following points

1. definition of a figure of merit for the algorithm (3.2.1);

2. optimisation of Γ with respect to this figure of merit;

3. definition of a new criterion associated only with the vector field $H(\theta_{n-1}, X_n)$ to evaluate the performance of (3.2.1) for optimal choices of Γ.

Clearly this second criterion should be used to compare the two phase-locked loops previously described.

3.2.2 A Figure of Merit for Algorithms of Decreasing Gain

We examine firstly the effect of the gain matrix Γ on the ODE. The effect of the gain Γ is to modify the ODE; with the usual notation, the result of this modification is given by,

$$\dot{\theta} = \Gamma h(\theta) \qquad (3.2.2)$$

The **primitive ODE** is defined to be the one which corresponds to the unmodified vector field; i.e.

$$\dot{\theta} = h(\theta) \qquad (3.2.3)$$

Of interest here is the modification of the stability properties of the ODE caused by the introduction of the gain matrix Γ. We know of no results which will allow us to carry over the global stability properties of (3.2.3) to (3.2.2); such properties must be examined case by case.

As far as local properties are concerned, the only point of note is that if θ_* is a stable equilibrium point of the primitive ODE (3.2.3), and if

$$\operatorname{Re} \lambda(\Gamma \cdot h_\theta(\theta_*)) < 0 \qquad (3.2.4)$$

then θ_* is also a stable equilibrium point of the modified ODE (3.2.2).

We assume therefore that θ_* is a stable equilibrium point of the primitive ODE. Theorem 3 provides the figure of merit in a natural way. We must restrict ourselves to so-called **admissible** gains, for which

$$\operatorname{Re} \lambda \left(\frac{I}{2} + \Gamma \cdot h_\theta(\theta_*) \right) < 0 \qquad (3.2.5)$$

Note that this condition implies (3.2.4). If we denote the trajectory of algorithm (3.2.1) by (θ_n^Γ), making the dependence on Γ explicit, the **asymptotic figure of merit** is given by

$$\operatorname{Tr}(P_\Gamma) = \lim_{n \to \infty} E \|\theta_n^\Gamma - \theta_*\|^2 \qquad (3.2.6)$$

where P_Γ is the symmetric (positive since Γ is admissible) solution of the Lyapunov equation

$$\left(\frac{I}{2} + \Gamma \cdot h_\theta(\theta_*) \right) P_\Gamma + P_\Gamma \left(\frac{I}{2} + \Gamma \cdot h_\theta(\theta_*) \right)^T + \Gamma R(\theta_*) \Gamma^T = 0 \qquad (3.2.7)$$

3.2.3 Optimal Choice of the Gain Matrix, and Introduction of the Intrinsic Quality Criterion Associated with the Random Vector Field

All the results are summarised in the following elementary proposition.

Proposition 4. *We have*

$$P_* = \min\{P_\Gamma : \Gamma \text{ admissible}\} = h_\theta^{-1}(\theta_*) \cdot R(\theta_*) \cdot h_\theta^{-T}(\theta_*) \qquad (3.2.8)$$

where the minimisation is with respect to the ordering of symmetric matrices. The corresponding optimal gain Γ_* *is given by*

$$\Gamma_* = -h_\theta^{-1}(\theta_*) \qquad (3.2.9)$$

Proof. Set

$$\tilde{P}_\Gamma := P_\Gamma - P_*, \qquad \tilde{\Gamma} := \Gamma - \Gamma_*$$

3.2 Algorithms with Decreasing Gain

and for simplicity, denote $R(\theta_*)$ by R and $h_\theta(\theta_*)$ by h_θ. It is easy to show that $\tilde{P}_\Gamma$ is a solution of the Lyapunov equation

$$\left(\frac{I}{2} + \Gamma h_\theta\right) \tilde{P}_\Gamma + \tilde{P}_\Gamma \left(\frac{I}{2} + \Gamma h_\theta\right)^T + \tilde{\Gamma} R \tilde{\Gamma}^T = 0$$

Since Γ is admissible, it follows that $\tilde{P}_\Gamma \geq 0$, with equality if and only if $\tilde{\Gamma} = 0$. This completes the proof of Proposition 4. □

Conclusion and Remarks.

1. Formula (3.2.8) leads naturally to the introduction of the **intrinsic quality criterion** associated with the vector field $H(\theta_{n-1}, X_n)$:

    $$\Delta := h_\theta^{-1}(\theta_*) \cdot R(\theta_*) \cdot h_\theta^{-T}(\theta_*) \qquad (3.2.10)$$

    Note that $\Gamma H(\theta_{n-1}, X_n)$ has the same quality criterion as $H(\theta_{n-1}, X_n)$, whence the term "intrinsic".

2. Note that when the **Fisher information matrix** $I(\theta)$ is defined (Dacunha-Castelle and Duflo 1983), the Cramer–Rao inequality ensures that

    $$\Delta \geq I^{-1}(\theta) \qquad (3.2.11)$$

3. Finally, note that when

    $$h = -\nabla J \qquad -h_\theta = \text{Hess } J$$

    (i.e. the mean field is a potential derivative), formula (3.2.9) shows that **the asymptotically optimal search directions are the Newtonian methods**: thus, for algorithms with decreasing gain, optimisation in the asymptotic sense is the same as optimisation of the direction of descent defining the ODE. Thus one should expect the stochastic quasi-Newtonian methods to be effective, both in the transient period (when the ODE describes the behaviour of the algorithm well), and in the asymptotic phase (where the behaviour is described by Theorem 3).

3.2.4 Examples

We shall consider two cases which give a good illustration of the use of the previous results.

3.2.4.1 Phase-Locked Loops

Exercise 3.4.1 concerns the calculation of the quality criterion Δ for the two phase-locked loops (1.2.25) and (1.2.26); this criterion should enable us to decide which of the loops is more effective under the given experimental conditions. Here we present just one of the results obtained. We assume that the channel causes a weak distortion (cf. formula (1.2.22)), so that

$$\sum_{k \neq 0}(s_k) \ll s_0 \qquad (3.2.12)$$

Then it can be shown that under these conditions, the two loops have approximately the same quality criterion, given by

$$\Delta = \frac{1}{2}\sum_{k>0}\frac{|s_k - \bar{s}_{-k}|^2}{s_0^2} + \frac{\sigma_\nu^2}{s_0^2} \qquad (3.2.13)$$

where $\bar{z}$ is the conjugate of z, and σ_ν^2 is the power of the noise ν.

There are two interesting things to note about (3.2.13). Firstly, as far as applications are concerned, this formula shows that the performance of the loop is not degraded by channels with Hermitian symmetry ($s_k = \bar{s}_{-k}$); this result was not previously known.

Moreover, as far as the theory is concerned, we have here an example of a vector field with **a quality criterion** Δ **which is independent of the equilibrium point** θ_*: this is clearly a highly favourable case in which everything (i.e. Δ and the optimal (scalar here) gain Γ_*) may be calculated before the algorithm is implemented.

3.2.4.2 Least Squares Algorithm

This is presented in a rather more general context than previously, namely that of regression problems. Let (y_n) be a stationary signal of the form

$$y_n = \Phi_n^T \cdot \theta_* + \nu_n \qquad (3.2.14)$$

where (Φ_n) is a regression vector of dimension p, and (ν_n) is a stationary sequence of variables with zero-mean, such that, for all n, ν_n and Φ_n are uncorrelated. The least squares algorithm to determine the unknown parameter θ_* is given by

$$\theta_n = \theta_{n-1} + \frac{1}{n}\Gamma\Phi_n e_n(\theta_{n-1})$$
$$e_n(\theta_{n-1}) := y_n - \Phi_n^T \cdot \theta_{n-1} \qquad (3.2.15)$$

for a chosen constant gain matrix Γ. We apply the previous results to this example. Recall now that we had

$$h(\theta) := E_\theta(\Phi_n \cdot e_n(\theta)) = -\nabla J(\theta)$$
$$J(\theta) = E_\theta(e_n^2(\theta)) \qquad (3.2.16)$$

The following formulae may be derived, where σ^2 is the variance of the noise ν_n in model (3.2.14).

(i) $\qquad h_\theta(\theta_*) = -\text{Hess } J(\theta_*) = -E_{\theta_*}(\Phi_n \Phi_n^T) := -\Sigma(\theta_*)$
(ii) $\qquad R(\theta_*) := \sum_{n \in \mathbb{Z}} E_{\theta_*}(\nu_n \Phi_n \Phi_0^T \nu_0) = \sigma^2 \Sigma(\theta_*) \qquad (3.2.17)$

3.2 Algorithms with Decreasing Gain

From Proposition 4 it follows that

(i) $\quad\quad\quad\quad\quad\quad\quad\quad\quad\quad \Gamma_* = -\Sigma^{-1}(\theta_*)$
(ii) $\quad\quad\quad\quad\quad\quad\quad\quad\quad\quad \Delta(\theta_*) = \sigma^2 \Sigma^{-1}(\theta_*)$ $\quad\quad\quad$ (3.2.18)

Comments.

1. $\Sigma(\theta_*)$ is effectively dependent on θ_*, except when the regression vector Φ_n is exogenous: in particular, this is the case when Φ_n^T is given by $\Phi_n^T = (y_{n-1}, \ldots, y_{n-p})$, corresponding to the Auto-Regressive model given in Paragraph 2.4.3.3. Consequently, in general, calculation of Δ and of the optimal gain Γ requires a priori knowledge of the parameter θ_* which is to be determined: this is clearly absurd. The recursive least squares (RLS) method given by

$$\theta_n = \theta_{n-1} + \frac{1}{n}\Gamma_n \Phi_n e_n(\theta_{n-1})$$
$$\Gamma_n^{-1} = \Gamma_{n-1}^{-1} + \frac{1}{n}(\Phi_n \Phi_n^T - \Gamma_{n-1}^{-1}) \quad\quad (3.2.19)$$

introduces a variable gain matrix Γ_n which converges to Γ_*; thus RLS provides a good practical way of realising an implementation of optimal gain asymptotically. We shall see how this procedure may be generalised.

2. Here, we demonstrated that the quality criterion is equal to the inverse of the Fisher information matrix: this again shows the (well-known) optimality of the least squares algorithm.

3.2.5 Optimal Choice of Search Direction; Effective Implementation

Here, we shall show how to derive an algorithm with asymptotically optimal gain, when the theoretical gain Γ_* cannot be calculated beforehand. In what follows, we shall assume that the function $H(., x)$ is differentiable, as usual, we denote

$$H_\theta = \frac{\partial}{\partial \theta} H$$

Consider the following algorithm

(i) $\quad\quad\quad\quad\quad\quad \theta_n = \theta_{n-1} + \frac{1}{n}\Gamma_n H(\theta_{n-1}, X_n)$
(ii) $\quad\quad\quad\quad\quad\quad \Gamma_n^{-1} = \Gamma_{n-1}^{-1} - \frac{1}{n}(H_\theta(\theta_{n-1}, X_n) + \Gamma_{n-1}^{-1}) \quad\quad (3.2.20)$

This augmented algorithm still has state vector (X_n); we also need to assume the regularity of the function $\theta \to H_\theta(\theta, X)$. If we denote

$$R_n := \Gamma_n^{-1}$$

then the ODE associated with algorithm (3.2.20) is

(i) $\qquad\qquad\qquad\qquad \dot{\theta} = R^{-1} h(\theta)$
(ii) $\qquad\qquad\qquad\qquad \dot{R} = -h_\theta(\theta) - R \qquad\qquad$ (3.2.21)

Note now that the 2-tuple

$$(\theta_*, R_*) \qquad R_* = -h_\theta(\theta_*) \qquad\qquad (3.2.22)$$

is a stable equilibrium point of the ODE (3.2.21), and thus $\Gamma_n \to \Gamma_*$.

3.3 Conclusions from Section 3.2

In Section 3.2, we chose to study the vector field $H(\theta_{n-1}, X_n)$ and the search direction Γ separately; these were separately introduced in formula (3.2.1) for the algorithm with step size $1/n$.

We characterised $H(\theta_{n-1}, X_n)$ by an intrinsic quality criterion giving a convenient generalisation of the Fisher information matrix for adaptive algorithms.

Then we calculated the optimal search direction Γ_*, and gave a method for realising this asymptotically, when it cannot be calculated beforehand.

Taken together, the results comprise all the elements needed for evaluation of the performance of adaptive algorithms with gain $1/n$, and for the optimal design of such algorithms.

3.4 Exercises

3.4.1 Calculation of the Quality Criterion for Phase-Locked Loops

The notation is as in Subsection 1.2.2, the algorithms correspond to formulae (1.2.25) and (1.2.26). This exercise depends on the work of (Benveniste, Vandamme and Joindot 1979).

Part 1. *Costas Loop, formula (1.2.26-i).*

1. Show that the equilibrium points ϕ_∞ of the Costas loop are given by

$$\phi_\infty = \phi_* + \frac{1}{4} \arg(\sum_{k=-\infty}^{+\infty} s_k^4) \mod \frac{\pi}{2}$$

where ϕ_* and (s_k) are defined in formula (1.2.22).

3.4 Exercises

2. Deduce the following approximation, given a weak distortion $\sum_{k\neq 0}|s_k| \ll s_0$

$$\phi_\infty \approx \phi_* + \sum_{k\neq 0} \frac{|s_k|^4}{s_0^4} \cdot \arg s_k \mod \frac{\pi}{2}$$

Note that the value of ϕ_∞ is independent of the noise and depends only on the intersymbol interference.

3. Calculate the derivative of the mean field at an equilibrium point: show that

$$\varepsilon'(\phi_\infty) = -4 |\sum_{k=-\infty}^{+\infty} s_k^4|$$

4. Show that the cumulative variance at the equilibrium point is given by

$$R := \sum_{k=-\infty}^{+\infty} E[\varepsilon_k(\phi_\infty)\varepsilon_0(\phi_\infty)]$$

$$= 16 |\sum_{-\infty}^{+\infty} s_k^4|^2 \sum_{-\infty}^{+\infty} E\left(\text{Im}\left(\frac{y_k^4}{Ey_k^4}\right) \cdot \text{Im}\left(\frac{y_0^4}{Ey_0^4}\right)\right)$$

5. Deduce the following formula for the quality criterion

$$\Delta = \frac{1}{16} \sum_{-\infty}^{+\infty} E\left(\text{Im}\left(\frac{y_k^4}{Ey_k^4}\right) \cdot \text{Im}\left(\frac{y_0^4}{Ey_0^4}\right)\right)$$

6. Given a weak distortion, and a high signal to noise ratio ($\sigma \ll s_0$ where σ^2 is the variance of the noise ν_n), show that

$$\Delta \approx \frac{1}{2} \sum_{k>0} \frac{|s_k - \bar{s}_{-k}|^2}{s_0^2} + \frac{\sigma^2}{s_0^2}$$

Part 2. *Decision-feedback loop, formula (1.2.26-ii).*

For these calculations, we assume for simplicity that $\hat{a}_n = a_n$; this allows us to derive results for a weak distortion and a high signal to noise ratio.

1. Show that the equilibrium points ϕ_∞ are now given by

$$\phi_\infty = \phi_* \mod \frac{\pi}{2}$$

2. Show that

$$\varepsilon'(\phi_\infty) = -s_0$$

3. Show that

$$\Delta = \frac{1}{2} \sum_{k>0} \frac{|s_k - \bar{s}_{-k}|^2}{s_0^2} + \frac{\sigma^2}{s_0^2}$$

3.4.2 Efficiency of Lattice Algorithms

This exercise is a continuation of Exercise 1.7.3.

1. Is the lattice gradient algorithm with $\gamma_n = \frac{1}{n}$ (formulae (1.7.14)) optimal in the sense of Subsection 3.2.3?

2. What about Burg's algorithm (formulae (1.7.15) of the same exercise)? What do you conclude from a comparison of these two algorithms?

3.4.3 Algorithms with a Sliding Window

Given the vector field $H(\theta_{n-1}, X_n)$ (with the usual notation), an algorithm with a sliding window is given by

$$\theta_n(N) = \text{solution of } \sum_{n-N+1}^{n} H(\theta, X_k) = 0$$

It is assumed that $\theta_n(N)$ lies in the neighbourhood of a stable equilibrium point θ_* of the ODE associated with the vector field $H(\theta_{n-1}, X_n)$. Using the heuristics given at the beginning of Chapter 3, show that for large N

$$N^{1/2}(\theta_n(N) - \theta_*) \approx N(0, P)$$
$$P = h_\theta^{-1}(\theta_*) \cdot R(\theta_*) \cdot h_\theta^{-T}(\theta_*)$$

where the notation is as in Theorem 3. What can you deduce from the comparison of this result with that of Theorem 3?

3.5 Comments on the Literature

General Comments.

The first important results on diffusion approximations are those due to (Khas'minskii 1966) concerning diffusion approximations in continuous time, with constant gain, for a finite horizon, with mixing properties which are weakened in (Borodin 1977); these results apply to vector fields which are stationary, mixing and smooth in θ. In the same vein, we note also the work of Kushner (Kushner and Clark 1978), and most importantly (Kushner and Huang 1981) concerning algorithms with constant gain. Moving away from diffusion approximations in a finite horizon, (Kushner 1984) gives a result analogous to Theorem 2 for algorithms with constant gain, in an infinite horizon.

The Markov setting was first considered by Ljung and Caines for algorithms with decreasing gain (Ljung and Caines 1979). The subject is in essence covered by the works of Kushner et al., referred to above; additionally, (Kushner and Huang 1979) gives a full treatment of algorithms with gain decreasing as $n^{-\alpha}$, for various values of α. Finally (Derevitskii and Fradkov 1974) gives an invariance principle for fields of Robbins–Monro type which admit discontinuities, thanks to a method relying on martingales. The present work takes in all these results, and at the same time also includes both the Markov setting and the discontinuities of the random vector field.

References Used in this Chapter.

This chapter is entirely original, apart from the example of the phase-locked loop which is taken from (Benveniste, Vandamme and Joindot 1979).

Chapter 4
Tracking Non-Stationary Parameters

Here we come to the very heart of the subject, namely an analysis of adaptive algorithms in the context in which they are mainly used (as described in Chapter 1). Throughout this chapter the "true" parameter θ_* will be considered to be time-varying.

The main purpose of this chapter is to provide a guide to the design of adaptive algorithms and to their optimisation for non-stationary systems. Thus, the ingredients of this chapter are the same as those of the previous one; definition of a figure of merit, definition of an intrinsic quality criterion to measure the tracking ability of an algorithm, and optimisation of a search direction (gain matrix). In our investigation of the tracking of non-stationary parameters, the algorithms to be studied will have constant gain.

This chapter is divided into three sections. The first is given over to a study of classical algorithms (with constant gain). The second considers a new class of *multistep algorithms*, introduced by the Soviet school: as we shall see, these algorithms are essential in tracking of non-stationary parameters. The third section is the user guide.

4.1 Tracking Ability of Algorithms with Constant Gain

The first subsection is concerned with expressing the problem properly; selecting a satisfactory setting and introducing the figure of merit. The second subsection gives the main results: the optimal gains and the intrinsic quality criterion. A third subsection is devoted to the application of the previous results.

4.1.1 Posing the Problem

We begin by examining the two examples already used in Chapter 3.

4.1.1.1 Two Examples

The Phase-Locked Loop.
See Subsection 1.2.2 for the description of this example. Let us return to formula (1.2.22), which we restate here in a slightly different form, making

4.1 Tracking Ability of Algorithms with Constant Gain

the dependence on ϕ_* explicit.

(i) $$y_n(\phi_*) = y_n \cdot e^{i\phi_*}$$
(ii) $$y_n := \sum_{k=-\infty}^{+\infty} s_k a_{n-k} + \nu_n \qquad (4.1.1)$$

If now we suppose that ϕ_* varies with time, and that (s_k) and the statistics of (ν_n) remain unchanged, we must replace (4.1.1-i) by

$$y_n(\phi_*(n)) = y_n \cdot e^{i\phi_*(n)} \qquad (4.1.2)$$

If now we take (4.1.2) into account, the temporal variation caused by ϕ_* has repercussions on the algorithm of (1.2.25) and (1.2.26) (we stress that **from the user's point of view, the algorithm is not modified!**). This gives

$$\phi_n = \phi_{n-1} + \gamma \varepsilon[\phi_{n-1}, \phi_*(n-1); y_n, \nu_n] \qquad (4.1.3)$$

with for the Costas and the decision-feedback loop respectively:

(i) $$\varepsilon(\phi, \phi_*; y_n, \nu_n) = -\mathrm{Im}[y_n^4(\phi_*) e^{-i4\phi}]$$
(ii) $$\varepsilon(\phi, \phi_*; y_n, \nu_n) = \mathrm{Im}[y_n(\phi_*) e^{-i\phi} \overline{\hat{a}}_n(\phi_*, \phi)] \qquad (4.1.4)$$

where $\hat{a}_n$ is defined by formula (1.2.27), using $y_n(\phi_*)$, and $\overline{z}$ denotes the conjugate of z.

The Least Squares Algorithm.

This algorithm was described in Paragraph 3.2.4.2. Here again, we begin by rewriting model (3.2.14) to make the dependence on the "true" parameter θ_* (which represents the system, and which we assume to be time-varying) explicit. We obtain

$$y_n(\theta_*(n)) = \Phi_n^T(\theta_*(n)) \theta_*(n) + \nu_n \qquad (4.1.5)$$

The least squares algorithm (3.2.15), of which we use a version with a constant gain matrix, must then be rewritten in the following form which brings out the role of θ_* (remember that we don't touch the algorithm!):

(i) $$\theta_n = \theta_{n-1} + \Gamma \Phi_n[\theta_*(n-1)] e_n[\theta_{n-1}, \theta_*(n-1)]$$
(ii) $$e_n(\theta, \theta_*) := y_n(\theta_*) - \Phi_n^T(\theta_*)\theta \qquad (4.1.6)$$

Conclusion: Repercussion on the Form of the Algorithms.

The analysis of these two examples allows us to derive the following convenient description of adaptive algorithms operating on time-varying systems:

$$\theta_n = \theta_{n-1} + \Gamma H(\theta_{n-1}, z_{n-1}; X_n) \qquad (4.1.7)$$

In (4.1.7), for simplicity, we have neglected the complementary term ε_n. Here (z_n) denotes the trajectory of the **true parameter** which θ_n is designed to identify and then track. The **state (X_n) has a Markov representation controlled** by the 2-tuple (θ, z), see Section 1.5. We stress that, unlike θ_n, the true parameter z_n is not observed; its effect is only made apparent through the state X_n. In fact, it would be more appropriate to describe z_n as a parameter governing the state X_n, rather than expressing it explicitly in the vector field $H(\ldots)$, as in (4.1.7). This would give the expression

$$\theta_n = \theta_{n-1} + \Gamma H(\theta_{n-1}; X_n)$$
$$P(\xi_n \in G \mid \xi_{n-1}, \ldots; \theta_{n-1}, \ldots; z_{n-1}, \ldots) = \int_G \pi_{\theta_{n-1}, z_{n-1}}(\xi_{n-1}, dx)$$
$$X_n = f(\xi_n)$$

However, we prefer formula (4.1.7), since this will enable us to use more compact notation in what follows.

4.1.1.2 Introduction of the Hypermodel

In order to study the behaviour of the algorithm thus modified, (4.1.7), we now need a recursive model to describe the evolution of the true system z_n.

We choose to describe z_n recursively by the following model:

$$z_n = z_{n-1} + K(z_{n-1}, \zeta_n) \tag{4.1.8}$$

where (ζ_n) is a semi-Markov state controlled by the parameter z. Thus the recursive model for (z_n) is autonomous, since updating the parameter θ by algorithm (4.1.7) has, logically, no effect on the true system z.

This algorithm is quite general in describing the evolution of a parameter which varies slowly with time. The two formulae (4.1.7) and (4.1.8) describe the {system,algorithm} behaviour.

As it is, this form is not yet directly appropriate for "asymptotic analysis". The model is lacking in a certain aspect; namely our postulation of slow variations of the true system. To include this assumption, we multiply the vector field $K(z, \zeta)$ by a small parameter μ. The asymptotic investigation may now proceed by letting μ tend to 0. In the third section of this chapter, we shall show how to select a model for (z_n). In parallel, we express the gain of the algorithm (4.1.7) in the form $\gamma \cdot \Gamma$, where the matrix Γ will always remain fixed throughout the asymptotic analysis (Γ characterises the search direction), and the scalar γ preceding it will tend to 0 with μ.

Finally, we describe the {system, algorithm} pair by the model

(i) $$z_n = z_{n-1} + \mu K(z_{n-1}, \zeta_n)$$
(ii) $$\theta_n = \theta_{n-1} + \gamma \Gamma H(\theta_{n-1}, z_{n-1}; X_n) \tag{4.1.9}$$

as commented on below.

4.1 Tracking Ability of Algorithms with Constant Gain

a) Equation (4.1.9-i) defines what we shall term the **hypermodel**: this is a stochastic approximation whose description is given in a form identical to that of the adaptive algorithms studied from the start. The state of the hypermodel is (ζ_n); it has a Markov representation controlled by the parameter z. We shall impose on the hypermodel all the assumptions necessary to enable us to apply the convergence theorems and the results on asymptotic normality, as required. Note that *none of the objects involved in the hypermodel (neither z_n nor ζ_n) is accessible to the user*. Only the form of the model (the function K and the conditional distribution of ζ_n) may be postulated and even identified in certain cases, as we shall see in Paragraph 4.1.3.3.

b) Equation (4.1.9-ii) describes the algorithm itself. We stress that from the user's point of view, this algorithm has not been modified: we have simply introduced an extra variable parameter z_n to represent the conditional distribution of the state X_n. In fact (X_n) now has a Markov representation controlled by the 2-tuple (θ, z), where only θ is accessible to the user.

c) We define the precise roles of the objects μ, γ and Γ

1. Γ is a constant matrix which defines the search direction;
2. γ is a scalar which defines the rate of progress of the algorithm, γ will vary as a function of μ, as we shall see in Subsection 4.1.2;
3. μ is not a gain: it is a small parameter used to express the fact that the true system (z_n) is slowly varying; **the asymptotic analysis will correspond to letting μ tend to 0**, this amounts to assuming smaller and smaller variations in the true system.

The product $\gamma \cdot \Gamma$ is the gain of the algorithm. In Subsection 4.1.3, we shall see a practical use of the results of the asymptotic analysis.

Examples of Hypermodels, and Counterexamples.

1. Constant drift: the true system evolves at a constant rate.
$$z_n = z_{n-1} + \mu \delta z \qquad (4.1.10)$$

2. Oscillations: it is not possible to describe oscillations by a hypermodel of the form (4.1.9-i), for this, a second order difference equation is required.

3. Random walk, linear model.
$$z_n = z_{n-1} + \mu(Az_{n-1} + W_n)$$
$$\operatorname{Re}\lambda(A) < 0 \quad \text{or} \quad A = 0 \qquad (4.1.11)$$

where W_n is a stationary white noise.

4. Jump process.
$$z_n = z_{n-1} + \mu \xi_n W_n \qquad (4.1.12)$$
where W_n is a stationary sequence of independent variables and (ξ_n) is a series of Bernoulli trials, i.e. a sequence of independent, variables having the identical distribution
$$P\{\xi_n = 1\} = \alpha \ll 1, \qquad P\{\xi_n = 0\} = 1 - \alpha$$
α may also depend upon z_{n-1}.

We summarise the assumptions which we shall use in the sequel.

Assumptions NS (Non-Stationary).

NS1 *The state (ζ_n) has a Markov representation controlled by z, and it has a stationary asymptotic behaviour for all z in the effective domain of the hypermodel. The state (X_n) has a Markov representation controlled by the pair (θ, z), and it has a stationary asymptotic behaviour for all (θ, z) in the effective domain of the algorithm.*

NS2 *The functions $K(z, \zeta)$ and $H(\theta, z; X)$ are sufficiently smooth that*

(i) $\qquad\qquad k(z) := \lim_{n \to \infty} E_z[K(z, \zeta_n)]$

(ii) $\qquad\qquad h(\theta, z) := \lim_{n \to \infty} E_{\theta, z}[H(\theta, z; X_n)] \qquad (4.1.13)$

are defined and suitably smooth (locally Lipschitz in this case).

NS3 *We assume that the true system may be identified by the algorithm, i.e. that for all z*
$$h(\theta, z) = 0 \Leftrightarrow \theta = z \qquad (4.1.14)$$

The following property of the partial derivatives of h may be deduced:

Consequence of Properties NS2 and NS3.

For all z,
$$h_\theta(z, z) = -h_z(z, z) \qquad (4.1.15)$$

Proof. We apply Taylor's formula to each variable separately.
$$\begin{aligned} h(\theta, z) &= h(z, z) + h_\theta(z, z)(\theta - z) + o(\|\theta - z\|) \\ h(\theta, z) &= h(\theta, \theta) + h_z(\theta, \theta)(z - \theta) + o(\|\theta - z\|) \end{aligned}$$
But then (4.1.14) and the continuity of $\theta \to h_z(\theta, \theta)$ give formula (4.1.15). $\square$

Comments. Note that condition NS3 means effectively that the model structures associated with the tunable parameter θ and with the true system z are the same. This does not require the model to be an exact representation of the true system.

4.1.1.3 Questions Arising

These are the following:

Problem 1. *Given the hypermodel, what is the best way to choose γ and Γ?*

Problem 2. *How should we evaluate the tracking ability of the vector field $H(\theta_{n-1}, z_{n-1}; X_n)$, with no prior knowledge of the hypermodel?*

Note that Problem 2 is particularly important, since it concerns an intrinsic property of the algorithm, whilst the hypermodel is often inappropriate in practice.

4.1.1.4 A Figure of Merit

We are interested in evaluating the performance of the algorithm once the transient phase ends: the trajectory θ_n of the algorithm is then in the vicinity of the true parameter z_n, and we would like to measure the difference between θ_n and z_n. Logically, we end up with the figure of merit

$$E\|\theta_n - z_n\|^2 \tag{4.1.16}$$

where n is sufficiently large that the transient phase is ended. Note that for the moment, (4.1.16) defines a family of criteria indexed by n.

4.1.2 Main Theorems

This subsection is divide into three parts. In the first part, we shall use the asymptotic normality results of Chapter 3 to reformulate the figure of merit of formula (4.1.16). This is of interest for two reasons. Firstly, we shall show that (4.1.16) may be written as {bias + variance} (cf. the model of the elastic string given in Subsection 1.4.4), so that increasing the gain γ will reduce the bias but increase the variance, and decreasing the gain will have the opposite effect.

We shall also highlight two cases which give rise to very different behaviours; namely the cases when the solution of the ODE of the hypermodel is degenerate (i.e. constant), and when it is not.

The second part of this subsection considers the first of these cases, which we shall call the "hypermodel without drift". The third part of the subsection considers the second case of the "hypermodel with drift". Lastly, by way of conclusion, we derive the appropriate intrinsic criterion.

4.1.2.1 Reformulation of the Figure of Merit

Assumptions and Notation.

We reinforce Assumptions NS by introducing the following condition:

NS4 *We assume that for each fixed z in the effective domain of the hypermodel, there exists a non-empty set of so-called **admissible** gain matrices Γ, such that*

$$\operatorname{Re} \lambda(\Gamma \cdot h_\theta(z, z)) < 0 \qquad (4.1.17)$$

Conditions NS3 and NS4 imply that for any fixed z, $\theta = z$ is a stable equilibrium point of the differential equation

$$\dot{\theta} = \Gamma h(\theta, z)$$

In other words, in the absence of non-stationarity in z, when suitably initialised, the algorithm with constant gain (4.1.9-ii) will converge to z.

We shall also require the following notation

$$Q(z) := \sum_{n=-\infty}^{+\infty} \operatorname{cov}_z[K(z, \zeta_n), K(z, \zeta_0)] \qquad (4.1.18)$$

$$R(\theta, z) := \sum_{n=-\infty}^{+\infty} \operatorname{cov}_{\theta,z}[H(\theta, z; X_n), H(\theta, z; X_0)] \qquad (4.1.19)$$

As previously, in formula (3.1.14), P_z (resp. $P_{\theta,z}$) denotes the asymptotic distribution under which the state ζ_n (resp. X_n) with the Markov representation controlled by z (resp. (θ, z)) is stationary, and cov_z (resp. $\operatorname{cov}_{\theta,z}$) denotes the covariance matrix corresponding to this distribution.

We now use the results of Chapters 2 and 3 to describe the behaviour of (4.1.9).

Derivation of the ODE.

We begin by deriving the ODE. The ODE associated with (4.1.9-i) and its solution, with the continuous time/discrete time relationship are given by

$$\dot{z} = k(z)$$
$$(z(s))_{s \geq 0}, \qquad s_n = \mu n \qquad (4.1.20)$$

The ODE for (4.1.9-ii) is more difficult to obtain. We suppose that

$$\eta := \frac{\mu}{\gamma} \leq C \qquad (4.1.21)$$

for some constant $C < \infty$. The approach given in Subsection 2.2.1 may be applied to (4.1.9-i), since z_n does not vary faster than θ_n, thanks to (4.1.21);

4.1 Tracking Ability of Algorithms with Constant Gain

thus we may associate (4.1.9-ii) with the following ODE and its solution with the continuous time/discrete time relationship

$$\dot{\theta} = \Gamma h(\theta, z)$$
$$(\theta(t))_{t \geq 0}, \qquad t_n = \gamma n \qquad (4.1.22)$$

Of course, in (4.1.22), the trajectory z is that given by the ODE (4.1.20). Note that since the gains γ and μ are not identical, we have had to introduce two continuous time parameters s and t for each ODE; these time parameters are effectively different as the continuous time/discrete time relationships show.

Derivation of the Gaussian Approximations.

We next consider the Gaussian approximations. Since the interaction between (4.1.9-i) and (4.1.9-ii) has a more complex effect than in the case of the ODE, we distinguish the two cases $\eta = 1$ and $\eta \to 0$ (the case in which η tends to a finite limit reduces to the first case, by inclusion of this limit in the gain matrix Γ).

Case where $\eta = 1$.

We consider the 2-tuple (z_n, θ_n) as a single adaptive algorithm to which we apply Theorems 1 and 2 of Chapter 3. Recalling that $s = t$, since $\mu = \gamma$, this gives

$$\begin{aligned}(i) \qquad & z_n = z(t_n) + \gamma^{\frac{1}{2}} \tilde{z}_{t_n} + o(\gamma^{\frac{1}{2}}), \quad t_n = \gamma n \\ (ii) \qquad & \theta_n = \theta(t_n) + \gamma^{\frac{1}{2}} \tilde{\theta}_{t_n} + o(\gamma^{\frac{1}{2}}) \end{aligned} \qquad (4.1.23)$$

where the 2-tuple $[z(t), \theta(t)]$ is a solution of the ODE

$$\begin{aligned}(i) \qquad & \dot{z} = k(z) \\ (ii) \qquad & \dot{\theta} = \Gamma h(\theta, z) \end{aligned} \qquad (4.1.24)$$

whilst the 2-tuple $(\tilde{z}_t, \tilde{\theta}_t)$ is a solution of the stochastic linear differential equation

$$\begin{pmatrix} d\tilde{z}_t \\ d\tilde{\theta}_t \end{pmatrix} = \begin{pmatrix} k_z & 0 \\ \Gamma h_z & \Gamma h_\theta \end{pmatrix} \begin{pmatrix} \tilde{z}_t \\ \tilde{\theta}_t \end{pmatrix} dt + \begin{pmatrix} Q^{\frac{1}{2}} & 0 \\ 0 & \Gamma \cdot R^{\frac{1}{2}} \end{pmatrix} dW_t \qquad (4.1.25)$$

where the functions k_z and Q, on the one hand, and h_z, h_θ and R on the other hand, are evaluated along the trajectories $z(t)$ and $(\theta(t), z(t))$ (respectively) of the ODE; also (W_t) is a standard Wiener process in an appropriate dimension. Formulae (4.1.23), (4.1.24), and (4.1.25) describe the behaviour of (4.1.9) when $\eta = 1$.

Case where $\eta \to 0$.

Here we must take into account the fact that the gains μ and γ are of different orders of magnitude.

We may describe the behaviour of the hypermodel directly:

(i) $\quad z_n = z(s_n) + \mu^{\frac{1}{2}} \tilde{z}_{s_n} + o(\mu^{\frac{1}{2}}), \quad s_n = n\mu$

(ii) $\quad \dot{z} = k(z), \text{ solution } z(s)$

(iii) $\quad d\tilde{z}_s = k \cdot \tilde{z}_s ds + Q^{\frac{1}{2}} \cdot dW_s^z \qquad (4.1.26)$

where as before, the functions k_z and Q are evaluated along the trajectory $z(s)$, and (W_s^z) is a Wiener process.

For the algorithm (4.1.9-ii) itself, things are rather more difficult: since $\mu \ll \gamma$, we may neglect the effect on θ_n of the fluctuations of $\mu^{\frac{1}{2}} \tilde{z}_{s_n}$ in comparison with the fluctuations of θ_n itself which are of the order of $\gamma^{\frac{1}{2}}$. Consequently we have

(i) $\quad \theta_n = \theta(t_n) + \gamma^{\frac{1}{2}} \tilde{\theta}_{t_n} + o(\gamma^{\frac{1}{2}}), \quad t_n = \gamma n$

(ii) $\quad \dot{\theta} = \Gamma h(\theta, z), \text{ solution } \theta(t)$

(iii) $\quad d\tilde{\theta}_t = \Gamma h_\theta \tilde{\theta}_t dt + \Gamma R^{\frac{1}{2}} dW_t^\theta \qquad (4.1.27)$

In (4.1.27-ii), the trajectory $z(t)$ is simply the solution of the ODE (4.1.26-ii) with a change in time; in (4.1.27-iii), the functions h_θ and R are evaluated along the trajectory $(\theta(t), z(t))$ and (W_t^θ) is a Wiener process **independent** of (W_s^z). Formulae (4.1.26) and (4.1.27) describe the behaviour of (4.1.9) when $\eta \to 0$.

Reformulation of the Criterion, and Initial Analysis.

Finally, we decompose the figure of merit into the sum of a bias and a variance:

$$E\|\theta_n - z_n\|^2 = \|\theta(t) - z(s)\|^2 \qquad \text{(bias)}$$
$$+ \gamma E \|\tilde{\theta}_t - \eta^{\frac{1}{2}} \tilde{z}_s\|^2 \qquad \text{(variance)}$$
$$+ o(\gamma)$$

$$\mu = \eta\gamma, \qquad n = \frac{s}{\mu} = \frac{t}{\gamma} \qquad (4.1.28)$$

where $\theta(t)$, $z(s)$, $\tilde{\theta}_t$ and $\tilde{z}_s$ are given by (4.1.23), (4.1.24) and (4.1.25) when $\eta = 1$ and by (4.1.26) and (4.1.27) when $\eta \to 0$.

Note. Equation (4.1.28) which provides the decomposition into bias + variance **does not use Assumption NS4**, and makes absolutely no assumptions on the proximity of θ_n to z_n: this formula is extremely powerful.

From now on, we shall use Assumption NS4. Formula (4.1.28) shows that the contribution from the variance is of the order of γ, for any value of the ratio η of the gains. Thus we must choose η so as to reduce the contribution from the bias; it is here that NS4 is used.

In fact, let us assume that the hypermodel is such that its ODE has a degenerate solution

$$z(s) \equiv z(0) := z \qquad \forall s \geq 0 \qquad (4.1.29)$$

4.1 Tracking Ability of Algorithms with Constant Gain

Then by simple elimination of the transient phase of the ODE of the algorithm (4.1.9-ii), NS4 ensures that the bias tends to zero, whatever the value of η. It is then clear that, for given μ, (4.1.28) is optimised by the choice $\eta = 1$. In other words, **for the hypermodel with drift zero, we should choose $\gamma = \mu$**. The case $z(s) \to z_*$ clearly reduces to the previous case, and will also be ranked amongst the class of models with drift zero.

Conversely, let us now suppose that the ODE of the hypermodel does not have a stable equilibrium point. In this case, ongoing effort is necessary to keep the bias small; this can only be done by choosing $\gamma \gg \mu$, i.e. $\eta \to 0$, so that the ODE of the algorithm is more rapid than that of the hypermodel. Condition NS4 then does the rest.

To conclude, we shall study the cases of zero (with $\eta = 1$) and non-zero drift (with $\eta \to 0$) separately.

4.1.2.2 Analysis when the Hypermodel has Drift Zero

We recall that this corresponds to the case in which the ODE of the hypermodel satisfies

$$\dot{z} = k(z) \; : \; z(t) \to z_*. \tag{4.1.30}$$

Examples (4.1.11) and (4.1.12) of Paragraph 4.1.1.2 belong to this class. As we have seen, the gain γ is chosen to be equal to μ. The model which we use to analyse formula (4.1.28) is based on the formulae (4.1.24) and (4.1.25).

Using (4.1.30) together with Assumption NS4, once the transient phase is ended, we may neglect the bias in (4.1.28) and consider the functions k_z, h_θ, h_z, Q and R in (4.1.25) as **constants** equal to

$$
\begin{aligned}
(i) &\qquad k_z \equiv k_z(z_*), \; Q \equiv Q(z_*) \\
(ii) &\qquad h_\theta = -h_z \equiv h_\theta(z_*, z_*) \\
(iii) &\qquad R \equiv R(z_*, z_*)
\end{aligned}
\tag{4.1.31}
$$

Note that equation (4.1.15) is used in (4.1.31-ii). We shall consider two extreme cases, which allow us to handle any other case, using an appropriate decomposition.

Case 1. k_z *is asymptotically stable.*

Taking into account (4.1.31), (4.1.25) may, in this case, be rewritten as follows : *the process $(\tilde{z}_t, \tilde{\theta}_t - \tilde{z}_t)$ is a solution of the following stochastic linear differential equation in which $\delta\tilde{\theta}_t$ denotes $\tilde{\theta}_t - \tilde{z}_t$*

$$\begin{pmatrix} d\tilde{z}_t \\ d(\delta\tilde{\theta}_t) \end{pmatrix} = \begin{pmatrix} k_z & 0 \\ -k_z & \Gamma h_\theta \end{pmatrix} \begin{pmatrix} \tilde{z}_t \\ \delta\tilde{\theta}_t \end{pmatrix} dt + \begin{pmatrix} Q^{\frac{1}{2}} & 0 \\ -Q^{\frac{1}{2}} & \Gamma R^{\frac{1}{2}} \end{pmatrix} dW_t \tag{4.1.32}$$

Since the gain Γ is admissible (cf. NS4), (4.1.32) defines a stationary process.

Thus the figure of merit (4.1.28) is in this case given by: for $\gamma = \mu$, small,

$$\lim_{n \to \infty} E\|\theta_n - z_n\|^2 \approx \gamma E\|\tilde{\theta}_t - \tilde{z}_t\|^2 \qquad (4.1.33)$$

where the right-hand side is given by the solution of the following augmented Lyapunov equation

(i) $\quad P := \mathrm{cov}(\tilde{z}_t, \tilde{\theta}_t - \tilde{z}_t)$

(ii) $\quad \begin{pmatrix} k_z & 0 \\ -k_z & \Gamma h_\theta \end{pmatrix} P + P \begin{pmatrix} k_z^T & -k_z^T \\ 0 & h_\theta^T \Gamma^T \end{pmatrix} + \begin{pmatrix} Q & -Q \\ -Q & Q + \Gamma R \Gamma^T \end{pmatrix} = 0$

$\qquad\qquad\qquad\qquad\qquad\qquad\qquad\qquad\qquad\qquad\qquad (4.1.34)$

Formulae (4.1.34) do not provide a simple expression for the optimal gain Γ_*. The following two points should be made.

Remarks.

1. The Lyapunov equation (4.1.34-ii) illustrates the *compromise between tracking and accuracy*. In fact, increasing Γ makes the matrix Γh_θ more stable, and thus contributes to the reduction of P; on the other hand, such an increase in Γ results in an increase in $\Gamma R \Gamma^T$, and thus contributes to an increase in P.

2. Suppose that Γ_0 is a fixed, asymptotically stable matrix, and normalise the gain Γ with the constraint

$$\Gamma h_\theta = -\Gamma_0 \qquad (4.1.35)$$

Then (4.1.34-ii) may be rewritten as a function of Γ_0:

(i) $\quad \begin{pmatrix} k_z & 0 \\ -k_z & -\Gamma_0 \end{pmatrix} P + P \begin{pmatrix} k_z^T & -k_z^T \\ 0 & -\Gamma_0^T \end{pmatrix} + \begin{pmatrix} Q & -Q \\ -Q & Q + \Gamma_0 \Delta \Gamma_0^T \end{pmatrix} = 0$

(ii) $\qquad\qquad\qquad \Delta := h_\theta^{-1} \cdot R \cdot h_\theta^{-T} \qquad (4.1.36)$

Given the constraint (4.1.35), it appears that the effect of the algorithm on the matrix P is entirely concentrated on the *intrinsic criterion* Δ, as previously introduced for algorithms with decreasing gain in formula (3.2.10); the smaller Δ, the better the tracking ability of the algorithm.

Case 2. $k_z = 0$

In this case, (4.1.32) reduces to

(i) $\qquad\qquad d\tilde{z}_t = (Q^{\frac{1}{2}} \; 0) dW_t$

(ii) $\qquad\qquad d(\delta \tilde{\theta}_t) = \Gamma h_\theta \cdot \delta \tilde{\theta}_t dt + (-Q^{\frac{1}{2}} \; \Gamma R^{\frac{1}{2}}) dW_t \qquad (4.1.37)$

4.1 Tracking Ability of Algorithms with Constant Gain

We then obtain

$$\lim_{n\to\infty} E\|\theta_n - z_n\|^2 \approx \gamma E\|\tilde{\theta}_t - \tilde{z}_t\|^2 = \gamma \operatorname{Tr} P \qquad (4.1.38)$$

where P is a solution of the Lyapunov equation

$$(\Gamma h_\theta)P + P(\Gamma h_\theta)^T + \Gamma R \Gamma^T + Q = 0 \qquad (4.1.39)$$

Remarks 1 and 2 of the previous case apply once again to equation (4.1.39). Moreover, we have the following result:

Theorem 1. Optimal choice of gain for "random walk" hypermodels.
The optimal 2-tuple (Γ_*, P_*) satisfying the Lyapunov equation (4.1.39) is given by

(i) $\qquad P_* \cdot \Delta^{-1} \cdot P_* = Q$

(ii) $\qquad \Gamma_* = -P_* \cdot h_\theta^T \cdot R^{-1} = Q^{\frac{1}{2}} \cdot R^{-\frac{1}{2}} \qquad (4.1.40)$

where the last equality holds for an appropriate choice of the square roots.

Proof. Set $\Gamma_0 = \Gamma h_\theta$, then (4.1.39) may be rewritten as

$$\Gamma_0 P + P \Gamma_0^T + \Gamma_0 \Delta \Gamma_0^T + Q = 0 \qquad (4.1.41)$$

Let $\tilde{\Gamma}_0 = \Gamma_0 - \Gamma_{0*}$ and $\tilde{P} = P - P_*$. Then we have

$$\Gamma_0 \tilde{P} + \tilde{P} \Gamma_0 + \tilde{\Gamma}_0 \Delta \tilde{\Gamma}_0^T = 0 \qquad (4.1.42)$$

which proves that $\tilde{P} \geq 0$, with equality if and only if $\tilde{\Gamma}_0 = 0$, since Γ_0 is asymptotically stable. $\square$

Remarks.

3. Models (4.1.11) and (4.1.12) are "random walks" when $A = 0$.

4. A comparison of (4.1.40-ii) with (3.2.9) shows that *the optimal search direction for tracking is in general not related to the optimal search direction for algorithms with decreasing gain*. Clearly this calls in question the established theory of "exponential forgetting methods" currently used for tracking, since they use the search direction (3.2.9) originating from algorithms with decreasing gain. See Paragraph 4.2.4.3 for details of exponential forgetting factor techniques.

4.1.2.3 Analysis when the Hypermodel has Non-Zero Drift

In this case, as we have seen, we must choose γ so that

$$\eta := \frac{\mu}{\gamma} \to 0 \qquad (4.1.43)$$

We thus choose the gain γ to be of the form

$$\gamma = \mu^\alpha, \qquad 0 < \alpha < 1 \tag{4.1.44}$$

The problem is then to choose the pair (α, Γ). Thus we use formulae (4.1.26), (4.1.27) and (4.1.28). Using (4.1.44), (4.1.28) may be rewritten as

(i) $\quad E\|\theta_n - z_n\|^2 = \|\theta(t) - z(s)\|^2 + \mu^\alpha E\|\tilde{\theta}_t\|^2 + o(\mu^\alpha)$

(ii) $\quad n = \dfrac{s}{\mu} = \dfrac{t}{\gamma} \tag{4.1.45}$

We have already established that *for a hypermodel with non-zero drift, the fluctuations z_s of the hypermodel do not play any part.*

The essential point here is the evaluation of the bias $(\theta(t) - z(s))$. For this, we shall use a formal calculus originating from the study of singularly perturbed differential equations, as described in the excellent article (Hoppenstaedt 1971).

We express the two ODEs (4.1.26-ii) and (4.1.27-ii) in terms of the slow timescale s associated with the hypermodel: after the temporal transformation $s \leftarrow t$ is applied to (4.1.27-ii), we obtain the following slow-fast differential equations

(i) $\quad \dot{z} = k(z)$

(ii) $\quad \eta \dot{\theta} = \Gamma h(\theta, z) \tag{4.1.46}$

Using Assumption NS4, we next try to obtain an expansion to the first order in η of the solution of this system, following Hoppenstaedt's method.

The zero order term in the expansion is given by setting $\eta = 0$ in the differential equation with the slow timescale (equations (4.1.46)). If we denote the zero order term by (z_0, θ_0), we then have

$$\begin{aligned} \dot{z}_0 &= k(z_0) \\ 0 &= \Gamma h(\theta_0, z_0) \end{aligned} \tag{4.1.47}$$

If we denote the solution of (4.1.46) by (z, θ), Assumption NS3 allows us to translate (4.1.47) into the form

$$\begin{aligned} z_0 &= z \\ \theta_0 &= z \end{aligned} \tag{4.1.48}$$

To obtain the first order term of the expansion in η, we subtract (4.1.47) from (4.1.46) and the return to the fast timescale $t = s/\eta$. Since we already know from (4.1.48) that $z_0 \equiv z$, this gives

$$\begin{aligned} (\theta - \theta_0) &= \Gamma h(\theta, z) - \eta k(z) \\ &= \Gamma h_\theta(z, z) \cdot (\theta - z) - \eta k(z) + o(\|\theta - z\|) \end{aligned} \tag{4.1.49}$$

4.1 Tracking Ability of Algorithms with Constant Gain

Next we calculate the equilibrium point of (4.1.49), considering z (which is slowly varying) as fixed. This gives the following first order expansion

$$\theta = z + \eta(\Gamma \cdot h_\theta(z,z))^{-1} \cdot k(z) + o(\eta) \qquad (4.1.50)$$

We stress that this formal calculus is supported by a precise mathematical formulation; in the expansion of formula (4.1.50), the underlying constant in the notation $o(\eta)$ is uniform for $s \geq s_0$ (slow timescale), for some arbitrary, fixed $s_0 > 0$.

We are now in a position to choose α. In fact, if we substitute (4.1.50) in (4.1.45), then using (4.1.44), we have

$$E\|\theta_n - z_n\|^2 = \mu^{2(1-\alpha)}\|(\Gamma.h_\theta)^{-1} \cdot k\|^2 + \mu^\alpha E\|\tilde{\theta}_s\|^2 + o(\max[\mu^\alpha, \mu^{2(1-\alpha)}]) \qquad (4.1.51)$$

where $\mu n = s \geq s_0$ (some arbitrary, fixed $s_0 > 0$), and the functions h_θ and k are given by

$$h_\theta := h_\theta(z(s), z(s)), \qquad k := k(z(s)) \qquad (4.1.52)$$

The optimal solution with respect to α is then given by

$$\alpha = 2(1 - \alpha)$$

whence

$$\alpha = \frac{2}{3} \qquad (4.1.53)$$

Using (4.1.27-iii), we now have an expression for the contribution of the variance in (4.1.51), and we have proved the following theorem:

Theorem 2. *Hypermodel with non-zero drift.*
(i) The optimal scalar gain γ is given by

$$\gamma = \mu^{\frac{2}{3}} \qquad (4.1.54)$$

(ii) Using the value in (4.1.54), we have the following formula, where the various functions are evaluated at the points (z, z) or z (respectively) where, $z = z(s_n)$, $s_n = n\mu \geq s_0$ (for some arbitrary, fixed $s_0 > 0$):

$$E\|\theta_n - z_n\|^2 = \mu^{\frac{2}{3}}(\|(\Gamma \cdot h_\theta)^{-1} \cdot k\|^2 + \operatorname{Tr} P) + o(\mu^{\frac{2}{3}}) \qquad (4.1.55)$$

where P is a solution of the Lyapunov equation

$$(\Gamma h_\theta) \cdot P + P(\Gamma h_\theta)^T + \Gamma R \Gamma^T = 0 \qquad (4.1.56)$$

Here too, Remarks 1 and 2 of Paragraph 4.1.2.2 may be repeated in respect of (4.1.55) and (4.1.56). We add the following remark

Remarks.

3. In the case of a hypermodel with non-zero drift, the algorithm is only concerned with tracking this drift, the fluctuations $\tilde{z}_s$ of the hypermodel may be totally ignored.

4.1.2.4 Conclusion: Quality Criterion

We review the important points, beginning with those which are common to all cases.

The Tracking/Accuracy Compromise.

In all cases, we have shown that the performance of the algorithm is a compromise between the tracking rate and accuracy. This compromise is seen in the decomposition of the figure of merit into a sum of bias and variance terms; an increase in the gain corresponds to a decrease in the bias, and an increase in the variance, and vice-versa.

The Intrinsic Quality Criterion.

Remark 2 applies to all cases, and we have the following result.

Theorem 3. *With the constraint*

$$\Gamma h_\theta \equiv -\Gamma_0 \qquad (4.1.57)$$

where Γ_0 is fixed and asymptotically stable, we have

$$E\|\theta_n - z_n\|^2 \approx \mu^\alpha \Psi(\Gamma_0, \Delta) \qquad (4.1.58)$$

where $\alpha = 1$ or $\alpha = \frac{2}{3}$, according as to whether the hypermodel has zero or non-zero drift; in (4.1.58), the **intrinsic quality criterion**

$$\Delta := h_\theta^{-1} R h_\theta^{-T} \qquad (4.1.59)$$

is evaluated at the point $z = z(s_n)$, $s_n = n\mu$; lastly, the function Ψ is such that the function

$$\Delta \to \Psi(\Gamma_0, \Delta) \qquad (4.1.60)$$

is **increasing** *for fixed Γ_0.*

The intrinsic quality criterion Δ, previously introduced in Chapter 3 for algorithms with decreasing gain, thus characterises the **tracking ability** of the vector field $H(\theta_{n-1}, X_n)$. Note that this tracking ability may depend on the current status of the true system z. We have seen that this is not the case for the phase-locked loop (formula (3.2.13)), but that it is the case for the least squares algorithm (formula (3.2.18)); in any event, the tracking ability depends only on the mean value of z, and this mean value is constant for hypermodels with drift zero.

The Role of the Hypermodel.

The drift in the hypermodel plays a critical role. If it is zero, the speed of the algorithm will be of the same order as that of the true system; if the drift is non-zero, the speed of the algorithm will be infinitely larger ($\gamma = \mu^{\frac{2}{3}}$). As far as the probabilistic model for z is concerned, its effect is described by a mean

4.1 Tracking Ability of Algorithms with Constant Gain

speed k and a noise covariance matrix Q. Long evening fireside discussions about the realism of such and such a hypermodel (him: "the true system is slowly varying, and yet you describe it with white noise?" me: "yes,...") are thus superfluous!

Optimisation of the Search Direction Γ.

It is clearly always possible to find an explicit formula when θ is scalar (cf. the phase-locked loop). In the general case, we have only been able to derive the optimal value Γ_* explicitly for random walk hypermodels: the results bear no relation to the Newtonian methods, contrary to many beliefs.

Notes on the Underlying Mathematics.

The informal calculations described above, do not in this instance have a rigorous counterpart in Part II of the book. A rigorous proof of formula (4.1.28) is not out of our range, however the proof of Theorem 2 is certainly harder. Note that none of the approximations made by us presuppose that the algorithm θ_n is effectively in the vicinity of the true system z_n; only the slowness of the variations of z_n is assumed, and in order to quantify the degree of slowness, we would need to express the constant underlying the $o(.)$ notation explicitly.

4.1.3 Effective Realisation

Backed by the theorems of the previous section, we now return to complete the guides to algorithm design and analysis introduced in Chapter 2.

We recall that formulae (4.1.9) constitute a model for the study of the system/algorithm interaction, and that the user only has access to the algorithm in the usual form

$$\theta_n = \theta_{n-1} + \Gamma_n H(\theta_{n-1}, X_n) \tag{4.1.61}$$

(compare the models in Paragraph 4.1.1.1 with the algorithms from which they originated). Thus the user must select

1. the vector field $H(\theta_{n-1}, X_n)$, and
2. the gains Γ_n for a given tracking application.

4.1.3.1. Reformulation of the Results of Subsection 4.1.2

These results were presented in the context of the asymptotic analysis using the small parameter μ, where the gain matrix was in the form of the product $\gamma \cdot \Gamma$. In practice, this will not be the case, since the algorithm is given in the form (4.1.61), and the hypermodel is in the form (4.1.8). Accordingly, we shall reformulate the results of Subsection 4.1.2 with the model

$$\begin{aligned} z_n &= z_{n-1} + K(z_{n-1}, \zeta_n) \\ \theta_n &= \theta_{n-1} + \Gamma H(\theta_{n-1}, z_{n-1}; X_n) \end{aligned} \tag{4.1.62}$$

where K is assumed to be small. This amounts to making the following changes of scale in the calculations of Subsection 4.1.2

(i) $\qquad\qquad\qquad\qquad \gamma\Gamma \leftarrow \Gamma$
(ii) $\qquad\qquad\qquad\qquad \mu k \leftarrow k, \quad \mu^2 Q \leftarrow Q \qquad\qquad (4.1.63)$

whilst h and R remain unchanged. Thus we obtain the following results, which we shall use in practice for algorithm synthesis.

Hypermodel with Drift Zero, k_z Asymptotically Stable.

Reformulation of (4.1.33) and (4.1.34) for the model (4.1.62) of the system/algorithm interaction gives (exercise for the reader):

(i) $\qquad\qquad\qquad \lim_{n\to\infty} E\|\theta_n - z_n\|^2 = \operatorname{Tr} P_{22}$

(ii) $\qquad\qquad\qquad \boldsymbol{P} := \begin{pmatrix} P_{11} & P_{12} \\ P_{21} & P_{22} \end{pmatrix}$ solution of

$$\begin{pmatrix} k_z & 0 \\ -k_z & \Gamma h_\theta \end{pmatrix} \boldsymbol{P} + \boldsymbol{P} \begin{pmatrix} k_z^T & -k_z^T \\ 0 & h_\theta^T \Gamma^T \end{pmatrix} + \begin{pmatrix} Q & -Q \\ -Q & Q + \Gamma R \Gamma^T \end{pmatrix} = 0 \qquad (4.1.64)$$

Hypermodel with Drift Zero, $k_z = 0$.

Reformulation of (4.1.38), (4.1.39) and (4.1.40) for model (4.1.62) gives

(i) $\qquad\qquad\qquad \lim_{n\to\infty} E\|\theta_n - z_n\|^2 = \operatorname{Tr} P$
(ii) $\qquad\qquad\qquad \Gamma h_\theta P + P(\Gamma h_\theta)^T + \Gamma R \Gamma^T + Q = 0 \qquad (4.1.65)$

where the optimal 2-tuple (Γ_*, P_*) is again given by

(i) $\qquad\qquad\qquad P_* \Delta^{-1} P_* = Q$
(ii) $\qquad\qquad\qquad \Gamma_* = -P_* h_\theta^T R^{-1} \qquad (= Q^{\frac{1}{2}} R^{-\frac{1}{2}}) \qquad (4.1.66)$

Hypermodel with Non-Zero Drift.

The reformulation of Theorem 3 for the model (4.1.62) follows directly from formulae (4.1.55) and (4.1.56):

(i) $\qquad\qquad\qquad E\|\theta_n - z_n\|^2 \approx \|(\Gamma \cdot h_\theta)^{-1} \cdot k\|^2 + \operatorname{Tr} P$
(ii) $\qquad\qquad\qquad (\Gamma h_\theta) \cdot P + P(\Gamma h_\theta)^T + \Gamma R \Gamma^T = 0 \qquad (4.1.67)$

where the various functions are evaluated for $z = z(s_n)$ with $s_n = n$. Note that the optimal scaling factor $\|\Gamma\|$ in (4.1.67) is once again given by $\|\Gamma\| \approx \|k\|^{\frac{2}{3}}$ for small k.

4.1.3.2 Guide to Algorithm Design for Tracking Applications

Stage 1. *Choice of the vector field $H(\theta_{n-1}, X_n)$.*

This choice may for example be made in accordance with the procedure established in Chapter 2:

1. choice of the functional to minimise,

2. calculation of the stochastic gradient.

By Theorem 3, the vector field must be chosen from those fields with the least possible intrinsic quality criterion Δ: such a vector field H will have the best tracking ability. As we have seen, algorithms which attain the Cramer-Rao bound asymptotically (Dacunha-Castelle and Duflo 1983) provide optimal vector fields for tracking; in any case they will always be associated with a gain matrix Γ which adjusts the search direction.

Stage 2. *Choice of the hypermodel.*

The hypermodel is only relevant via

1. its mean behaviour or "drift" (given by its ODE),

2. the matrix Q described in (4.1.18).

Moreover the "drift zero" and "non-zero drift" cases must be treated in different ways.

Thus we may make the following recommendations.

- If there are no physical reasons which force the selection of a particular hypermodel, the hypermodel should be simple. The hypermodel should be chosen either to be deterministic, or to have drift zero (purely random), since a combination of these two features is difficult to handle.

- Except in special cases, the standard expressions for hypermodels are (1) (formula (4.1.10)) or (3) (formula (4.1.11)). For a robust method, it may be best simply to choose the hypermodel structure and to determine some of its parameters on-line.

Stage 3. *Practical tuning of the gain Γ.*

The results of Paragraph 4.1.3.1 allow us to calculate Γ_* in advance (once the hypermodel is chosen), when Δ is independent of the true system z, as is for example the case for the phase-locked loop (formula (3.2.12)). Otherwise, we must identify z and Γ_* simultaneously, according to a procedure analogous to that of Subsection 3.2.5.

4.1.3.3 Examples

The Phase-Locked Loop with Frequency Offset.

Frequency offset corresponds to the hypermodel

$$z_n = z_{n-1} + \omega \tag{4.1.68}$$

where z is the true channel phase, and ω is the frequency offset, assumed known. Thus this is a purely deterministic model. We seek to identify z using

$$\phi_n = \phi_{n-1} + \gamma \varepsilon_n \tag{4.1.69}$$

where ε_n corresponds either to the Costas loop (1.2.26-i) or to the decision-feedback loop (1.2.26-ii), both of which have the same tracking ability for weak distortion (cf. formula (3.2.13)).

We denote the mean field by $\varepsilon(\phi)$ and its derivative by ε_ϕ. If we set

$$\gamma_0 = -\gamma \varepsilon_\phi, \tag{4.1.70}$$

then, using (4.1.67), we obtain

$$E(\phi_n - z_n)^2 \approx \frac{\omega^2}{\gamma_0^2} + \gamma_0 \Delta \tag{4.1.71}$$

which gives the optimal value of γ (here again we have the fateful $\frac{2}{3}$):

$$\gamma_* = -\frac{1}{\varepsilon_\phi}\left(\frac{\Delta}{2\omega^2}\right)^{\frac{1}{3}} \tag{4.1.72}$$

or (with an explicit expression of ε_ϕ, the derivative of the mean field at equilibrium) (cf. Chapter 3, Exercise 3.4.1)

(i) $$\gamma_{\text{Costas}} = -\frac{1}{4|\sum_k s_k^4|}\left(\frac{\Delta}{2\omega^2}\right)^{\frac{1}{3}}$$

(ii) $$\gamma_{\text{feedback}} = -\frac{1}{s_0}\left(\frac{\Delta}{2\omega^2}\right)^{\frac{1}{3}} \tag{4.1.73}$$

These formulae generalise the known formulae for the case in which there is no channel dispersion effect ($s_k = 0$ for $k \neq 0$). Note that when the offset ω is unknown, it may be replaced by an estimate ω_n given, for example, by

$$\omega_n = f(q^{-1}) \cdot (\phi_n - \phi_{n-1}) \tag{4.1.74}$$

where $f(q^{-1})$ is an elementary lowpass filter to be selected. The three formulae (4.1.69), (4.1.73) and (4.1.74) thus constitute a robust phase-locked loop for tracking an unknown offset.

4.1 Tracking Ability of Algorithms with Constant Gain

Least Squares Algorithm and Kalman Filter.

With the now customary notation, the least squares algorithm is

$$\theta_n = \theta_{n-1} + \Gamma \Phi_n e_n \tag{4.1.75}$$

We shall use a random walk hypermodel

$$z_n = z_{n-1} + W_n, \quad \text{cov}(W_n) = Q \tag{4.1.76}$$

where (W_n) is a white noise with covariance matrix Q assumed small. Using (3.2.17) and (3.2.18), formulae (4.1.66) now give

(i) $$P_* \cdot \Sigma \cdot P_* = \sigma^2 \cdot Q$$

(ii) $$\Gamma_* = \frac{P_*}{\sigma^2} \tag{4.1.77}$$

where

$$\begin{aligned} y_n &= z_{n-1}^T \Phi_n + \nu_n \\ \sigma^2 &= E\nu_n^2 \\ \Sigma &= E_z(\Phi_n \Phi_n^T) \end{aligned} \tag{4.1.78}$$

where P_z is the distribution corresponding to $z = E(z_n)$. This allows us to calculate Γ_* in advance, once z has been selected.

Next we show that **the commonly used Kalman filter corresponding to the model**

$$\begin{aligned} z_n &= z_{n-1} + W_n, \quad \text{cov}(W_n) = Q, \quad z_0 = z \\ y_n &= z_{n-1}^T \Phi_n + e_n, \quad E e_n^2 = \sigma^2 \end{aligned} \tag{4.1.79}$$

where

$$\sigma^{-2} Q \ll I \tag{4.1.80}$$

conforms to our approach. The Kalman formulae for the estimator $\theta_n := \hat{z}_{n|n-1}$ are

(i) $$\theta_{n+1} = \theta_n + \Gamma_{n+1} \Phi_n (y_n - \Phi_n^T \theta_n)$$

(ii) $$\Gamma_{n+1} = \frac{P_n}{1 + \Phi_n^T P_n \Phi_n}$$

(iii) $$P_{n+1} = P_n - \frac{P_n \Phi_n \Phi_n^T P_n}{1 + \Phi_n^T P_n \Phi_n} + \sigma^{-2} Q$$

(iv) $$\text{cov}(\theta_n - z_n) = \sigma^2 P_n \tag{4.1.81}$$

We shall analyse the asymptotic behaviour of (4.1.81-iii) using (4.1.80). For n sufficiently large, P_n is small and varies slowly with respect to Φ_n. Using the same methods as in our discussion of the ODE, we shall:

1. neglect $\Phi_n^T P_n \Phi_n$ with respect to 1,
2. replace $\Phi_n \Phi_n^T$ by its mean Σ. This will allow us to replace (4.1.81-iii) by

$$P_{n+1} = P_n - P_n \Sigma P_n + \sigma^{-2} Q \qquad (4.1.82)$$

which, when (4.1.81-iv) is taken into account, gives precisely (4.1.77-i) as equilibrium point; thus (4.1.81-ii) corresponds to (4.1.7-ii).

To summarise, the **use of the Kalman filter in the model (4.1.79) provides asymptotically the optimal gain Γ_* for a random walk hypermodel.**

When it is desired to determine Q, an "adaptive Kalman filter", which determines the state z and the covariance matrix Q simultaneously, might be used. Such a procedure is used in (Bohlin 1976) and in (Gersh and Kitagawa 1985); the reader should refer to Exercise 4.4.2 for details of the adaptive Kalman filter.

Least Squares Algorithm with Exponential Forgetting Factor.

We return to the least squares algorithm (4.1.75), choosing for the hypermodel (4.1.76)

$$Q = \gamma^2 \cdot \sigma^2 \cdot \Sigma^{-1}, \quad \gamma \ll 1 \qquad (4.1.83)$$

Formulae (4.1.77) then give

(i) $\qquad\qquad\qquad P_* = \gamma \sigma^2 \Sigma^{-1}$
(ii) $\qquad\qquad\qquad \Gamma_* = \gamma \Sigma^{-1}$ $\qquad\qquad (4.1.84)$

We shall show that the gain Γ_* given in (4.1.84-ii) may be attained asymptotically via the **least squares algorithm with exponential forgetting factor** below:

$$\theta_n := \arg\min_\theta \sum_{k=1}^n (1-\gamma)^{n-k}(y_k - \Phi_k^T \theta)^2 \qquad (4.1.85)$$

In fact, if we set

$$R_n := \gamma \sum_{k=1}^n \lambda^{n-k} \Phi_k \Phi_k^T \qquad (4.1.86)$$

where $\lambda := 1 - \gamma$ is the **forgetting factor** in the algorithm, it may be shown that, if the reasoning of Paragraph 2.4.3.3 for the RLS method is copied mutatis mutandis, then (4.1.85) may be rewritten recursively in the form

(i) $\qquad\qquad\qquad \theta_n = \theta_{n-1} + \gamma R_n^{-1} \Phi_n e_n$
(ii) $\qquad\qquad\qquad R_n = R_{n-1} + \gamma(\Phi_n \Phi_n^T - R_{n-1})$ $\qquad (4.1.87)$

once n is sufficiently large for R_n to be invertible. Then it is clear in the expression (4.1.87) that this algorithm has a gain matrix which converges

4.1 Tracking Ability of Algorithms with Constant Gain

towards the value given in (4.1.84). To illustrate the commonality and the differences between (4.1.87) and the Kalman filter of formulae (4.1.81), we shall rewrite (4.1.87) using the Riccati equation to give a direct expression in terms of

$$\Gamma_n := \gamma R_n^{-1} \tag{4.1.88}$$

Applying the formula

$$(A + BCD)^{-1} = A^{-1} - A^{-1}B(DA^{-1}B + C^{-1})^{-1}DA^{-1} \tag{4.1.89}$$

with $A = \gamma^{-1}\lambda R_{n-1}$, $B = \Phi_n$, $C = 1$, $D = \Phi_n^T$, (4.1.87) may be rewritten in the form

$$\begin{aligned}(i) \qquad & \theta_n = \theta_{n-1} + \Gamma_n \Phi_n e_n \\ (ii) \qquad & \Gamma_n = \frac{1}{\lambda}\left(\Gamma_{n-1} - \frac{\Gamma_{n-1}\Phi_n\Phi_n^T\Gamma_{n-1}}{\lambda + \Phi_n^T\Gamma_{n-1}\Phi_n}\right)\end{aligned} \tag{4.1.90}$$

In summary, the least squares algorithm with exponential forgetting factor corresponds to the hypermodel given by (4.1.83), that is to say to a random walk with noise covariance matrix proportional to the inverse ratio of the covariance matrix of the regression vector Φ_n, and the observation noise variance σ^2: it is good to know this when using this algorithm.

On the other hand, choosing

$$Q = \gamma^2 \sigma^2 \Sigma, \quad \gamma \ll 1 \tag{4.1.91}$$

for the hypermodel gives

$$\begin{aligned}(i) \qquad & P_* = \gamma \sigma^2 I \\ (ii) \qquad & \Gamma_* = \gamma I\end{aligned} \tag{4.1.92}$$

Thus the optimal tracking algorithm given by the hypermodel (4.1.76) and (4.1.91) is the simple stochastic gradient

$$\theta_n = \theta_{n-1} + \gamma \Phi_n e_n \tag{4.1.93}$$

These two examples were designed to alert the user to the following points.

1. The least squares algorithm with exponential forgetting factor is recommended without hesitation for the **transient phase**, where its relationship to the maximum likelihood method on the one hand, and its Newtonian ODE on the other, guarantee that it performs well.

2. However, in tracking problems proper, the algorithm may still be used, although it has no particular optimality property (since the choice of (4.1.84) is quite arbitrary); thus it should not be surprising that the simple stochastic gradient is preferred in data transmission (cf. the examples of Chapter 1) for long periods of tracking the non-stationary parameters.

3. Often, for least squares algorithms, tracking ability is measured on the basis of the prediction error: for an example of this see Exercise 4.4.1.

4.1.4 Conclusions on the Design of Algorithms with Constant Gain to Track Non-Stationary Systems

We have described some tools of general importance in the analysis of the system/algorithm interaction, and we have illustrated the use of these tools with several examples. In the simple cases (such as the phase-locked loop), these tools enable us to determine completely the optimal design of adaptive tracking algorithms. Be that as it may, these tools should serve as a guide to the designer in each particular case. The main point to remember is that the design of adaptive algorithms is not simply an extrapolation of the usual pair {fixed system, algorithm with decreasing gain}.

4.2 Multistep Algorithms

The idea of filtering the vector field $H(\theta_{n-1}, X_n)$ is quite natural. In certain cases in particular, this idea is quite old. Telecommunications engineers have used loop filters in analogue phase-locked loops for some time now (see for example the basic work (Lindsey and Simon 1973)). Although less classical for digital loops, the use of loop filters can to all intents and purposes be considered as part of the folklore of transmission systems. This example apart, this topic has only been studied in a more general context by the Soviet school ((Tsypkin 1971), (Shil'man and Yastrebov 1976,1978) and (Korostelev 1981)) under the label of **multistep algorithms**. In fact, there are at least two different approaches to the study of multistep algorithms. The first of these is, like the word "multistep" itself, directly imported from the numerical analysis of ordinary differential equations, where techniques of this type allow one to decrease the order of magnitude of the discretisation error (cf. the classical book (Henrici 1963)). This approach is taken, for example, by Shil'man and Yastrebov. As Exercise 4.4.4 shows, it is of no use for adaptive algorithms. However, there is a second asymptotic approach, used in particular by Korostelev, which proves very successful when applied to tracking problems: it is this second approach which we shall describe.

This section is divided into four parts. The first introduces multistep algorithms. The second defines an adequate framework for tracking non-stationary parameters using multistep algorithms. The third presents the main results, and the last describes a practical means of implementing multistep algorithms to track non-stationary systems.

4.2.1 Construction of Multistep Algorithms

Here, we are given a vector field $H(\theta_{n-1}, X_n)$, together with its mean field $h(\theta)$, satisfying the usual assumptions. Let us begin by modifying the ODE

$$\dot{\theta} = h(\theta) \tag{4.2.1}$$

by inserting the transfer function (Laplace transform notation from control theory (Kailath 1980), the various matrices being of appropriate dimensions)

$$\Gamma(\sigma) := D + C(\sigma I - A)^{-1}B, \quad \sigma \in \mathbb{C} \tag{4.2.2}$$

as follows

$$\dot{\theta} = \Gamma(\sigma) \cdot h(\theta) \tag{4.2.3}$$

Using (4.2.2), it is well known that (4.2.3) may be written in the **state space description form**

$$\begin{aligned} \dot{\Theta} &= A\Theta + Bh(\theta) \\ \dot{\theta} &= C\Theta + Dh(\theta) \end{aligned} \tag{4.2.4}$$

where Θ is the state corresponding to the transfer function Γ. Let us now consider the following algorithm

$$\begin{aligned} \Theta_n &= \Theta_{n-1} + \gamma(A\Theta_{n-1} + BH(\theta_{n-1}, X_n)) \\ \theta_n &= \theta_{n-1} + \gamma(C\Theta_{n-1} + DH(\theta_{n-1}, X_n)) \end{aligned} \tag{4.2.5}$$

or equivalently

$$\begin{pmatrix} \Theta_n \\ \theta_n \end{pmatrix} = \begin{pmatrix} \Theta_{n-1} \\ \theta_{n-1} \end{pmatrix} + \gamma \begin{pmatrix} A & B \\ C & D \end{pmatrix} \begin{pmatrix} \Theta_{n-1} \\ H(\theta_{n-1}, X_n) \end{pmatrix} \tag{4.2.6}$$

Equation (4.2.6) clearly shows that we have constructed an adaptive algorithm relative to the 2-tuple

$$[\theta]_n := \begin{pmatrix} \Theta_n \\ \theta_n \end{pmatrix} \tag{4.2.7}$$

The gain matrix of this algorithm is

$$[\Gamma] := \gamma \begin{pmatrix} A & B \\ C & D \end{pmatrix} \tag{4.2.8}$$

whilst its ODE is simply (4.2.4). Note that all the Assumptions (A) relating to the original vector field $H(\theta_{n-1}, X_n)$ are carried over to the augmented vector field of the algorithm (4.2.6); note also that the state (X_n) is also the state of the new algorithm.

Now we can equally well rewrite (4.2.6) eliminating Θ in the following way

(i) $$\theta_n = \theta_{n-1} + \gamma \Gamma_\gamma(q^{-1}) \cdot H(\theta_{n-1}, X_n)$$
(ii) $$\Gamma_\gamma(q^{-1}) = D + \gamma C(qI - (I + \gamma A))^{-1}B \tag{4.2.9}$$

where, as usual, q^{-1} denotes the delay operator defined by

$$q^{-1} \cdot u_n = u_{n-1} \qquad (4.2.10)$$

Multistep algorithms of the form given in (4.2.9) have been studied by Korostelev, for the case in which the original filter in continuous time $\Gamma(\sigma)$ is all-pole (auto-regressive). In the case in which it is required to identify a parameter θ_* corresponding to a stable equilibrium point of the ODE (4.2.1), the condition

$$\operatorname{Re} \lambda \begin{pmatrix} A & Ch_\theta(\theta_*) \\ B & Dh_\theta(\theta_*) \end{pmatrix} < 0 \qquad (4.2.11)$$

implies that $[\Gamma]$ (cf. (4.2.8)) is invertible, and consequently guarantees that

$$[\theta]_* := \begin{pmatrix} 0 \\ \theta_* \end{pmatrix} \qquad (4.2.12)$$

is a stable equilibrium point of the augmented ODE (4.2.4). We shall frequently use condition (4.2.11). Note that when $\Gamma(\sigma)$ is a stable filter (i.e. $\operatorname{Re} \lambda(A) < 0$), condition (4.2.11) is equivalent to

$$\operatorname{Re} \lambda[\Gamma(0) \cdot h_\theta(\theta_*)] < 0 \qquad (4.2.13)$$

4.2.2 An Adequate Framework for Tracking Non-Stationary Parameters Using Multistep Algorithms

In our presentation of certain examples of hypermodels in Paragraph 4.1.1.2, we mentioned in (2), the case of oscillatory behaviours which cannot be described by a stochastic approximation such as (4.1.8).

We now broaden the class of hypermodels given in (4.1.8), by introducing hypermodels of the form

$$\begin{pmatrix} Z_n \\ z_n \end{pmatrix} = \begin{pmatrix} Z_{n-1} \\ z_{n-1} \end{pmatrix} + \begin{pmatrix} A_* & B_* \\ C_* & D_* \end{pmatrix} \begin{pmatrix} Z_{n-1} \\ k(z_{n-1}, \zeta_n) \end{pmatrix} \qquad (4.2.14)$$

where the matrices we introduce here are of appropriate dimensions, and the vector field $k(z_{n-1}, \zeta_n)$ is as in (4.1.8).

4.2.2.1 Examples of Hypermodels

To counterbalance the flexibility introduced by increasing the dimension of the hypermodel (4.2.14), in practice, we shall use a restricted class of vector fields $k(z_{n-1}, \zeta_n)$:

$$\begin{pmatrix} Z_n \\ z_n \end{pmatrix} = \begin{pmatrix} Z_{n-1} \\ z_{n-1} \end{pmatrix} + \begin{pmatrix} A_* & B_* \\ C_* & D_* \end{pmatrix} \begin{pmatrix} Z_{n-1} \\ W_n \end{pmatrix} \qquad (4.2.15)$$

4.2 Multistep Algorithms

where (W_n) is a standard white noise. As we shall see in later examples, model (4.2.15) will allow us to give a satisfactory representation of a large number of interesting behaviours: oscillation, quasi-drift, and combinations of these two factors. In common with the random walk hypermodel which we have already used for one-step algorithms, model (4.2.15) provides the optimal tracking filter $\Gamma(\sigma)$ explicitly.

4.2.2.2 Introduction to the Asymptotic Approach

We again follow the approach taken in Section 4.1. The system/algorithm interaction will once again be made evident by the explicit specification of the effect of the true system z on the vector field of the algorithm, giving the expression

$$H(\theta_{n-1}, z_{n-1}; X_n)$$

which we met previously.

Then we introduce a small parameter μ into the hypermodel (4.2.14). We shall let μ tend to zero, corresponding to a "vanishing" of the variations in the system. Finally, we obtain a global description of the system/algorithm interaction in the following form:

$$(i) \quad \begin{pmatrix} Z_n \\ z_n \end{pmatrix} = \begin{pmatrix} Z_{n-1} \\ z_{n-1} \end{pmatrix} + \mu \begin{pmatrix} A_* & B_* \\ C_* & D_* \end{pmatrix} \begin{pmatrix} Z_{n-1} \\ k(z_{n-1}, \zeta_n) \end{pmatrix}$$

$$(ii) \quad \begin{pmatrix} \Theta_n \\ \theta_n \end{pmatrix} = \begin{pmatrix} \Theta_{n-1} \\ \theta_{n-1} \end{pmatrix} + \gamma \begin{pmatrix} A & B \\ C & D \end{pmatrix} \begin{pmatrix} \Theta_{n-1} \\ H(\theta_{n-1}, z_{n-1}; X_n) \end{pmatrix} \quad (4.2.16)$$

where, for the moment, we make no particular assumptions about the respective dimensions of Z_n and Θ_n. We shall use two more concise forms of (4.2.16). The first uses fresh notation to express (4.2.16-i) and (4.2.16-ii) in terms of stochastic approximations:

$(i) \quad [z]_n = [z]_{n-1} + \mu[\Gamma_*] \cdot [k]([z]_{n-1}, \zeta_n)$

$(ii) \quad [\theta]_n = [\theta]_{n-1} + \gamma[\Gamma] \cdot [H]([\theta]_{n-1}, [z]_{n-1}; X_n)$

$(iii) \quad [z] := \begin{pmatrix} Z \\ z \end{pmatrix}, \quad [\theta] := \begin{pmatrix} \Theta \\ \theta \end{pmatrix}$

$(iv) \quad [k]([z]_{n-1}, \zeta_n) := \begin{pmatrix} Z_{n-1} \\ k(z_{n-1}, \zeta_n) \end{pmatrix}$

$(v) \quad [H]([\theta]_{n-1}, [z]_{n-1}; X_n) := \begin{pmatrix} \Theta_{n-1} \\ H(\theta_{n-1}, z_{n-1}; X_n) \end{pmatrix} \quad (4.2.17)$

The second uses notation relating to the filters we met in (4.2.9):

$(i) \quad z_n = z_{n-1} + \mu \Gamma_{*\mu}(q^{-1}) \cdot k(z_{n-1}, \zeta_n)$

$(ii) \quad \theta_n = \theta_{n-1} + \gamma \Gamma_\gamma(q^{-1}) \cdot H(\theta_{n-1}, z_{n-1}; X_n) \quad (4.2.18)$

where the filters are given by

(i) $\quad \Gamma_{*\mu}(q^{-1}) := D_* + \mu C_*[qI - (I + \mu A_*)]^{-1} B_*$
(ii) $\quad \Gamma_{\gamma}(q^{-1}) := D + \gamma C[qI - (I + \gamma A)]^{-1} B$ (4.2.19)

In the following section, we shall study this model when $\mu \to 0$. The points to examine are

1. the determination of the relative order of magnitude of γ and μ,

2. the choice of $\Gamma(\sigma)$ once Γ_* and the field $k(z_{n-1}, \zeta_n)$ are known.

$H(\theta_{n-1}, z_{n-1}; X_n)$ will satisfy Assumptions NS1 to NS3: see the previous section.

Note that these assumptions concern the vector field $H(\theta_{n-1}, z_{n-1}; X_n)$.

As for the filter $\Gamma(\sigma)$, we replace the previous assumption NS4 by:

NS4′ *For any given element z in the effective domain of the hypermodel, there is a non-empty set of* **admissible** *filters*

$$\Gamma(\sigma) := D + C(\sigma I - A)^{-1} B \qquad (4.2.20)$$

such that

(i) $\quad \operatorname{Re} \lambda(A) < 0$
(ii) $\quad \operatorname{Re} \lambda(\Gamma(0) \cdot h_\theta(z, z)) < 0$ (4.2.21)

The arguments given in (4.2.11), (4.2.12) and (4.2.13), together with the conditions (4.2.21), guarantee that for any given z, the ODE of the algorithm (4.2.16-ii) admits z as a stable equilibrium point.

Warning. Contrary to what one might expect from (4.2.17), the study of (4.2.16) does not lead back to the work of the previous section, even when Z_n and Θ_n are of the same dimension! In fact, it is assumed that only the "one-step" field $h(\theta, z)$ satisfies the Assumption NS3; moreover, this assumption is never satisfied by the augmented mean field

$$[h]([\theta], [z]) = \begin{pmatrix} \Theta \\ h(\theta, z) \end{pmatrix} \qquad (4.2.22)$$

In fact, even if Z and Θ have the same dimension

$$[h]_{[\theta]}([z], [z]) = \begin{pmatrix} I \\ h_\theta(z, z) \end{pmatrix} \qquad (4.2.23)$$

whilst

$$[h]_{[z]}([z], [z]) = \begin{pmatrix} 0 \\ h_z(z, z) \end{pmatrix} \qquad (4.2.24)$$

It follows from the above that the relation (4.2.15) is not satisfied by $[h]$. Thus we have to develop a specific theory for multistep algorithms. This is the aim of the next paragraph.

4.2.2.3 A Figure of Merit

Just as for classical one-step algorithms, we use

$$E\|\theta_n - z_n\|^2 \qquad (4.2.25)$$

as a figure of merit, where n is sufficiently large that the transient phase is completed.

4.2.3 Tracking Ability of Multistep Algorithms: Main Results

The questions which we seek to answer are the same as those relating to one-step algorithms. Thus we follow the plan of Subsection 4.1.2, to which the reader should refer for any notation which is not defined after this point.

4.2.3.1 Reformulation of the Figure of Merit

As the results of the corresponding Paragraph 4.1.2.1 did not use Assumption NS4, we may apply them directly to the formula for the system/algorithm, interaction (4.2.17).

The condition which we impose upon γ is once again

$$\eta := \frac{\mu}{\gamma} \leq C \qquad (4.2.26)$$

for some finite constant C.

The ODEs with the continuous time/discrete time relationship are given by (using the filter notation)

(i) $\qquad \dot{z} = \Gamma_*(\sigma) \cdot k(z); \qquad (z(s))_{s \geq 0}, \qquad s_n = \mu n$

(ii) $\qquad \dot{\theta} = \Gamma(\sigma) \cdot h(\theta, z); \qquad (\theta(t))_{t \geq 0}, \qquad t_n = \gamma n \qquad (4.2.27)$

where in (4.2.27-ii), z denotes the trajectory defined by equation (4.2.27-i).

We now consider the Gaussian approximations, and once again distinguish the cases where $\mu = \gamma$ from those where $\mu \ll \gamma$.

Cases where $\eta = 1$.

The diffusion approximation given in (4.1.25) is replaced here by the corresponding approximation for the multistep algorithm

(i) $\qquad \begin{pmatrix} d\widetilde{Z}_t \\ d\widetilde{z}_t \end{pmatrix} = \begin{pmatrix} A_* & B_* k_z \\ C_* & D_* k_z \end{pmatrix} \begin{pmatrix} \widetilde{Z}_t \\ \widetilde{z}_t \end{pmatrix} dt + \begin{pmatrix} B_* \\ D_* \end{pmatrix} Q^{\frac{1}{2}} dW_t^z$

(ii) $\qquad \begin{pmatrix} d\widetilde{\Theta}_t \\ d\widetilde{\theta}_t \end{pmatrix} = \begin{pmatrix} A & Bh_\theta \\ C & Dh_\theta \end{pmatrix} dt + \begin{pmatrix} Bh_z \\ Dh_z \end{pmatrix} \widetilde{z}_t dt + \begin{pmatrix} B \\ D \end{pmatrix} R^{\frac{1}{2}} dW_t^\theta \qquad (4.2.28)$

where (W_t^z) and (W_t^θ) are two independent Wiener processes; the various functions h_θ, h_z, Q and R, are evaluated as in (4.1.25) along the trajectories of the ODEs. Globally, equations (4.2.28) are the analogue of (4.1.25): we have split them into two parts for improved readability. To obtain these equations, it is sufficient to calculate the usual objects for the stochastic approximation (4.2.17), where $\mu = \gamma$, namely:

1. the derivative of the mean field,
2. the cumulative covariance, as in formulae (4.1.18) and (4.1.19).

This is not a difficult calculation, but it would take up an excessive amount of space on this page.

Cases in which $\eta \to 0$.

The approximation error $[\tilde{z}]_s$ satisfies equation (4.2.28-i), where the time index s replaces t. The dynamic equation for $[\tilde{\theta}]_t$ is

$$\begin{pmatrix} d\tilde{\Theta}_t \\ d\tilde{\theta}_t \end{pmatrix} = \begin{pmatrix} A & Bh_\theta \\ C & Dh_\theta \end{pmatrix} \begin{pmatrix} \tilde{\Theta}_t \\ \tilde{\theta}_t \end{pmatrix} dt + \begin{pmatrix} 0 \\ R^{\frac{1}{2}} \end{pmatrix} dW_t^\theta \qquad (4.2.29)$$

which is derived from (4.2.28-ii) by ignoring the (relatively negligible) contribution of $\tilde{z}$.

With these modifications, formula (4.1.28) remains entirely valid, giving in all cases

$$\begin{aligned} E\|\theta_n - z_n\|^2 &= \|\theta(t) - z(s)\|^2 & \text{(bias)} \\ &\quad + \gamma E\|\tilde{\theta}_t - \eta^{\frac{1}{2}}\tilde{z}_s\|^2 & \text{(variance)} \\ &\quad + o(\gamma) \\ \mu &= \eta\gamma, \qquad n = \frac{s}{\mu} = \frac{t}{\gamma} \end{aligned} \qquad (4.2.30)$$

where $\theta(t), z(s), \tilde{\theta}_t$ and $\tilde{z}_s$ are given by (4.2.27), and (4.2.28) when $\eta = 1$, and by (4.2.27), (4.2.28-i) with s replacing t, and (4.2.29), when $\eta \to 0$.

4.2.3.2 Theory when the Hypermodel has Drift Zero

This corresponds to the case in which the ODE (4.2.27-i) of the hypermodel has an attractor z_*. In this section, we shall calculate an optimal filter $\Gamma(\sigma)$. Among the invariants of this filter is its Smith–McMillan degree (Kailath 1980), i.e. the minimal dimension of Z for which there exists a state variable representation of the filter Γ_* of the ODE (4.2.27-i). We shall therefore require that in the ODE (4.2.4) of the algorithm we have

$$\dim \Theta \geq \text{Smith–McMillan degree of } \Gamma_* \qquad (4.2.31)$$

But then, **provided that we add components which are neither observable nor controllable to the state Z, we may assume for the remainder of this section that Θ and Z have the same dimension.**

4.2 Multistep Algorithms

In keeping with formulae (4.1.31), in all that follows, all the functions k_z, h_θ $h_z = -h_\theta, Q$ and R are evaluated at the equilibrium point z_*. We shall not treat this case in its full generality, but we shall content ourselves to giving the generalisation of Theorem 1, i.e. to treating the case in which

$$k_z = 0 \qquad (4.2.32)$$

This case contains models of the form given in (4.2.15). Following the steps given in Paragraph 4.1.2.2, in the case of drift zero, a convenient form of (4.2.30) is

$$\lim_{n\to\infty} E\|\theta_n - z_n\|^2 \approx \gamma E\|\tilde{\theta}_t - \tilde{z}_t\|^2 \qquad (4.2.33)$$

and thus we must ensure that the process $(\tilde{\theta}_t - \tilde{z}_t)$ is indeed stationary, and calculate its covariance. Using (4.2.32) and Assumption NS3, (4.2.28) may be rewritten in the form

$$(i) \qquad \begin{pmatrix} d\tilde{Z}_t \\ d\tilde{z}_t \end{pmatrix} = \begin{pmatrix} A_* & 0 \\ C_* & 0 \end{pmatrix} dt \begin{pmatrix} \tilde{Z}_t \\ \tilde{z}_t \end{pmatrix} + \begin{pmatrix} B_* \\ D_* \end{pmatrix} Q^{\frac{1}{2}} dW_t^z$$

$$(ii) \qquad \begin{pmatrix} d\tilde{\Theta}_t \\ d\tilde{\theta}_t \end{pmatrix} = \begin{pmatrix} A & Bh_\theta \\ C & Dh_\theta \end{pmatrix} \begin{pmatrix} \tilde{\Theta}_t \\ \tilde{\theta}_t - \tilde{z}_t \end{pmatrix} dt + \begin{pmatrix} B \\ D \end{pmatrix} R^{\frac{1}{2}} dW_t^\theta \qquad (4.2.34)$$

But then, if we set

$$\delta\Theta_t = \tilde{\Theta}_t - \tilde{Z}_t, \qquad \delta\theta_t = \tilde{\theta}_t - \tilde{z}_t \qquad (4.2.35)$$

the following equation describing the dynamics of $(\delta\Theta_t, \delta\theta_t)$ may be deduced from (4.2.34):

$$\begin{pmatrix} d(\delta\Theta_t) \\ d(\delta\theta_t) \end{pmatrix} = \begin{pmatrix} A & Bh_\theta \\ C & Dh_\theta \end{pmatrix} \begin{pmatrix} \delta\Theta_t \\ \delta\theta_t \end{pmatrix} dt + \begin{pmatrix} A - A_* \\ C - C_* \end{pmatrix} Z_t dt$$
$$+ \begin{pmatrix} B \\ D \end{pmatrix} R^{\frac{1}{2}} dW_t^\theta - \begin{pmatrix} B_* \\ D_* \end{pmatrix} Q^{\frac{1}{2}} dW_t^z \qquad (4.2.36)$$

From now on, we shall work from equations (4.2.34-i) and (4.2.36). We shall suppose from now on that

$$\text{Re}\,\lambda(A_*) < 0 \qquad (4.2.37)$$

and that the filter used $\Gamma(\sigma)$ is admissible; this implies condition (4.2.11), together with the **stationarity** of the joint process (self-explanatory notation):

$$\begin{pmatrix} [\tilde{z}]_t \\ [\delta\theta]_t \end{pmatrix} \qquad (4.2.38)$$

From this, we deduce that the covariance matrix Π of this joint process is the solution of the (very) composite Lyapunov equation

$$\boldsymbol{F}\Pi + \Pi\boldsymbol{F}^T + \boldsymbol{Q} = 0$$

$$\boldsymbol{F} := \begin{pmatrix} A_* & 0 & 0 & 0 \\ C_* & 0 & 0 & 0 \\ A - A_* & 0 & A & Bh_\theta \\ C - C_* & 0 & C & Dh_\theta \end{pmatrix}$$

$$\boldsymbol{Q} :=$$

$$\begin{pmatrix} \begin{pmatrix} B_* \\ D_* \end{pmatrix} Q[B_*^T, D_*^T] & -\begin{pmatrix} B_* \\ D_* \end{pmatrix} Q[B_*^T, D_*^T] \\ -\begin{pmatrix} B_* \\ D_* \end{pmatrix} Q[B_*^T, D_*^T] & \begin{pmatrix} B_* \\ D_* \end{pmatrix} Q[B_*^T, D_*^T] + \begin{pmatrix} Bh_\theta \\ Dh_\theta \end{pmatrix} \Delta[(Bh_\theta)^T, (Dh_\theta)^T] \end{pmatrix}$$
(4.2.39)

We are now in a position to present the analogue of Theorem 1, which describes the optimal filter relative to the figure of merit (4.2.33).

Theorem 4. An optimal filter $\Gamma_{opt}(\sigma)$ for "filtered random walk" hypermodels. The notation is as in formulae (4.2.16).
(i) *Suppose* $\dim \Theta$ *is fixed and satisfies (4.2.31). Let*

$$\Delta := h_\theta^{-1} R h_\theta^{-T} \quad (4.2.40)$$

denote the intrinsic quality criterion of the random field $H(\theta_{n-1}, X_n)$. *The following formula*

$$\Gamma_{opt}(\sigma) := D_{opt} + C_*(\sigma I - A_*)^{-1} B_{opt}$$
$$B_{opt} = P_{12} \cdot h_\theta^T \cdot R^{-1}, \quad D_{opt} = P_{22} \cdot h_\theta^T \cdot R^{-1}$$
$$\boldsymbol{P}_{opt} := \begin{pmatrix} P_{11} & P_{12} \\ P_{21} & P_{22} \end{pmatrix} \quad (4.2.41)$$

where $\boldsymbol{P}_{opt}$ *is the unique symmetric, non-negative solution of the algebraic Riccati equation*

$$P\begin{pmatrix} 0 & 0 \\ 0 & \Delta^{-1} \end{pmatrix} P = \begin{pmatrix} A_* & 0 \\ C_* & 0 \end{pmatrix} P + P\begin{pmatrix} A_*^T & C_*^T \\ 0 & 0 \end{pmatrix} + \begin{pmatrix} B_* \\ D_* \end{pmatrix} Q(B_*^T D_*^T)$$
(4.2.42)

define an optimal filter $\Gamma_{opt}(\sigma)$, *in the sense that*

$$\boldsymbol{P}_{opt} = \min_{\Gamma(\sigma)} \operatorname{cov}([\delta\theta]_t) \approx \min_{\Gamma(\sigma)} \lim_{n \to \infty} \gamma \operatorname{cov}([\theta]_n - [z]_n) \quad (4.2.43)$$

the minimum (not necessarily unique) being effectively realised by Γ_{opt}.
(ii) *Moreover it is possible to realise*

$$\min_{\dim \Theta} \min_{\Gamma} \operatorname{cov}(\delta\theta_t) \approx \min_{\dim \Theta} \min_{\Gamma} \lim_{n \to \infty} \gamma \operatorname{cov}(\theta_n - z_n) \quad (4.2.44)$$

as follows:

4.2 Multistep Algorithms

1. *choose a minimal realisation of* $\Gamma_{opt}(\sigma)$, *whence*

$$\dim \Theta = \dim Z = \text{Smith–McMillan degree of } \Gamma_* \qquad (4.2.45)$$

2. *calculate* Γ_{opt} *according to (4.2.41) and (4.2.42).*

Comments. Point (i) is a result for $\dim \Theta$ fixed, the minimisation depending on the joint covariance of the 2-tuple $(\delta\Theta, \delta\theta)$, this is a superfluously strong result, since only $\delta\theta$ interests us. Point (ii) gives the full solution: it improves upon (i) in that it gives an indication of the right dimension to choose for Θ. The general conclusion is quite simple: in order to obtain the optimal filter Γ_{opt}, the formulae of Point (i) of the theorem should be applied to the minimal realisation Γ_*.

Proof of Theorem 4. We begin with Point (i). Let us consider the following dynamical system:

(i) $$dX_t = \begin{pmatrix} A_* & 0 \\ C_* & 0 \end{pmatrix} X_t dt + \begin{pmatrix} B_* \\ D_* \end{pmatrix} Q^{\frac{1}{2}} dV_t$$

(ii) $$dY_T = (0 \ h_\theta) X_t dt + R^{\frac{1}{2}} dW_t \qquad (4.2.46)$$

where (V_t) and (W_t) denote two independent Wiener processes. Let us also introduce the dynamical system

$$dT_t = \begin{pmatrix} A & Bh_\theta \\ C & Dh_\theta \end{pmatrix} T_t dt - \begin{pmatrix} B \\ D \end{pmatrix} dY_t \qquad (4.2.47)$$

If we express dY_t as in (4.2.46-i), we obtain

$$dT_t = \begin{pmatrix} A & Bh_\theta \\ C & Dh_\theta \end{pmatrix} T_t dt - \begin{pmatrix} 0 & Bh_\theta \\ 0 & Dh_\theta \end{pmatrix} X_t dt + \begin{pmatrix} B \\ D \end{pmatrix} R^{\frac{1}{2}} dW_t \qquad (4.2.48)$$

But then, since $h_z = -h_\theta$, it becomes clear that, with a suitable change of notation, **equations (4.2.46-i) and (4.2.48) coincide with equations (4.2.28)**. On the other hand, (4.2.47) shows that T_t is $\mathcal{Y}_t$-measurable, where $\mathcal{Y}_t$ is the σ-algebra of the history of Y prior to the instant t, defined by

$$\mathcal{Y}_t := \sigma\{Y_s - Y_u : u \leq s \leq t\} \qquad (4.2.49)$$

Consequently, the problem of optimising the gain block matrix $[\Gamma]$ is equivalent to the problem

$$\min_{[\Gamma]} E\|T_t - X_t\|^2 \qquad (4.2.50)$$

the solution of which is known to be given by the Kalman filter which allows us to calculate the dynamical system with conditional expectation

$$T_t^{opt} = \hat{X}_t := E(X_t \mid \mathcal{Y}_t) \qquad (4.2.51)$$

The algebraic form of the Kalman filter equations for (4.2.46) which provides the matrices A, B, C, D, which together with (4.2.47) give a description of $\hat{X}_t$ in the stationary case, comprises exactly the equations (4.2.41) and (4.2.42). This proves Point (i).

We now consider Point (ii). Let us suppose that

$$\dim \Theta > \text{Smith–McMillan degree of } \Gamma_* \qquad (4.2.52)$$

Then let

$$\Gamma_* = D_* + C_+(\sigma I - A_+)^{-1} B_+ \qquad (4.2.53)$$

be a minimal realisation of Γ_* with consequently

$$\dim A_+ = \text{Smith–McMillan degree of } \Gamma_*$$

Now let (for example)

$$A_* = \begin{pmatrix} -I_r & 0 \\ 0 & A_+ \end{pmatrix}, \quad B_* = \begin{pmatrix} 0 \\ B_+ \end{pmatrix}, \quad C_* = (0 \; C_+) \qquad (4.2.54)$$

where the identity matrix I_r is chosen such that $\dim A_+ = \dim \Theta$. The results of Point (i) apply to this construction, and it now becomes clear that the parasitic components of Z introduced here are perfectly tracked; then the filter Γ_{opt} is of the same form as (4.2.54), and the unnecessary components of Θ are eliminated. On the other hand, it is impossible to achieve optimal performance with

$$\dim \Theta < \text{degree of } \Gamma_*$$

In fact, if we choose Θ of dimension equal to the degree of Γ_*, the pairs $(C_* A_*)$ and $(A_* B_* Q^{\frac{1}{2}})$ are respectively observable and controllable. But then, the observability and controllability matrices of the pairs

$$\left\{ (0, I), \begin{pmatrix} A_* & 0 \\ C_* & 0 \end{pmatrix} \right\} \text{ and } \left\{ \begin{pmatrix} A_* & 0 \\ C_* & 0 \end{pmatrix}, \begin{pmatrix} B_* \\ D_* \end{pmatrix} Q^{\frac{1}{2}} \right\}$$

are given, respectively, by

$$\begin{pmatrix} 0 & I \\ C_* & 0 \\ C_* A_* & 0 \\ \cdots & \cdots \\ \cdots & \cdots \\ C_* A_*^d & 0 \end{pmatrix} \quad \text{and}$$

$$\begin{pmatrix} B_* Q^{\frac{1}{2}} & A_* B_* Q^{\frac{1}{2}} & A_*^2 B_* Q^{\frac{1}{2}} & \cdots & A_*^{d+1} B_* Q^{\frac{1}{2}} \\ D_* Q^{\frac{1}{2}} & C_* B_* Q^{\frac{1}{2}} & C_* A_* B_* Q^{\frac{1}{2}} & \cdots & C_* A_*^d B_* Q^{\frac{1}{2}} \end{pmatrix}$$

4.2 Multistep Algorithms

From the observability of (C_*A_*), and the controllability of (A_*B_*), and using the fact that $D_*Q^{\frac{1}{2}}_*$ is invertible, we verify that these matrices are of full rank. But then, it is known (cf. (Goodwin and Sin 1984)) that P_{opt}, the solution of (4.2.42), is **positive definite**, and the filter Γ_{opt} cannot then be reduced without loss of performance. This completes the proof of Theorem 4. □

The use of multistep algorithms is thus effectively required to improve the tracking ability of hypermodels having complex dynamics.

4.2.3.3 Theory when the Hypermodel has Non-Zero Drift

Using the techniques of the previous section (increasing $\dim \Theta$ or $\dim Z$), we may restrict ourselves to the case in which Θ and Z are of the same dimension. We may then safely copy the calculations of Paragraph 4.1.2.3, applying them to the stochastic approximations written in the form (4.2.17), since these calculations do not rely on Assumption NS3 (recall that NS3 is the only assumption which is not satisfied by the augmented vector field of (4.2.17-ii)).

We now let $\eta \to 0$. To obtain the bias, we apply formula (4.1.50), making the substitutions

$$\theta \leftarrow \begin{pmatrix} \Theta \\ 0 \end{pmatrix}, \qquad z \leftarrow \begin{pmatrix} Z \\ z \end{pmatrix}$$

$$\Gamma \leftarrow \begin{pmatrix} A & B \\ C & D \end{pmatrix}, \qquad h_\theta \leftarrow \begin{pmatrix} I & 0 \\ 0 & h_\theta \end{pmatrix}$$

$$k(z) \leftarrow \begin{pmatrix} Z \\ k(z) \end{pmatrix}$$

Next we apply the formula for the inversion of partitioned matrices

$$\begin{pmatrix} A & Bh_\theta \\ C & Dh_\theta \end{pmatrix}^{-1} = \begin{pmatrix} A^+ & -A^{-1}BD^+ \\ -D^+CA^{-1} & D^+ \end{pmatrix}$$

$$A^+ := (A - BD^{-1}C)^{-1}$$

$$D^+ := (Dh_\theta - CA^{-1}Bh_\theta)^{-1} = (\Gamma(0)h_\theta)^{-1} \quad (4.2.55)$$

in order, finally to obtain the following expression for the bias

$$\theta = z + \eta(\Gamma(0)h_\theta(z;z))^{-1}(k(z) - CA^{-1}Z) + o(\eta) \quad (4.2.56)$$

Note that the bias depends on the 2-tuple (z, Z).

To obtain the variance, we use formula (4.2.28-ii) which gives $E\|\tilde{\theta}_t\|^2$ via the solution of an extended Lyapunov equation analogous to (4.1.56). Let us summarise the results we have obtained.

Theorem 5. Hypermodel with non-zero drift.
(i) The optimal order of magnitude of the scalar gain γ is given by

$$\gamma = \mu^{\frac{2}{3}} \tag{4.2.57}$$

(ii) Using (4.2.48), we obtain the following formula, where the various functions are evaluated along the trajectory $(Z(s_n), z(s_n))$ of the solution of the hypermodel for $s_n = n\mu \geq s_0$ ($s_0 > 0$ arbitrary but fixed):

$$E\|\theta_n - z_n\|^2 = \mu^{\frac{2}{3}}(\|(\Gamma(0) \cdot h_\theta)^{-1}(k - CA^{-1}Z)\|^2 + \text{Tr } P_{22}) + o(\mu^{\frac{2}{3}})$$
(4.2.58)

where

$$P := \begin{pmatrix} P_{11} & P_{12} \\ P_{21} & P_{22} \end{pmatrix} \tag{4.2.59}$$

is a solution of the Lyapunov equation

$$\begin{pmatrix} A & Bh_\theta \\ C & Dh_\theta \end{pmatrix} P + P \begin{pmatrix} A & Bh_\theta \\ C & Dh_\theta \end{pmatrix}^T + \begin{pmatrix} Bh_\theta \\ Dh_\theta \end{pmatrix} \Delta((Bh_\theta)^T (Dh_\theta)^T) = 0$$
(4.2.60)

4.2.3.4 Conclusion

Finally, subject to changes in detail, and with the exception of the case we did not investigate, we restate the conclusions of Paragraph 4.1.2.4.

- The tracking/accuracy compromise again comes to the fore.
- The validity of the **intrinsic quality criterion** Δ, introduced for one-step algorithms, extends without restriction to multistep algorithms, as is shown by formulae ((4.2.42) (4.2.43)) and ((4.2.58), (4.2.59), (4.2.60)).
- The role of drift in the hypermodel is again critical, it determines the choice of the ratio $\eta = \mu/\gamma$.
- The new point is that multistep algorithms give a real improvement as soon as the hypermodel has non-trivial dynamics. This point will be examined in greater detail in Subsection 4.2.4. On the other hand, multistep algorithms are of no particular value as algorithms of decreasing gain for the identification of a fixed system; this problem was examined in Chapter 3. The last point is not however surprising, since almost all the known algorithms which attain the Cramer-Rao bound (optimality property in the sense of Chapter 3, Proposition 4) are one-step.

4.2 Multistep Algorithms

4.2.4 Effective Realisation

Once again, we recall that the user only has access to the algorithm in the form (4.2.6) or (4.2.9), and that expressions (4.2.16) to (4.2.19) are simply models designed for the study of the system/algorithm interaction. We follow now, in somewhat less detail, the plan of Subsection 4.1.3.

4.2.4.1 Reformulation of the Results of Subsection 4.2.3

We proceed to eliminate the parameter μ which is then included in the vector field $K(z_{n-1}, \zeta_n)$, which we assume to be small. The results are similar to those of Paragraph 4.1.3.1, and may be summarised as follows: γ **is omitted from formulae (4.2.43) and (4.2.44) of Theorem 4, and** μ **is omitted from formula (4.2.58) of Theorem 5**, γ is thus included in the augmented gain $[\Gamma]$ defined in (4.2.8).

Note that in the derivation of algorithm (4.2.9), the replacement of $\gamma[\Gamma]$ by $[\Gamma]$, now corresponds to the omission of the parameter γ, both in algorithm (4.2.9-i) and in the expression for the filter Γ_γ given in (4.2.9-ii).

4.2.4.2 Guide to the Design of Multistep Algorithms

Stage 1. *Choice of the vector field* $H(\theta_{n-1}, X_n)$.

The comments of Paragraph 4.1.3.2 regarding one-step algorithms are still valid, since the intrinsic quality criterion is valid for multistep algorithms.

Stage 2. *Choice of the hypermodel.*

The general comments for one-step algorithms are again valid. To illustrate the additional flexibility introduced by multistep algorithms, we shall examine, via a simple example, some games that may be played with the hypermodel.

Analysis of an Example.

We consider the following two hypermodels in which, for increased understanding, the small parameter μ is made explicit. The first hypermodel has already been seen; we recall formula (4.1.11)

$$\begin{aligned} z_n &= z_{n-1} + \mu(A z_{n-1} + W_n) \\ \operatorname{Re} \lambda(A) &< 0 \end{aligned} \qquad (4.2.61)$$

We consider secondly the new hypermodel given by

$$\begin{aligned} Z_n &= Z_{n-1} + \mu(z_{n-1} - Z_{n-1}) \\ z_n &= z_{n-1} + \mu(A Z_{n-1} + W_n) \end{aligned} \qquad (4.2.62)$$

where A is the same matrix as in (4.2.61). For μ small, these two hypermodels behave in the same way; since A is stable, the difference $z_n - Z_n$ becomes small compared with z_n for n sufficiently large. The advantage of (4.2.62) is that it corresponds to a "filtered random walk hypermodel", as in Theorem 4: we know how to design the optimal multistep algorithm for this model. Moreover, Theorem 4 tells us that a multistep algorithm is actually necessary.

This example is a good illustration of a general rule: if a given behaviour can be described using a filtered random walk, this is preferable to the use of a more complex field $k(z_{n-1}, \zeta_n)$.

Stage 3. *Calculation of the filter $\Gamma(\sigma)$.*

The comments on one-step algorithms are applicable in their entirety.

4.2.4.3 Examples

The Phase-Locked Loop.

So-called "higher order" phase-locked loops are frequently used (cf. (Lindsey and Simon 1973)), at least for analogue loops. Second order loops correspond to the use of a filter $\Gamma(\sigma)$ of Smith–McMillan degree 1 (i.e. a rational function of degree 1), third order loops correspond to $d^0\Gamma = 2$, filters of higher order are rarely used. Consider the hypermodel

$$\begin{aligned} Z_n &= Z_{n-1} - \mu(\alpha Z_{n-1} + \beta W_n) \\ z_n &= z_{n-1} + \mu(Z_{n-1} + W_n) \end{aligned} \qquad (4.2.63)$$

where we have $0 < \alpha \ll 1$, $0 < \mu \ll 1$, and (W_n) is a standard white noise. The coefficient β is chosen so that

$$EZ_n^2 \approx \frac{\mu}{2\alpha}\beta^2 := \sigma^2 \qquad (4.2.64)$$

is of an order of magnitude specified in advance by the user. With the given choices of μ and α (4.2.63) is a model of drift of (z_n); this drift is slowly varying, since $\alpha \ll 1$. But then the modification of the formulae of Theorem 4 proposed in Paragraph 4.2.4.2 ensures that the best tracking algorithm for model (4.2.63) is a second order loop of the form

$$\phi_n = \phi_{n-1} + \frac{\gamma_0 + \gamma_1 q^{-1}}{1 - (1-\alpha\mu)q^{-1}}\varepsilon_n \qquad (4.2.65)$$

where ε_n is the usual error signal of the loop, whilst the coefficients γ_0 and γ_1 are obtained by numerical solution of the Riccati equation associated with this example, as in Theorem 4. Our proposed method allows the user to choose the parameters of the hypermodel (4.2.63), namely α, μ and σ^2, rather than α, γ_0 and γ_1 directly, since, in particular, $\frac{\gamma_0}{\gamma_1}$ is difficult to guess.

There is better to come: similarly, we may investigate the case of **jitter** corresponding to oscillatory behaviour. The best way to proceed is as follows. We begin with **a model of the jitter in continuous time**, where, to avoid

4.2 Multistep Algorithms

any confusion with the variance, we use s to denote the variable of the Laplace transform

$$\dot{z} = \frac{\beta s}{s^2 + \alpha s + \omega^2} \nu \qquad (4.2.66)$$

where ν is a white noise, ω is the frequency of the jitter, and α is a small positive parameter which will ensure that the model is stable, and where β is chosen so that

$$\frac{\beta^2}{\alpha \omega^2} = \text{amplitude of the jitter} \qquad (4.2.67)$$

Additionally, we denote the sampling period by μ, where it is assumed that

$$\mu \omega \ll 1 \qquad (4.2.68)$$

This latter equation reflects the fact that, after sampling, the variations of the true system are slow. Finally, we use μ and the matrices

$$A_* := \begin{pmatrix} 0 & 1 \\ -\omega^2 & -\alpha \end{pmatrix}, \quad B_* := \begin{pmatrix} 0 \\ \beta \end{pmatrix}$$
$$C_* := (0\ 1), \quad D_* = 0 \qquad (4.2.69)$$

which, together with the quality criterion Δ of the loops, allow us to calculate the coefficients γ_0, γ_1 and γ_2 in

$$\Gamma_{opt}(s) = \frac{\gamma_0 + \gamma_1 s + \gamma_2 s^2}{s^2 + \alpha s + \omega^2} \qquad (4.2.70)$$

where the additional scalar gain γ must be chosen to be equal to the sampling period μ.

Least Squares Algorithm and the Kalman filter.

We consider the following model of a linear regression with respect to a time-varying parameter

$$\begin{aligned} Z_n &= Z_{n-1} + A_* Z_{n-1} + B_* W_n \\ z_n &= z_{n-1} + C_* Z_{n-1} + D_* W_n \\ y_n &= \Phi_n^T z_{n-1} + \nu_n, \quad E(\nu_n^2) = \sigma^2 \end{aligned} \qquad (4.2.71)$$

where (W_n) is a standard white noise of the same dimension as z_n, and where the matrices A_*, B_*, C_* and D_* are small, with $\operatorname{Re} \lambda(A_*) < 0$. Exercise 4.4.5 leads the user to show that the **Kalman filter equations for model (4.2.71) constitute an effective means of approximating the optimal (in the sense of Theorem 4) multistep tracking algorithm** for a filtered random walk hypermodel.

Once again, adaptive Kalman filters enable us to make simultaneous estimates of certain hypermodel parameters, however this problem becomes frankly ill-conditioned for multistep algorithms.

4.3 Conclusions

In this chapter, we have studied the system/algorithm interaction, very generally, in relation to tracking non-stationary systems. The first lesson we learnt was that this problem is qualitatively different from the problem of optimising the rate of convergence of algorithms with decreasing gain, as described in Chapter 3. In particular, we have shown the arbitrary nature (from the asymptotic point of view) of the widely used algorithms with an exponential forgetting factor.

The theorems stated form a guide to the optimisation of adaptive tracking algorithms.

Stage 1. *Choice of the vector field $H(\theta_{n-1}, X_n)$.*
This is easy, using the intrinsic quality criterion Δ: the choice is similar to that of a functional to be minimised, and may be made without prior knowledge of the non-stationary parameters.

Stage 2. *Choice of the hypermodel.*

This choice should only take into account reliable information about the variations in the system. In the absence of such very simple information, variations, leading to one-step algorithms which are easier to implement, should be assumed. We also recommend the use of "filtered random walk" models, since these models provide an effective means of deriving optimal algorithms.

Stage 3. *Optimisation of the gains.*

We have given many illustrations of the use of the results of this chapter for such calculations. The operation may be more or less difficult, depending on the instance in hand.

In any case, after reading this chapter, the reader has the wherewithal for a better understanding of tracking problems.

4.4 Exercises

4.4.1 Average Excess Mean Square Error and Misadjustment in the Least Squares Algorithm

This exercise is inspired by the article (Widrow, McCool, Larimore and Johnson 1976), and follows (Benveniste 1984). The notation is as before: the true system is

$$y_n = \Phi_n^T z_{n-1} + \nu_n$$

where Φ_n is the regression vector, and ν_n is the noise (independent of Φ_n). The true system is represented by z which is time-varying. Let θ_n be the least

4.4 Exercises

squares estimator

$$\theta_n = \theta_{n-1} + \Gamma \Phi_n e_n$$
$$e_n = y_n - \Phi_n^T \theta_{n-1}$$

Define the average excess mean square error (AEMSE) by

$$AEMSE = \frac{Ee_n^2}{E\nu_n^2}$$

1. Prove that

$$AEMSE = 1 + \frac{E(e_n - \nu_n)^2}{E\nu_n^2}$$
$$= 1 + \frac{1}{E\nu_n^2} E((\theta_n - z_n)^T \Phi_n \Phi_n^T (\theta_n - z_n))$$

2. Suppose that z_n is given by model (4.1.79), with condition (4.1.80). Follow the reasoning used in the analysis of formulae (4.1.81) to show that

$$AEMSE \approx 1 + \frac{1}{\sigma^2} E((\theta_n - z_n)^T \Sigma (\theta_n - z_n))$$

(cf. formulae (4.1.78) for any undefined notation).

3. Using formula (4.1.65-ii), show that, when the gain Γ is constrained to be of the form γI for some scalar γ, then

$$AEMSE \approx 1 + \frac{1}{2}(\gamma \operatorname{Tr}\Sigma + \frac{1}{\gamma \sigma^2} \operatorname{Tr} Q)$$

Determine the scalar gain which minimises the AEMSE.

4. Prove that, in all cases, minimisation of the AEMSE is equivalent to minimisation of the figure of merit used in this chapter.

4.4.2 Examples of Adaptive Tuning of the Gain in Least Squares Algorithms, when the Model of the Variation of the True System is Unknown

We consider the tracking problem with the following system/algorithm interaction

$$\begin{aligned} z_n &= z_{n-1} + W_n, & \operatorname{cov} W_n &= Q \\ \theta_n &= \Phi_n^T \theta_{n-1} + \Gamma \phi_n e_n \\ e_n &= y_n - \Phi_n^T \theta_{n-1} \\ y_n &= \Phi_n^T z_{n-1} + \nu_n, & \nu_n &\perp W_n \end{aligned} \quad (4.4.1)$$

We shall examine several ideas for dealing with the case in which Q is unknown.

Let G be a family of gains Γ, specified in advance. we attempt to choose Γ in G, using the AEMSE defined in the previous exercise; this leads to the criterion:

$$J(\Gamma) = E(e_n^2(\Gamma))$$

where $e_n(\Gamma)$ is the prediction error for the gain Γ in (4.4.1).

1. **Stochastic gradient adaptive gain.**
   Let G be the set of scalar gains γ. Show that a stochastic gradient method applied to $J(\gamma)$ leads to the following algorithm, where $\nu > 0$ is a small gain:

$$\begin{aligned} \theta_n &= \theta_{n-1} + \gamma_n \Phi_n e_n \\ e_n &= y_n - \Phi_n^T \theta_{n-1} \\ \gamma_n &= \gamma_{n-1} + \nu \Phi_n^T \Psi_{n-1} e_n \\ \Psi_n &= (I - \gamma_n \Phi_n \Phi_n^T)\Psi_{n-1} + \Phi_n e_n \end{aligned}$$

   Extend this algorithm to the case where $\Gamma = \gamma \Gamma_0$, for some previously selected gain matrix Γ_0.

2. **Least squares with adaptive forgetting factor.**
   Here the algorithm is

$$\begin{aligned} \theta_n &= \theta_{n-1} + \gamma R_n^{-1} \Phi_n e_n \\ R_n &= R_{n-1} + \gamma(\Phi_n \Phi_n^T - R_{n-1}) \\ e_n &= y_n - \Phi_n^T \theta_{n-1} \end{aligned}$$

   and the forgetting factor $1 - \gamma$ is chosen to minimise the criterion

$$J(\gamma) = E(e_n^2(\gamma))$$

   Show that a stochastic gradient method leads to the following algorithm, where $\nu > 0$ is a small gain:

$$\begin{aligned} \theta_n &= \theta_{n-1} + \gamma_n R_n^{-1} \Phi_n e_n \\ R_n &= R_{n-1} + \gamma_n(\Phi_n \Phi_n^T - R_{n-1}) \\ e_n &= y_n - \Phi_n^T \theta_{n-1} \\ \gamma_n &= \gamma_{n-1} - \nu \Phi_n^T \Psi_{n-1} e_n \\ \Psi_n &= (I - \gamma_n R_n^{-1} \Phi_n \Phi_n^T)\Psi_{n-1} + (I - \gamma_n R_n^{-1} S_n R_n^{-1})\Phi_n e_n \\ S_n &= (1 - \gamma_n)S_{n-1} + \Phi_n \Phi_n^T - R_{n-1} \end{aligned}$$

3. Try to determine a common philosophy behind 1. and 2.

4.4.3 Sliding Window Algorithms

The notation is as in Exercise 3.4.3, here the aim is to study the tracking performance of this algorithm.

1. Suppose, for the time being, that θ_* is fixed, and let $\tilde{\theta}_n = \theta_n(N) - \theta_*$, $\gamma = \frac{1}{N}$ (N is assumed to be large and fixed). Prove that $\tilde{\theta}_n$ is determined by the following algorithm for $n \geq N$

$$\tilde{\theta}_n = \tilde{\theta}_{n-1} + \gamma R_n^{-1}\{[H(\theta_*, X_n) - H(\theta_*, X_{n-N})]$$
$$- [H_\theta(\theta_*, X_n) - H_\theta(\theta_*, X_{n-N})]\tilde{\theta}_{n-1}\} + \gamma^2 \varepsilon_n$$
$$R_n = R_{n-1} + \gamma(H_\theta(\theta_*, X_n) - H_\theta(\theta_*, X_{n-N}))$$
$$R_N = \frac{1}{N}\sum_1^N H_\theta(\theta_*, X_k)$$

2. Suppose that N is sufficiently large that the variables X_n and X_{n-N} may be considered to be independent (cf. Chapters 1 and 3 for temporal properties of the state vector X_n). Show, by applying Theorems 1 and 2 of Chapter 3, that for large N, the sliding window algorithm behaves like the following algorithm with exponential forgetting factor:

$$\theta'_n = \text{solution of } \mu \sum_1^n (1-\mu)^{n-k} H(\theta, X_k) = 0$$

with $\mu = 2\gamma$.

3. What can you deduce about the tracking behaviour of sliding window algorithms?

4.4.4 Another Type of Multistep Algorithm

This version of the algorithm is most commonly found in Soviet literature ((Tsypkin 1971), (Shil'man and Yastrebov 1976, 1978)). It is closely related to multistep schemes for the solution of ordinary differential equations (Henrici 1963). The exercise follows (Benveniste 1984). We use the notation q^{-1} for the delay operator. We examine the following algorithms

$$\theta_n = \theta_{n-1} + \gamma \Gamma(q^{-1}) H(\theta, X_n)$$

where $\Gamma(q^{-1})$ is a discrete-time transfer function operating on the vector field $H(\theta_{n-1}, X_n)$; note here that $\Gamma(q^{-1})$ is independent of γ.

1. Show that the associated ODE is

$$0 = \Gamma(1) h(\theta)$$

2. Show that
$$R(\theta) := \sum_{n\in\mathbb{Z}} \text{cov}_\theta(H(\theta, X_n), H(\theta, X_0))$$

and

$$R_\Gamma(\theta) := \sum_{n\in\mathbb{Z}} \text{cov}_\theta(\Gamma(q^{-1})H(\theta, X_n), \Gamma(q^{-1})H(\theta, X_0))$$

satisfy the formula

$$R_\Gamma(\theta) = \Gamma(1)^{-1} R(\theta) \Gamma(1)^{-T}$$

3. Deduce that this type of multistep scheme is of no interest either for the identification of a fixed system θ_* (with $\gamma_n \approx \frac{1}{n}$), or for the improvement of tracking performance.

4.4.5 Multistep Algorithms and the Kalman Filter

This exercise analyses the Kalman filter corresponding to formula (4.2.71). By applying the reasoning used in the case of the least squares approximation for one-step algorithms, prove that when $\text{cov}\, W_n \ll \sigma^2$, the Kalman filter corresponding to the system (4.2.71) stabilises around the optimal tracking filter given by the equations of Theorem 4 as modified by the rules given in Paragraph 4.2.4.1.

4.4.6 Joint Equalisation and Phase Recovery

This follows on from Exercise 2.7.10, the notation is as in that exercise.

In practice, the objective of algorithm (2.7.4) (with error signal $e_n(1)$ or $e_n(2)$) is to tune the rates of variation of the two estimators θ_n and ϕ_n, to take into account that the channel phase is more rapidly varying than the intersymbol interference. In place of (2.7.4), we shall use

$$\begin{aligned}
\theta_n &= \theta_{n-1} + \mu \overline{Y}_n e^{i\phi_{n-1}} e_n \\
\phi_n &= \phi_{n-1} + \gamma \,\text{Im}(c_n \bar{e}_n) \\
\mu &\ll \gamma
\end{aligned} \qquad (4.4.2)$$

1. Suppose, for the time being, that the true parameters θ_* and ϕ_* are fixed. Applying the analysis of stochastic approximations with two timescales, given in Chapter 4, determine the ODE and the diffusion approximation corresponding to (4.4.2). Be sure to distinguish the cases $\gamma \ll \eta^{\frac{2}{3}}$, $\gamma \approx \eta^{\frac{2}{3}}$, $\gamma \gg \eta^{\frac{2}{3}}$.

2. Next suppose that the true parameters θ_* and ϕ_* vary according to the models

$$\theta_{*_n} = \theta_{*_{n-1}} + W_n$$
$$\phi_{*_n} = \phi_{*_{n-1}} + \Delta\phi$$

where W_n is a white noise with covariance matrix Q, and $\Delta\phi$ is a drift with deterministic phase (frequency offset); suppose also that $\operatorname{Tr} Q \ll \Delta\phi$. How should the gains μ and γ be chosen in (4.4.2)?

4.5 Comments on the Literature

General Comments.

The first attempt to study an adaptive algorithm is found in (Widrow, McCool, Larimore and Johnson 1976), where the least squares algorithm for independent regression vectors is studied from this viewpoint; the investigation is asymptotic, and in much the same spirit as here, although it is more rudimentary. Then, there are a number of articles which study a tracking algorithm for systems with a bounded rate of variation (Farden and Sayood 1980), and most importantly (Eweda and Macchi 1983, 1985) for the least squares algorithm with m-dependent regression vectors. The latter articles analyse this example in full. Also noteworthy is (Kushner and Huang 1981), where the asymptotic investigation, whilst exotic, is closer to our approach to the study of tracking problems. The present approach is a generalisation and development of ideas found in (Benveniste and Ruget 1982).

As far as pragmatic methods of gain adjustment to track non-stationary systems are concerned; there have been numerous recent attempts at this, which in our opinion are uninteresting, since they are too ad hoc. We mention only (Bohlin 1976) and (Gersh and Kitagawa 1985) who use Kalman filters, in more or less adaptive forms, to track systems with linear Gaussian dynamics.

The introduction of multistep algorithms is due to (Tsypkin 1971), based on ideas from numerical analysis for the improvement of approximations to ordinary differential equations (Henrici 1963). Historically, we mention (Shil'man and Yastrebov 1976, 1978), for the description and study of multistep schemes based (too) narrowly on those of numerical analysis. These schemes are of little interest, except perhaps to improve transient behaviour; this type of algorithm is analysed in Exercise 4.4.4. The algorithms described here are a generalisation of those described and studied from the point of view of "large deviations" in the excellent article (Korostelev 1981)! On the other hand (Gersh and Kitagawa 1985) also use multistep methods, without saying so.

References Used in this Chapter.

(Benveniste 1984) formed the basis for this chapter. Additionally, the results and the asymptotic description were improved, and errors in the proof of the key theorem concerning the derivation of the optimal filter for multistep algorithms were corrected. The sections concerning the optimal choice of gains for least squares algorithms were introduced to quash the all too common belief (see for example (Ljung and Soderström 1983) etc.) that forgetting factors are a universal panacea for tracking non-stationary parameters.

Exercise 4.4.1 takes up and clarifies the results of (Widrow, McCool, Larimore and Johnson 1976). The other exercises are original.

Chapter 5
Sequential Detection;
Model Validation

In this chapter we continue our analysis of non-stationary systems. After having shown in Chapter 4 how adaptive algorithms may be used to track non-stationary parameters, we shall now investigate ways of proceeding when the underlying system is subject to abrupt change.

As far as adaptive techniques for system identification are concerned, the detection of abrupt changes is an indispensable complement to tracking methods: it allows **active**, not just passive intervention by the algorithm (reinitialisation, or even compensation for the estimated jump). In fact, change detection is a much more fundamental technique than the tracking of non-stationary parameters. Far from simply improving the identification process, this may become the central consideration in applications such as the segmentation of a signal into homogeneous bands, or the detection of faults in control systems. We do not intend to study this subject exhaustively: the reader is referred to (Basseville and Benveniste 1986) for further details.

The aim of this introduction to change detection is modest. Once again, we leave aside the problem of selecting the model to be used; in this respect, the same comments as for system identification apply here. We simply wish to show that **there is a very general procedure by which any adaptive algorithm may be directly associated with a sequential method of change detection.** In so doing, we will be forced to leave numerous very strong methods on the touchline, but, broken in by system identification, the reader will see that, with a reasonable amount of effort, he will be well armed to look at sequential change detection in signals and dynamical systems.

The chapter is divided into five sections. In the first section, we introduce the problem and give a number of examples. The second section will describe the basic techniques for the solution of the simplest problems. The third section is devoted to a central limit theorem upon which the general methods will depend. The fourth section describes the local tests which may be generally associated with an adaptive algorithm, and how these tests may be implemented. Lastly, in the fifth section, we shall see how and why exactly the same techniques may be applied, with some modification, to model validation.

5.1 Introduction and Description of the Problem

The description which we shall give will meet the objectives below. We consider a dynamical system which is assumed to represent a physical system whose characteristics change abruptly at some unknown instant: the objectives are:

1. to detect this change,
2. to estimate the change time,
3. to identify the nature and magnitude of the change.

We begin with a number of examples.

5.1.1 Examples

5.1.1.1 Jumps in the Mean Value of a Signal

Suppose we have a signal of the form

$$y_n = \theta_*(n) + \nu_n \qquad (5.1.1)$$

where (ν_n) is a stationary sequence of independent variables, identically distributed (with distribution μ), with mean zero, and where θ_* is a piecewise constant function (cf. Fig. 22).

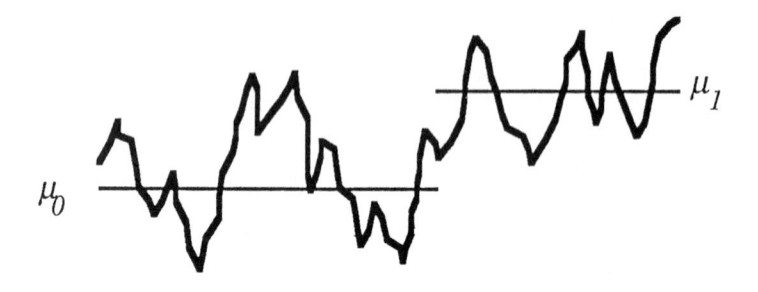

Figure 22

The aim is to detect the instants at which the mean of θ_* changes, and to estimate the mean value between jumps.

The distribution of the variable y_n is thus the image under the mapping $y \to y + \theta_*$ of the distribution μ.

5.1.1.2 Changes in an Auto-Regressive Signal

This is clearly a much harder problem: this is one of the ways of addressing in practice the detection of changes in the spectral characteristics of a signal.

5.1 Introduction and Description of the Problem

Consider an AR signal of the form

$$y_n = \sum_{i=1}^{p} a_i y_{n-i} + \sigma \nu_n \qquad (5.1.2)$$

where (ν_n) is a sequence of independent, identically distributed variables, with mean zero and variance one. Model (5.1.2) may be summarised by the vector

$$\theta_*^T := (a_1, \ldots, a_p; \sigma) \qquad (5.1.3)$$

If we set

$$\Phi_{n-1}^T := (y_{n-1}, \ldots, y_{n-p}) \qquad (5.1.4)$$

then (5.1.2) may be written in the state variable form

(i) $\quad \Phi_n = A(\theta_*)\Phi_{n-1} + B(\theta_*)\nu_n$

(ii) $\quad y_n = (1, 0, \ldots, 0)\Phi_n$

(iii) $\quad A(\theta_*) = \begin{pmatrix} a_1 & \cdots & & a_p \\ 1 & 0 & & \\ & \ddots & \ddots & \\ & & 1 & 0 \end{pmatrix}, \quad B(\theta_*) = \begin{pmatrix} \sigma \\ 0 \\ \vdots \\ 0 \end{pmatrix} \qquad (5.1.5)$

in which, in the definition of the matrix $A(\theta_*)$, the unspecified elements are zero. Formulae (5.1.5) now express the fact that (Φ_n) is a **Markov chain controlled by** θ_*. If we now assume that θ_* is piecewise constant, we thus need to

1. detect the jumps,
2. estimate the instants at which the jumps occur,
3. determine the magnitude of the jumps and the interim values of θ_*.

5.1.2 Change Detection in a Markov Process

The two examples described above illustrate once again the appropriateness of the Markov approach. In fact, we shall use Markov processes as the general setting for detection problems. We shall not consider the problems associated with multiple successive detection: since the observations are taken on the fly, it is in general possible to detect these jumps successively, so that at a given moment, attention may be concentrated solely on the detection of the next jump.

Thus let (X_n) be a process with a Markov representation controlled by the parameter θ_*, which satisfies

$$P\{\xi_n \in G | \xi_{n-1}, \xi_{n-2}, \ldots\} = \int_G \pi_{\theta_*}(\xi_{n-1}, dx)$$
$$X_n = f(\xi_n) \qquad (5.1.6)$$

where $\pi_{\theta_*}(\xi, dx)$ is the transition probability (parameterised by θ_*) of the Markov chain (ξ_n). Note that this definition is much simpler than (2.1.2) since θ_n has disappeared from the expression. Model (5.1.6) is a model of the **true system**. The problem of **sequential change detection** in the system (5.1.6) may be formulated as follows:

[**SD**]. *There exists a time* $r : 0 < r \leq +\infty$, *such that* (X_n) *is controlled by*

$$\theta_* = \theta_0 \text{ for } n < r$$
$$\theta_* = \theta_1 \text{ for } n \geq r$$

Thus the problems to be resolved, given the observations $X_1, \ldots, X_n$ up to time n, are the following:

1. **Detection.** Decide if

$$n < r \quad \text{(there has been no change)}$$

or if

$$r \leq n \quad \text{(there has been a change)}$$

2. **Estimation.** If it has been decided that $r \leq n$, estimate the moment at which r changed.

3. **Identification.** If one or other of these parameters is unknown, identify θ_0, and if needed θ_1.

Of course, we may be interested only in some of these problems. For example, in fault detection problems where no diagnosis is required, it is sufficient to solve 1.

5.1.3 The Problems to be Solved

As before, we leave aside the modelling problem, namely the choice of the model for the Markov representation used to describe the behaviour of the observations. We assume that the transition probability $\pi_\theta(\xi, dx)$ and the function f of model (5.1.6) are given.

Then three major questions remain to be examined (for a satisfactory solution to [SD]).

5.1.3.1 Design of the Test

Detection is generally based upon the **stopping rule** (cf. (Shiryaev 1978))

$$\nu = \min\{n : l_n(X_1, \ldots, X_n) \geq \lambda\} \tag{5.1.7}$$

5.1 Introduction and Description of the Problem

where λ is a threshold to be chosen, and l_n is a function of the n coordinates $X_1, \ldots, X_n$. **Thus at time ν, a change in θ_* is detected.**

Note that it is sufficient to observe the sample $X_1, \ldots, X_n$ up to the instant n, to know if $\nu \leq n$. In probabilistic terms, ν is a stopping time with respect to the family of σ-algebras generated by the observations. This property justifies the term **sequential detection** for procedures of type (5.1.7). The detector design incorporates both the choice of the family of functions l_n, and the choice of threshold (this problem is often resolved experimentally).

Furthermore, we must construct an **estimator** $\hat{r}$ of the time of the jump. In conformity with the principle of sequential methods, $\hat{r}$ will be a function $\hat{r}(X_1, \ldots, X_\nu)$ of the observations up to the time of detection, and clearly $\hat{r} \leq \nu$.

The last point concerns the identification of θ_0 and θ_1, the values of θ_* before and after the jump. We shall return to this point later.

5.1.3.2 Performance Evaluation

Since in the course of detecting changes, there are a number of tasks to be accomplished (tasks 1 to 3 as listed in our introduction to the sequential detection problem [SD]), performance evaluation may stress one or other of the attendant aspects, namely

- the quality of the detection itself:
- the quality of the estimate of the time r of the jump, and of the unknown parameters θ_0 and θ_1.

For example, the second aspect is more pertinent for signal segmentation, whilst the first is often preferred for fault detection in control systems.

We have chosen to evaluate performance from the more classical point of view (cf. (Shiryaev 1978)) of the quality of detection. The quality of the detection may be measured in two ways, which differ mathematically, but which are in fact closely related.

The Sequential Point of View.

This may be summarised by the set {mean time between false alarms, detection delay}. We now consider the stopping rule (5.1.7), our main interest being to study its essential elements, namely the family of functions l_n. If we decrease the threshold λ, we accelerate the occurrence of ν, which may correspond either to a decrease in the **detection delay** if $r < \nu$, or to a decrease in the elapsed time between **false alarms** otherwise. Usually, both these factors are studied simultaneously.

The **mean time between false alarms** is defined by

$$\alpha := E_{\theta_0}(\nu) \tag{5.1.8}$$

where P_{θ_0} is the distribution of X_n when $\theta_* = \theta_0$; in other words P_{θ_0} defines the distribution of X_n assuming that there is no jump, i.e. $r = +\infty$. The **detection delay** is harder to define. For any finite value r, P_r will denote the distribution of the process with Markov representation (X_n) controlled by θ_0 up to time r, and by θ_1 from time r onwards. Now we set

$$\beta(r) := \operatorname{ess\,sup}[E_r(\nu - r + 1 | X_1, \ldots, X_{r-1}) \cdot 1_{\{\nu \geq r\}}] \tag{5.1.9}$$

where $1_A(\omega) = 1$ if $\omega \in A$, $= 0$ if $\omega \notin A$. In (5.1.9), the expression in square brackets defines a random variable (which depends only on the first $r - 1$ observations), and "ess sup" denotes the supremum ignoring sets of probability zero. Since (5.1.9) does not provide a unique criterion, we may choose one of the following two formulae as a final measure of the detection delay.

(i)
$$\beta^* := \max_{r > 0} \beta(r)$$

or

(ii)
$$\beta^\infty := \limsup_{r \to \infty} \beta(r) \tag{5.1.10}$$

according (respectively) as to whether we are interested in the behaviour for all r, or in the behaviour for r large (corresponding to rare changes).

The following procedure gives a fair comparison of different tests in the sequential case

(i) choose the mean time between false alarms α,

(ii) determine the threshold λ corresponding to this α,

(iii) evaluate the detection delay for this λ.

The global viewpoint.

This applies less naturally to the essentially recursive methods studied in this book: we shall however have recourse to it. The key here is the pair {level, power}.

For each fixed n, we seek to test the hypothesis $\{n < r\}$ (there has not been a jump) against the hypothesis $\{r \leq n\}$ (there has been a jump), using the following decision rule

$$\begin{aligned} \{n < r\} & \quad \text{is decided if} \quad \{n < \nu\} \\ \{r \leq n\} & \quad \text{is decided if} \quad \{\nu \leq n\} \end{aligned} \tag{5.1.11}$$

where ν is given in (5.1.7).

The **level** is then defined by

$$\alpha'(n) := P_{\theta_0}\{\nu \leq n\} \tag{5.1.12}$$

5.1 Introduction and Description of the Problem

This is the probability of false alarm, since P_{θ_0} is the distribution corresponding to the absence of a jump.

The **power** is more complicated to define. Firstly, for fixed $r \leq n$, we let P_r denote the distribution of X_n for the case in which the jump from θ_0 to θ_1 takes place at time r. Next we set

$$\beta'(n,r) := P_r\{\nu \leq n\} \tag{5.1.13}$$

For fixed n, the following procedure is used to compare tests:

(i) fix the level $\alpha'(n)$,
(ii) determine the least threshold λ corresponding to this level,
(iii) evaluate the function $\beta'(n,r)$ for this choice of threshold.

For example, if after this procedure, two tests l_1 and l_2 have $\beta'_1(n,r) \geq \beta'_2(n,r)$ for all $r \leq n$, then we say that l_1 is **uniformly more powerful** than l_2: this is a very strong property. Since it is desired to eliminate the parameter n, the behaviour of the pair {level, power} is generally studied as n tends to infinity. Global performance evaluation is more difficult.

5.1.3.3 Effective Realisation

Except in very special cases, change detection is a difficult problem. The theoretical procedures which we shall describe, together with corresponding techniques for performance evaluation provide a basic methodology. After successive simplifications, the operational realisation may be quite different from the original theory. Certain types of simplification are to be recommended, we shall use examples to describe this point. The reader is referred to (Basseville and Benveniste 1986) for more details on this.

5.2 Two Elementary Problems, and their Solution

These two problems illustrate our case. But, more to the point, they constitute two important building blocks for the more complex constructions which we shall describe later.

5.2.1 The Page–Hinkley Stopping Rule

This is the simplest detection problem imaginable. Suppose we have two **known** distributions μ_0 and μ_1 on $\mathbb{R}$. We consider a sequence of independent variables $(x_n)_{n \geq 1}$ such that x_n has distribution μ_0 up to time $r-1$, then distribution μ_1 from time r, where $r \leq +\infty$ is unknown. Then the two problems to be resolved are:

1. detection of the change,
2. estimation of the change time $\hat{r}$.

The solution is given by the Page–Hinkley stopping rule. The most intuitive way to derive this rule is the following. Consider a fixed sample $x_1, \ldots, x_n$ of size n. For r between 1 and n, consider the following two hypotheses

H_0 : x_k has distribution μ_0 for $k = 1, \ldots, n$

H_1 : x_k has distribution μ_0 for $k = 1, \ldots, r-1$, and μ_1 for $k = r, \ldots, n$.

The log-likelihood ratio of these two hypotheses is then given by

$$\log \boldsymbol{L}_{H_1/H_0} = \sum_{k=r}^{n} \log \frac{d\mu_1}{d\mu_0}(x_k) := S_r^n(\mu_0, \mu_1) \tag{5.2.1}$$

and is denoted simply by S_r^n; in this formula, $\frac{d\mu_1}{d\mu_0}$ is a version of the Radon–Nikodym derivative of μ_1 with respect to μ_0.

To obtain the most favourable value of r for H_1, we set

(i) $\qquad G_n := \max_{1 \leq r \leq n} S_r^n = S_1^n - \min_{1 \leq r \leq n} S_1^r$

(ii) $\qquad \hat{r}_n := \arg\max_{1 \leq r \leq n} S_r^n \tag{5.2.2}$

The Page–Hinkley stopping rule is then:

(i) $\qquad \nu = \min\{n : G_n \geq \lambda\}$

(ii) $\qquad \hat{r} := \hat{r}_\nu \tag{5.2.3}$

Note that G_n may be calculated recursively using the formula

$$G_n = \left(G_{n-1} + \log \frac{d\mu_1}{d\mu_0}(x_n) \right)_+ \tag{5.2.4}$$

where $x_+ = \max(x, 0)$. The behaviour of this test is shown in Fig. 23.

5.2 Two Elementary Problems, and their Solution

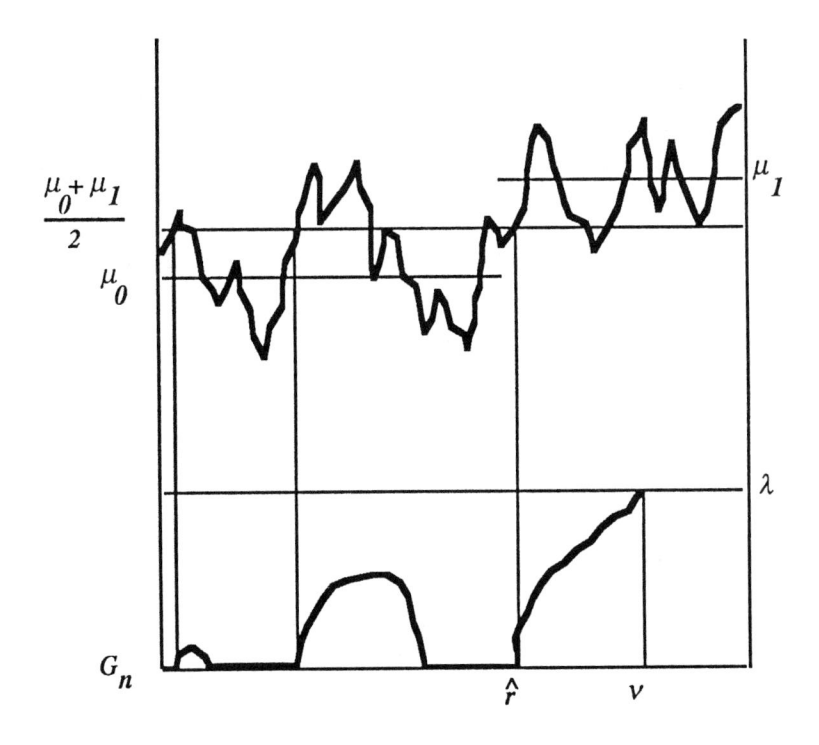

Figure 23. Page's test, behaviour.

The statistic S_r^n introduced in (5.2.1) is called the **cumulative sum**.

5.2.1.1 Gaussian Case

This corresponds to the distributions

$$\begin{aligned}\mu_0 &:= N(\theta_0, \sigma^2)\\ \mu_1 &:= N(\theta_1, \sigma^2).\end{aligned} \qquad (5.2.5)$$

Easy calculations give the following expression for the cumulative sum S_r^n:

$$S_r^n = \frac{\theta_1 - \theta_0}{\sigma^2} \sum_{k=r}^{n} \left(x_k - \frac{\theta_0 + \theta_1}{2} \right) \qquad (5.2.6)$$

In particular, (5.2.6) explains the behaviour of the cumulative sum S_r^n in Fig. 23.

5.2.1.2 Bernoulli Case

This corresponds to

$$\begin{aligned}\mu_0\{1\} &= \theta_0, & \mu_0\{-1\} &= 1 - \theta_0 \\ \mu_1\{1\} &= \theta_1, & \mu_1\{-1\} &= 1 - \theta_1\end{aligned} \qquad (5.2.7)$$

Here, it is easy to calculate G_n directly using (5.2.4):

$$G_n = \left(G_{n-1} + 1_{\{x_n=1\}} \cdot \log \frac{\theta_1}{\theta_0} + 1_{\{x_n=-1\}} \cdot \log \frac{1-\theta_1}{1-\theta_0} \right)_+ \qquad (5.2.8)$$

5.2.1.3 Properties of the Test

This very simple test is strongly recommended: it performs excellently in practice. Its theoretical properties are evaluated from the global viewpoint in (Deshayes and Picard 1986); the results are complicated, which is not surprising, since, as we have seen, the global approach is difficult. On the other hand, (Moustakides 1986) showed that Page's rule is optimal from the sequential point of view, and that taking the detection delay β^* (cf. (5.1.10)): for an arbitrary false alarm rate ν fixed in advance, as given by (5.2.1), (5.2.2) and (5.2.3), provides the shortest detection delay.

5.2.2 Detection of a Change in the Mean of a Sequence of Independent Gaussian Variables when the Mean after the Jump is Unknown

The new point here is that the mean after the jump is assumed to be unknown; the mean before the jump is known, and for simplicity , we assume that it is zero. Thus we have a multi-dimensional sequence (Y_n) of independent Gaussian variables, with the identical covariance matrix R, with mean zero

5.2 Two Elementary Problems, and their Solution

up to time $r-1$ and mean θ from time r onwards, where θ is unknown. This complicates the problem. In this case a Generalised Likelihood Ratio (GLR) test is recommended. This may be derived as follows.

First we fix r and θ. The log-likelihood ratio of the hypothesis "no jump to time n" against "jump of magnitude θ at time r", given the sample $Y_1, \ldots, Y_n$ is:

$$\begin{aligned} S_r^n(\theta) &= \sum_{k=r}^n Y_k^T R^{-1} Y_k - \sum_{k=r}^n (Y_k - \theta)^T R^{-1}(Y_k - \theta) \\ &= 2 \sum_{k=r}^n Y_k^T R^{-1} \theta - (n-r+1)\theta^T R^{-1}\theta \end{aligned} \quad (5.2.9)$$

If we select the most likely value of θ, still for r fixed, we have

$$S_r^n := \max_\theta S_r^n(\theta) = (\Delta_r^n)^T R^{-1} \Delta_r^n \quad (5.2.10)$$

where

$$\Delta_r^n := (n-r+1)^{-\frac{1}{2}} \sum_{k=r}^n Y_k \quad (5.2.11)$$

whilst

$$\hat{\theta}(n,r) := \arg \max_\theta S_r^n(\theta) = (n-r+1)^{-\frac{1}{2}} \Delta_r^n \quad (5.2.12)$$

Maximising (5.2.10) with respect to r, we have

(i) $\qquad G_n := \max_r S_r^n$

(ii) $\qquad \hat{r}_n = \arg \max_r S_r^n \quad (5.2.13)$

Thus the stopping rule in this case is given by

(i) $\qquad \nu = \min\{n : G_n \geq \lambda\}$

(ii) $\qquad \hat{r} = \hat{r}_\nu, \qquad \hat{\theta}_1 = \hat{\theta}(\nu, \hat{r}) \quad (5.2.14)$

The approach is quite natural: we attempt to compare the hypothesis of no jump with the hypothesis of a jump with parameters (r,θ), where the parameters (r,θ) are chosen in the most favourable way, this is the GLR principle. Thanks to the Gaussian assumptions, maximisation with respect to θ is **explicit**, this provides for a considerable simplification. Note that, regrettably, we no longer have a recursive expression for G_n analogous to (5.2.4) (see (Willsky and Jones 1976) on this subject).

Performance of the GLR.

We take the sequential view. Taking into account the independence of the Y_k, and that Δ_r^n has covariance matrix R, it follows that if there is no jump, the doubly indexed processes $(S_r^n)_{0 \leq r \leq n}$ have distribution independent of R. In particular, $(n-r+1)S_r^n$ is χ^2 with mean zero and $d(n-r+1)$ degrees of freedom, where $d = \dim Y_k$. We can already see that **the behaviour of the statistic G_n is independent of the covariance matrix R.** Thus the threshold $\lambda(\alpha)$ in (5.2.14) may be chosen solely as a function of the desired false alarm rate α, independently of R.

It is more difficult to determine the detection delay. We content ourselves with the following heuristic reasoning. If the (Y_k) have mean θ, then the mean of S_r^n is given by

$$E_\theta(S_r^n) = (n - r + 1)\theta^T R^{-1}\theta \qquad (5.2.15)$$

It is now clear that **the smaller R is, the smaller the detection delay.** We shall use this result in the sequel.

As far as the optimality properties of GLR are concerned, there is no simple result. To our knowledge, there are no results from the sequential viewpoint, as taken by us. (Deshayes and Picard 1986) contains a result of "optimality in the sense of large deviations", which this test satisfies; however, a description of this result would go beyond the scope of this book.

5.3 Central Limit Theorem and the Asymptotic Local Viewpoint

This section is the core of this chapter. We shall describe a tool which is capable, in all generality, of associating any adaptive algorithm with a method of change detection, and which relies on a general-purpose procedure.

5.3.1 The Problem

The following point is common both to the problem of change detection and to the tracking problem: since the true system is subject to variation, it must be expressed independently of the parameters accessed by user. Thus we follow once again the procedure described in Chapter 4. Subsection 4.1.1 allowed us to determine the form of the following vector field which was introduced in formula (4.1.7) to describe adaptive algorithms for non-stationary systems.

$$H(\theta, z_n; X_n) \qquad (5.3.1)$$

In (5.3.1), we recall that:

- θ effectively denotes the parameter which the user has at his disposal;
- z denotes a parameter representing the true system: the user does not have access to z;
- X_n represents the state at time n, it is a random variable, whose dynamic behaviour is represented by a Markov process controlled by the pair (θ, z).

5.3 Central Limit Theorem and the Asymptotic Local Viewpoint

We refer the reader to the important remark at the end of Paragraph 4.1.1.1 concerning the meaning of the notation (5.3.1), this point will play an important role in the proofs which follow.

We shall use (5.3.1) in a way different to that given in Chapter 4. In the sequel,

$$\theta = \theta_0$$

will represent the (fixed) **nominal model** chosen by the user. We shall try to detect small deviations of the true system z from the nominal model, by simple observation of the vector field $H(\theta, z; X_n)$.

To this end, we use a subset of Assumptions NS of Chapter 4: which, for clarity, we restate here.

Assumptions NS'.

NS'1. *The state X_n has a Markov representation controlled by the pair (θ, z). It is asymptotically stationary for any fixed (θ, z) in the effective domain of the vector field $H(\theta, z; X_n)$.*

NS'2. *The function $H(\theta, z; X)$ is sufficiently regular that*

$$h(\theta, z) := \lim_{n \to \infty} E_{\theta; z}(H(\theta, z; X_n)) \qquad (5.3.2)$$

is defined and locally Lipschitz.

NS'3. *We assume that the true system may be identified by the algorithm; this means that for all z*

$$h(\theta, z) = 0 \Leftrightarrow \theta = z \qquad (5.3.3)$$

Consequence. We have the following formula

$$h_\theta(z, z) = -h_z(z, z) \qquad (5.3.4)$$

All properties relating to the hypermodel used in Chapter 4 have been cancelled, as has Assumption NS4, which was specific to tracking problems.

In the next paragraph, we shall examine the following two problems. We have

- a nominal model θ_0 chosen by the user,
- an n element sample $X_1, \ldots, X_n$ of the state vector,

and we wish to distinguish the following situations by examining the vector field $[H(\theta_0, z; X_k)]_{1 \leq k \leq n}$:

- H_0 : $z \equiv \theta_0$;
- H_1 : $z = \theta_0 + \frac{\theta}{\sqrt{n}}$ where $\theta \neq 0$ is a fixed unknown jump;
- H_1' : there exists $\tau \in]0, 1[$, such that
  $z = \theta_0$ for times $k < \tau n$
  $z = \theta_0 + \frac{\theta}{\sqrt{n}}$ for $\tau n \leq k \leq n$, where $\theta \neq 0$ is a fixed, unknown jump.

Hypothesis H_0 says simply that the system conforms to the nominal model, Hypothesis H_1 corresponds to a constant difference of order $n^{-\frac{1}{2}}$ between the system and the nominal model, and Hypothesis H_1' corresponds to the occurrence at an unknown instant of a difference of order $n^{-\frac{1}{2}}$. The normalisation by $\sqrt{n}$ is clearly unnatural: it is carried out because it simplifies the results considerably, we shall see a convenient practical use of this. In fact, the procedure is very similar to the introduction in Chapter 4 of the small parameter μ for tracking problems. To statisticians, this approach is classical, it is known as the **asymptotic local approach**. We refer the curious reader to (Deshayes and Picard 1986) for a presentation of this approach in comparison with others.

5.3.2 Fundamental Results

We fix a nominal model θ_0, and consider the following cumulative sum, where $m \leq n$

$$(i) \qquad D_{n,m}(\theta_0, \theta) := n^{-\frac{1}{2}} \sum_{k=1}^{m} H(\theta_0, \theta_0 + \frac{\theta}{\sqrt{n}}; X_k)$$

$$(ii) \qquad D_n(\theta_0, \theta) := D_{n,n}(\theta_0, \theta) \qquad (5.3.5)$$

We shall describe the behaviour of this cumulative sum for each of the three hypotheses H_0, H_1 and H_1', under Assumptions NS'.

5.3.2.1 Behaviour of the Cumulative Sum when there is No Jump

The behaviour is governed by the following theorem:

Theorem 1.
(i) Limit of the marginal distribution: we have

$$D_n(\theta_0, 0) \xrightarrow[n \to \infty]{} N(0, R(\theta_0)) \tag{5.3.6}$$

where $R(\theta)$ is defined by formula (4.1.19) with $\theta = z = \theta_0$, which gives

$$R(\theta_0) := \sum_{n=-\infty}^{+\infty} \text{cov}_{\theta_0, \theta_0}(H(\theta_0, \theta_0; X_n), H(\theta_0, \theta_0; X_0)) \tag{5.3.7}$$

with the usual notation.
(ii) Behaviour of the trajectories: for $t \in [0, 1]$, set

$$D_{n,t}(\theta_0, \theta) := D_{n,m}(\theta_0, \theta) \text{ where } m = [nt] \tag{5.3.8}$$

Then

$$\{R(\theta_0)^{-\frac{1}{2}} \cdot D_{n,t}(\theta_0, 0)\}_{0 \le t \le 1} \xrightarrow[n \to \infty]{} (W_t)_{0 \le t \le 1} \tag{5.3.9}$$

where (W_t) is a Wiener process, and $\to$ denotes weak convergence of the processes.

Clearly (i) is implied by (ii), but we have retained (i) for clarity.

Proof. See Section 5.7 of this chapter. □

5.3.2.2 Behaviour of the Cumulative Sum when a Jump Occurs

We investigate the instance of H'_1, of which H_1 is a special case. We suppose that $\tau \in [0, 1[$, and introduce the following cumulative sum for $m \le n$

$$D_{n,m}(\theta_0, \theta, \tau) := n^{-\frac{1}{2}} \sum_{k=1}^{\min(m, [n\tau])} H(\theta_0, \theta_0; X_k)$$

$$+ n^{-\frac{1}{2}} \sum_{k=\min(m, [n\tau])+1}^{m} H(\theta_0, \theta_0 + \frac{\theta}{\sqrt{n}}; X_k) \tag{5.3.10}$$

Note that this cumulative sum expresses the effect on the vector field H of a difference of order $n^{-\frac{1}{2}}$ from the nominal model at time $n\tau$. The behaviour of the cumulative sum is governed by the following theorem:

Theorem 2. Behaviour of the cumulative sum under Hypothesis H_1'.
For $\tau \in [0,1]$, we set

$$D_{n,t}(\theta_0, \theta, \tau) := D_{n,m}(\theta_0, \theta, \tau) \text{ where } m = [nt] \qquad (5.3.11)$$

Then the process $[D_{n,t}(\theta_0, \theta, \tau)]_{0 \leq t \leq 1}$ converges weakly as n tends to infinity to the process $[D_t(\theta_0, \theta, \tau)]_{0 \leq t \leq 1}$, which is a solution of the stochastic linear differential equation

$$dD_t = -1_{\{t \geq \tau\}} \cdot h_\theta \cdot \theta.dt + R^{\frac{1}{2}}(\theta_0) \cdot dW_t \qquad (5.3.12)$$

where $R(\theta_0)$ is given by (5.3.7), whilst

$$h_\theta := h_\theta(\theta_0, \theta_0)$$

is as defined in (5.3.2).

Corollary 3. Hypothesis H_1

$$D_n(\theta_0, \theta) \xrightarrow[n \to \infty]{} N[-h_\theta \cdot \theta, R(\theta_0)] \qquad (5.3.13)$$

The corollary is derived from Theorem 2 by setting $\tau = 0$.

Proof. See Section 5.7 of this chapter. □

In conclusion, for a large sample and a small jump, Theorems 1 and 2 may be used to reduce the decision between Hypotheses H_0 (no change) and H_1' (change occurs in the given sample) to a problem of change detection in a Gaussian setting, analogous in continuous time to the problem examined in Subsection 5.2.2

5.4 Local Methods of Change Detection

In this section we shall show how the theorems of the previous section may be used to associate a method of change detection with any adaptive algorithm.

5.4.1 The Local Test

We return to the cumulative sums $D_{n,m}$ given in formulae (5.3.5) or (5.3.10); from the user's point of view, these are identical, since they differ only in a change in the true parameter z, which by definition is not observed. This common expression from the user's point of view is simply obtained by omitting from the formulae any reference to the dependence of the vector field on the true parameter z :

(i) $$D_{n,m}(\theta_0) = n^{-\frac{1}{2}} \sum_{k=1}^{m} Y_k(\theta_0)$$

(ii) $$Y_k(\theta_0) := H(\theta_0; X_k) \qquad (5.4.1)$$

5.4 Local Methods of Change Detection

Theorem 2 may be interpreted as follows: suppose for the moment that the variables $Y_k(\theta_0)$ are independent with distribution:

$$
\begin{aligned}
&(i) \qquad\qquad Y_k(\theta_0) \approx N[0, R(\theta_0)], \qquad k < r \\
&(ii) \qquad\qquad Y_k(\theta_0) \approx N[-h_\theta(\theta_0)\cdot\theta, R(\theta_0)], \qquad k \geq r
\end{aligned} \qquad (5.4.2)
$$

where r is the unknown change time, and θ is the unknown magnitude of the jump. Then formulae (5.4.1) and (5.4.2) again yield the behaviour described by formulae (5.3.12) in Theorem 2, which effectively governs the cumulative sum (5.4.1) given the occurrence of small jumps. It is thus as though *the test of H_0 against H_1' reduces to the detection of changes of mean in the variables $Y_k(\theta_0)$, according to the formulae (5.4.1) and (5.4.2)*. Thus we may use the procedure of Subsection 5.2.2 to investigate the case in which the direction of change θ is totally unknown; the reader is referred to Exercises 5.8.1 and 5.8.3 for instances in which a priori knowledge about θ is available.

With this simplification, the log-likelihood ratio, for fixed r and θ, of Hypotheses H_1' and H_0 is

$$S_r^n(\theta) = -2 \sum_{k=r}^n Y_k^T R^{-1} h_\theta \cdot \theta - (n - r + 1)\theta^T \cdot h_\theta^T R^{-1} h_\theta \cdot \theta \qquad (5.4.3)$$

where, for simplicity, we have omitted θ_0. Maximising with respect to θ, for fixed r gives

$$
\begin{aligned}
&(i) \qquad\qquad S_r^n := \max_\theta S_r^n(\theta) = (\Delta_r^n)^T R^{-1} \Delta_r^n \\
&(ii) \qquad\qquad \Delta_r^n := (n - r + 1)^{-\frac{1}{2}} \sum_{k=r}^n Y_k \\
&(iii) \qquad\qquad \hat\theta(n, r) := \arg\max_\theta S_r^n(\theta) = -(n - r + 1)^{-\frac{1}{2}} h_\theta^{-1} \Delta_r^n
\end{aligned} \qquad (5.4.4)
$$

The stopping rule and the estimators of the change time and of the jump are given by

$$
\begin{aligned}
&(i) \qquad\qquad G_n := \max_r S_r^n, \qquad \nu = \min\{n : G_n \geq \lambda\} \\
&(ii) \qquad\qquad r_n = \arg\max_r S_r^n, \qquad \hat r = \hat r_\nu, \qquad \hat\theta = \hat\theta(\nu, \hat r)
\end{aligned} \qquad (5.4.5)
$$

Note that (5.4.5-i) is sufficient if we are only interested in detecting changes. The local test is then given by formulae (5.4.1-ii), (5.4.4) and (5.4.5). One hint in setting the threshold λ is that, in the absence of changes,

$$E_{\theta_0}(S_r^n) = d \qquad (5.4.6)$$

since $(n - r + 1)S_r^n$ is χ^2 with mean zero and $d(n - r + 1)$ degrees of freedom.

Remarks.

Alternatively we could have followed the procedure below:

1. establish the adaptive algorithm with constant gain

$$\theta_n = \theta_{n-1} + \gamma \Gamma H(\theta_{n-1}, X_n) \qquad (5.4.7)$$

2. use a χ^2 test of the form

$$(\theta_n - \theta_0)^T \Sigma^{-1} (\theta_n - \theta_0) \geq \lambda \qquad (5.4.8)$$

for an appropriate matrix Σ, taking into account that $\theta_n - \theta_0$ is approximately Gaussian with mean zero when γ is small and no changes occur.

This intuitive procedure is in fact less effective than our own method. The results of Chapter 3 show that the dynamic behaviour of $\theta_n - \theta_0$ is quite complicated: it is a Gaussian, first order, continuous-time Markov process. The change detection test should be based on the innovation of the process; this is harder to calculate. In other words, test (5.4.8) is certainly naive and ineffective compared with our test. We shall see later that our procedure is simply an extrapolation of the so-called **local likelihood** methods which possess asymptotic optimality properties for small changes.

5.4.2 Examples

5.4.2.1 Change detection in AR Processes

The problem is to detect a change in the parameter θ of the system

$$\begin{aligned} y_n &= \Phi_n^T \theta + \nu_n \\ \Phi_n^T &= (y_{n-1}, \ldots, y_{n-p}) \end{aligned} \qquad (5.4.9)$$

We apply the previous method, with initially

(i) $\qquad\qquad H(\theta, y_n, \Phi_n) := \Phi_n e_n(\theta)$

(ii) $\qquad\qquad e_n(\theta) := y_n - \Phi_n^T \theta \qquad (5.4.10)$

We have already calculated the matrix $R(\theta_0)$ corresponding to (5.3.7) from formula (3.2.17):

$$R(\theta_0) = E_{\theta_0}(\nu_n \Phi_n \Phi_n^T \nu_n) = \sigma^2 \Sigma(\theta_0) \qquad (5.4.11)$$

where σ^2 is the variance of ν_n, and $\Sigma(\theta_0)$ is the covariance of the regression vector Φ_n for the nominal model θ_0. Thus here we have

$$\Delta_r^n(\theta_0) = (n - r + 1)^{-\frac{1}{2}} \sum_{k=r}^{n} \Phi_k e_k(\theta_0) \qquad (5.4.12)$$

5.4 Local Methods of Change Detection

Note here that $\sigma^{-2}\Delta_0^n(\theta_0)$ is the derivative with respect to θ of the log-likelihood of the sample $y_1, \ldots, y_n$ under θ_0, whilst $\sigma^{-2}\Sigma(\theta_0)$ is the Fisher information matrix. Comparison with (Nikiforov 1986), and with (Davies 1973), shows that the test based on formulae (5.4.1-ii), (5.4.4), (5.4.5), (5.4.11), and (5.4.12) is just the **local likelihood test** which is the well established way of detecting small changes in the parameters of an AR process.

5.4.2.2 Detection of Changes in the Poles of an ARMA Signal by the Instrumental Test

Here we begin with an ARMA signal of the form

$$y_n = \sum_{i=1}^{p} a_i y_{n-i} + \sum_{j=1}^{q} b_j \nu_{n-j} + \nu_n \qquad (5.4.13)$$

where (ν_n) is a white noise. The goal is to detect changes of the AR parameters $(a_i)_{1 \leq i \leq p}$, without reference to the MA parameters $(b_j)_{1 \leq j \leq q}$. We begin with an identification method which has the desired characteristics: the method of **instrumental variables**. We recall that this method was described in Exercise 2.7.7. We set

(i) $\quad \theta^T := (a_1, \ldots, a_p)$
(ii) $\quad \Phi_n^T := (y_{n-1}, \ldots, y_{n-p})$
(iii) $\quad \Psi_n^T := (y_{n-q-1}, \ldots, y_{n-q-p})$ $\qquad (5.4.14)$

where Ψ_n is known as the **instrumental variable**, and the true system θ_* is characterised by the equation

$$E[\Psi_n(y_n - \Phi_n^T \theta_*)] = 0 \qquad (5.4.15)$$

In fact, if θ_* is the true parameter, model (5.4.13) implies that

$$y_n - \Phi_n^T \theta_* = \sum_{j=1}^{q} b_j \nu_{n-j} + \nu_n \qquad (5.4.16)$$

is the MA part of the signal, which is consequently independent of the variables $y_{n-q-1}, y_{n-q-2}, \ldots$. Whence we have equation (5.4.15), which we shall use to determine θ_*. Thus the method of instrumental variables is given by

(i) $\quad \theta_n = \theta_{n-1} + \dfrac{1}{n}\Gamma_n^{-1}\Psi_n(y_n - \Phi_n^T \theta_{n-1})$

(ii) $\quad \Gamma_n = \Gamma_{n-1} + \dfrac{1}{n}(\Psi_n \Phi_n^T - \Gamma_{n-1})$ $\qquad (5.4.17)$

(this is the version with decreasing gain). Hence we begin with the vector field

$$H(\theta_0; \Psi_n, \Phi_n, y_n) = \Psi(y_n - \Phi_n^T \theta_0) := Y_n(\theta_0) \qquad (5.4.18)$$

to which we apply the procedure previously described. In so doing, we calculate

(i) $$R(\theta_0) := \sum_{n \in \mathbb{Z}} E_0(Y_n(\theta_0) \cdot Y_0^T(\theta_0))$$
$$= \sum_{n=-q}^{+q} E_0(\Psi_n \Psi_0^T (y_n - \Phi_n^T \theta_0)(y_0 - \Phi_0^T \theta_0))$$

(ii) $$h_\theta(\theta_0) = -E_0(\Psi_n \Phi_n^T) \qquad (5.4.19)$$

where E_0 denotes that θ_0 is assumed to be the true parameter. The reduction to a finite sum in (5.4.19-i) arises because in this case $y_n - \Phi_n^T \theta_0$ is a moving average, and thus $y_n - \Phi_n^T \theta_0$ and $y_0 - \Phi_0^T \theta_0$ are decorrelated once $|n| > q$. The **instrumental test** is thus given by applying formulae (5.4.1-ii), (5.4.4), (5.4.5), (5.4.18) and (5.4.19). This test exhibits good robustness properties with respect to the MA coefficients of the process, such properties are studied and used in (Moustakides and Benveniste 1986). This test is original and does not correspond to any known local likelihood method.

5.4.3 Implementation

Various implementations are possible. We shall do no more than describe the problems associated with implementation, and propose some solutions. For more details, we refer the reader to (Basseville and Benveniste 1986).

5.4.3.1 The problems

These are of two types. The first type is associated with the derivation of the nominal model θ_0, and with the calculation of $R(\theta_0)$ and if necessary of $h_\theta(\theta_0)$. In fact, for on-line (or sequential) use of the test, the nominal model θ_0 is in general unknown a priori. Thus, in order for the test to work, the model must be first estimated during an initialisation period.

The other type of problem concerns the computational load associated with the maximisation over r according to formulae (5.4.4) and (5.4.5). In certain problems, detection may be fast: this will allow us a priori to restrict the search for the maximum over r of S_r^n to a window $r \in [n - l, n]$, where l is sufficiently small that the computation of all the S_r^n and the maximisation is not too expensive. Otherwise this procedure may be too demanding.

5.4.3.2 Some solutions

Identification of θ_0 and $R(\theta_0)$.

The test procedure begins at time 0 (this may also be the instant preceding the jump). There are two possible solutions.

5.4 Local Methods of Change Detection

The first is to identify the nominal model (for example using the algorithm associated with the test) for a finite data record $X_1, \ldots, X_N$; the test proper only begins after this, with θ_0 now fixed.

In the second realisation, the test begins when the learning phase for the first N observations is complete; but we continue to update the nominal model. In other words, for the calculation of Δ_r^n in (5.4.4) we use the variables $Y_k(\theta_k)$, where θ_k is calculated using the adaptive algorithm associated with the test (a version with gain in $\frac{1}{n}$ is preferable since for the identification of the nominal model it is better to avoid too great a sensitivity to the non-stationarities of the system). Similarly R (and if need be h_θ) may be continually updated.

Maximisation over r

If necessary, this may be avoided, thanks to the following observation. Fix an integer n_0. If r is the true change time, then if we observe the sequence $[S_{n-n_0}]_{n \geq n_0}$, after a certain time the instants r and $n - n_0$ coincide. Thus we would expect the stopping rule

$$\nu = \min\{n : S_{n-n_0}^n \geq \lambda\} \tag{5.4.20}$$

to be satisfactory, in the sense that whilst it increases the detection delay slightly, it does not produce too many spurious jumps (for this, n_0 should not be chosen too small; for example $n_0 \approx 100$ is a reasonable order of magnitude for auto-regressive processes). Note that the stopping rule (5.4.20) does not provide a very satisfactory estimator for the change time r: in general, we can hope for $n - n_0 \leq r \leq n$. But then, at the moment of detection (with respect to the simplified rule (5.4.20)), it is possible (cf. (Appel and Brandt 1983) for a two pass approach of this type) to maximise S_r^ν over r, thus giving $\hat{r}$.

5.5 Model Validation by Local Methods

Model validation is a completely inherent and essential part of identification. Suppose that we have a model which is said to represent a system, then it is natural to ask the following question: **given new observations of the system, how can we verify that the given model actually provides a good description of the system?**

The term "model validation" refers specifically to this problem, viewed as a complement to the identification phase: thus the nominal model results from an initial identification phase, and validation is carried out using another sample of observations of the system. In this way, it is possible to verify that the chosen model structure provides an appropriate description of the system (Soderström and Stoica 1989). In other applications, model validation may simply be an off-line method of proving that the system has evolved considerably between two distant records; this method was, for example, applied in (Basseville, Benveniste and Moustakides 1986) and

(Basseville, Benveniste, Moustakides and Rougée 1987) to monitoring the vibration of offshore structures.

In this section, we shall again show that, with an asymptotic local approach, it is possible to associate any adaptive algorithm with a method of model validation.

5.5.1 Model Validation for Systems with a Markov Representation

We begin by returning to our previous starting point in Subsection 5.1.2. Let (X_n) be a process with a Markov representation controlled by the parameter θ_* according to formulae (5.1.6)

$$P\{\xi_n \in G | \xi_{n-1}, \xi_{n-2}, \ldots\} = \int_G \pi_{\theta_*}(\xi_{n-1}, dx)$$
$$X_n = f(\xi_n) \tag{5.5.1}$$

where $\pi_{\theta_*}(\xi, dx)$ is the transition probability, parameterised by θ_*, of the Markov chain (ξ_n). The model validation problem is then simply formulated as the following hypothesis testing problem:

[MV]. *Given a nominal model θ_0, decide between the following hypotheses*
  $\theta_* = \theta_0$: *the model is valid*
  $\theta_* \neq \theta_0$: *the model is not valid.*

Here, θ_0 is the model to be validated (nominal model), whilst θ_* is the (unknown) true parameter corresponding to the test sample $X_1, \ldots X_N$ used in validation. Note that the [MV] problem is easier than the [SD] problem of Subsection 5.1.2.

5.5.2 Validation Method Associated with an Adaptive Algorithm

We begin by returning to the previous procedure of Subsection 5.3.1. The starting point is once again the formalism previously used to investigate algorithm behaviour for non-stationary systems. The vector field is

$$H(\theta, z; X_n) \tag{5.5.2}$$

where

- θ denotes the nominal model chosen by the user;
- z is a parameter representing the "true system", which is not available to the user;
- X_n represents the state at time n, it is a semi-Markov random variable controlled by the pair (θ, z).

Once again, we suppose that Assumptions NS' of Subsection 5.3.1 hold.

5.5 Model Validation by Local Methods

The model validation problem is then the following: we are given

- a nominal model θ_0; and
- an n long record $X_1, \ldots, X_n$ of the state vector,

and it is desired to decide between the hypotheses

$$H_0 : z = \theta_0$$
$$H_1 : z = \theta_0 + \frac{\theta}{\sqrt{n}} \text{ where } \theta \neq 0 \text{ is a fixed but unknown separation}$$

by monitoring the random vector field $(H(\theta_0, z; X_k))_{1 \leq k \leq n}$. Choosing H_0 is equivalent to accepting the validity of the nominal model θ_0, whilst choosing H_1 is equivalent to rejecting θ_0.

If we refer back to Subsection 5.3.1, it is now clear that, in fact, we have already solved this problem, thanks to the theoretical results given in Section 5.3.

Now we follow the procedure of Subsection 5.4.1. Actually, we may now simplify this procedure, since we do not need to estimate the change time. We form the statistic

(i) $$D_n(\theta_0) := \frac{1}{n} \sum_{k=1}^{n} Y_k(\theta_0)$$

(ii) $$Y_k(\theta_0) := H(\theta_0, X_k) \tag{5.5.3}$$

The test to validate the model now amounts to testing the hypotheses

(i) $$Y_k(\theta_0) \approx N[0, R(\theta_0)]$$
(ii) $$Y_k(\theta_0) \approx N[-h_\theta(\theta_0) \cdot \theta, R(\theta_0)], \; \theta \neq 0 \tag{5.5.4}$$

against each other, where here h_θ and R are the usual matrices associated with the algorithm (see for example Subsection 5.3.1).

Finally, the test is given by setting $r = 0$ and by omitting the maximisation over r in formulae (5.4.3) and (5.4.4). This gives the following formulae, where reference to θ_0 is omitted for simplicity. For a fixed θ, the log-likelihood ratio between Hypotheses H_1 and H_0 is given by

$$S_n(\theta) = -2 \sum_{k=1}^{n} Y_k^T R^{-1} h_\theta \cdot \theta - n\theta^T \cdot h_\theta^T R^{-1} h_\theta \cdot \theta \tag{5.5.5}$$

This gives rise to the following χ^2 test

(i) $$\Delta_n^T R^{-1} \Delta_n \geq \lambda$$

(ii) $$\Delta_n = n^{-\frac{1}{2}} \sum_{k=1}^{n} Y_k \tag{5.5.6}$$

Note that under Hypothesis H_0 (validity of the nominal model), $n\Delta_n^T R^{-1} \Delta_n$ is χ^2 with n degrees of freedom; this enables us to choose the threshold λ appropriately. We refer the reader to the exercises of this chapter for several examples of applications of this method.

5.6 Conclusion

In this chapter, we have described a systematic means of associating a method of change detection and a method of model validation with any adaptive algorithm.

Our approach is far from the only possible method, nor is it the most effective in every case: it has the advantage of being systematically based. We refer any interested readers to (Basseville and Benveniste 1986). We have used two reputedly non-simple problems to illustrate this approach; change detection in AR processes, and change detection in the poles of an ARMA process. We refer the reader to the exercises for further examples and illustrations of model validation; see also (Benveniste, Basseville and Moustakides 1987) in which details of an extension of this method to the problem of diagnosis are given.

Lastly, we have only touched upon the topic of performance evaluation for these tests; this is a very difficult mathematical problem, and one might usefully refer to (Deshayes and Picard 1986), (Nikiforov 1986) and (Benveniste 1986).

5.7 Annex: Proofs of Theorems 1 and 2

The proofs of Theorems 1 and 2 of this chapter cannot be derived from the theoretical results in Part II of the book. We shall see that the central limit theorems which we need relate to "triangular arrays". Since this is the last chapter of Part I, we decided to ignore the principle of completeness which has prevailed thus far; the requisite mathematical results will not be described in detail in Part II. The basic notion which we require is that of *mixingales*. Mixingales were introduced by McLeish and investigated in (McLeish 1975a,b, 1977), see also (Hall and Heyde 1980).

5.7.1 Mixingales: a Theorem due to McLeish

Mixingales

Mixingales were introduced in (McLeish 1975a) and are a particularly large class of dependent variables. Let $\{\Omega, \mathcal{F}, \mathcal{F}_n, P\}$ be a probability space endowed with an increasing family of σ-algebras; the conditional expectation $E(U|\mathcal{F}_n)$ will be denoted in short by $E_n U$, and $\|.\|_2$ will denote the L^2 norm.

Definition A.1. *The sequence of random variables Z_n is a* **mixingale** *if there exist sequences of finite, non-negative constants, c_n and ψ_m, where $\psi_m \to 0$ as $m \to \infty$, for which we have, for all $n \geq 1, m \geq 0$*

(i) $\qquad\qquad\qquad \|E_{n-m} Z_n\|_2 \leq \psi_m c_n$

(ii) $\qquad\qquad\qquad \|Z_n - E_{n+m} Z_n\|_2 \leq \psi_{m+1} c_n \qquad\qquad (5.7.1)$

5.7 Annex: Proofs of Theorems 1 and 2

For example, $\psi_m \equiv 0$ corresponds to martingales. Strong laws and invariance principles for mixingales with exponentially vanishing ψ_m are given in (McLeish 1975a,b). The class of mixingales contains all the classes of mixing processes, and also all the functions of mixing processes introduced by (Billingsley 1968). However, we shall need stronger results, namely invariance principles for *triangular arrays of mixingales*: such results are found in (McLeish 1977) and restated below.

Triangular arrays of mixingales.

Let $\{Z_{n,i}\}_{i=1,2,\ldots\; n=1,2,\ldots}$ be an array of zero-mean random variables defined on a probability space $\{\Omega, \mathcal{F}, P\}$. Let $k_n(t)$ be a sequence of non-random, integer valued, increasing, right-continuous functions on $[0, \infty)$. We form the random function

$$W_n(t) = \sum_{i=1}^{k_n(t)} Z_{n,i} \qquad (5.7.2)$$

and wish to examine the convergence of $W_n(t)$ to a standard Wiener process W. Suppose

$$\sigma_{n,i}^2 = \sum_j E(Z_{n,i} Z_{n,j}) \qquad (5.7.3)$$

exists and is such that the following conditions hold for all $T < \infty$:

(i) $\quad \sup_{s<t<T} \limsup_{n \to \infty} \dfrac{\sum_{k_n(s)}^{k_n(t)} \sigma_{n,i}^2}{t - s} < \infty$

(ii) $\quad \left\{ \dfrac{Z_{n,i}^2}{\sigma_{n,i}^2} \right\}_{n=1,2,\ldots\; i \leq k_n(T)}$ is uniformly integrable

(iii) $\quad \max_{i \leq k_n(T)} \sigma_{n,i} \to 0$ as $n \to \infty$ $\qquad (5.7.4)$

These conditions guarantee that the Lindeberg condition holds for the given array. We also require that the sequence $\{Z_{n,i}\}_{i=1,2,\ldots}$ constitutes a mixingale with respect to the array of σ-algebras $\mathcal{F}_{n,i}$, in the sense that

$$\|E(Z_{n,i} \mid \mathcal{F}_{n,i-k})\|_2 \leq \psi_k \sigma_{n,i}$$
$$\|Z_{n,i} - E(Z_{n,i} \mid \mathcal{F}_{n,i+k})\|_2 \leq \psi_{k+1} \sigma_{n,i} \qquad (5.7.5)$$

Theorem (McLeish 1977). *Suppose conditions (5.7.4) and (5.7.5) hold, that $\psi_k \downarrow 0$ exponentially fast, and that for all $s < t < u$*

$$E \left| E \left\{ \left(\sum_{i=k_n(s)}^{k_n(t)} Z_{n,i} \right)^2 \middle| \mathcal{F}_{n,k_n(u)} \right\} - (u-t) \right| \to 0 \qquad (5.7.6)$$

as $n \to \infty$; then W_n is weakly convergent to a standard Wiener process.

5.7.2 Assumptions for Theorems 1 and 2

To apply McLeish's Theorem, we choose

$$\begin{aligned}
z_n &= \theta_0 + n^{-\frac{1}{2}}\theta \\
Z_{n,i} &= n^{-\frac{1}{2}}\lambda^T[R(\theta_0, z_n)]^{-\frac{1}{2}}(H(\theta_0, X_i) - h(\theta_0, z_n)) \\
\mathcal{F}_{n,i} &= \sigma\{\xi_k : k \leq i\} \\
P &= P_{\theta_0, z_n} \\
T &= 1, \quad k_n(t) = t
\end{aligned} \qquad (5.7.7)$$

where λ is an arbitrary unit vector, and

$$R(\theta, z) := \sum_{n=-\infty}^{+\infty} \text{cov}_{\theta, z}(H(\theta; X_n), H(\theta; X_0))$$

The conditions (5.7.4) are obviously satisfied, at least when the stationary regime of state X_n is reached. We shall furthermore assume that conditions (5.7.5) and (5.7.6) hold with constants ψ_k that are uniform in λ and n large enough, and that the function $z \to R(\theta_0, z)$ is continuous and bounded from below. These are very weak conditions: discontinuities of the random vector field $\theta \to H(\theta, X_n)$ are allowed, and the mixingale condition is satisfied for every reasonable Markov chain. See for example (Benveniste, Goursat and Ruget 1980a,b) for non-trivial cases in which the stronger ϕ-mixing property is satisfied (for example the transversal blind equaliser).

5.7.3 Proofs

Proof of Theorem 1. Using the notation of (5.3.8), McLeish's Theorem implies that

$$\lambda^T[R(\theta_0)]^{-\frac{1}{2}} \cdot \{D_{n,t}(\theta_0, 0)\}_{0 \leq t \leq 1} \to \lambda^T(W_t)_{0 \leq t \leq 1} \qquad (5.7.8)$$

Since λ is arbitrary, this proves Theorem 1. $\square$

In fact, a simple invariance principle for mixingales would have sufficed for Theorem 1, as would one of the invariance principles of Part II, Chapter 4.

Proof of Theorem 2. The proof of Theorem 2 requires the stronger result of asymptotic normality for triangular arrays. It is clearly sufficient to prove Theorem 2 for $\tau = 0$. We shall write $D_{n,t}(\theta)$, for short, instead of $D_{n,t}(\theta_0, \theta, \tau)$, and omit the dependence on θ_0 when no confusion can occur. Thus McLeish's Theorem now implies that

$$\lambda^T[R(\theta_0)]^{-\frac{1}{2}} \left(D_{n,t}(\theta) - n^{-\frac{1}{2}} \sum_{k=1}^{[nt]} h(\theta_0, z_n) \right)_{0 \leq t \leq 1} \to \lambda^T(W_t)_{0 \leq t \leq 1} \qquad (5.7.9)$$

5.7 Annex: Proofs of Theorems 1 and 2

Finally, a first-order Taylor expansion
$$h(\theta_0, z_n) \approx h(\theta_0, \theta_0) + n^{-\frac{1}{2}} h_z(\theta_0, \theta_0) \cdot \theta$$
$$= -n^{-\frac{1}{2}} h_\theta(\theta_0, \theta_0) \cdot \theta$$
completes the proof of Theorem 2. □

5.8 Exercises

5.8.1 Changes in the Mean of Gaussian Variables: Complements

Throughout this exercise, we have a sequence (X_n) of independent vector variables with distributions $N(\theta, R)$.

1. Write down the log-likelihood ratio test to detect a change from θ_0 to θ_1 (both known).

2. Suppose we wish to detect a change from $\theta = 0$ to $\theta = A\mu$, $\mu \neq 0$, where A is a matrix of full column rank. Prove that the GLR log-likelihood ratio test is given by

$$\Delta_r^n := (n-r+1)^{-\frac{1}{2}} \sum_{k=r}^{n} X_k$$
$$S_r^n := \Delta_r^{nT} \cdot R^{-1} A (A^T R^{-1} A)^{-1} A^T R^{-1} \cdot \Delta_r^n$$
$$\hat{\mu}(n,r) = (A^T R^{-1} A)^{-1} A^T R^{-1} \frac{\Delta_r^n}{(n-r+1)^{\frac{1}{2}}}$$
$$G_n = \max_r S_r^n, \quad \hat{r}_n = \arg\max_r S_r^n$$

and that the stopping rule is

$$\nu = \min\{n : G_n \geq \lambda\}$$
$$\hat{r} = \hat{r}_\nu, \quad \hat{\mu} = \hat{\mu}(\nu, \hat{r})$$

How should this test be modified when A does not have full column rank?

3. Use a likelihood method to detect a change in θ from the condition $\theta^T U < -\varepsilon$ to $\theta^T U > \varepsilon$ where U is a given unit vector. Using local methods, deduce a procedure for detecting change from $\theta^T U < 0$ to $\theta^T U > 0$.

4. Suppose we wish to detect a change from $\theta = 0$ to $\theta \neq 0$, but that we wish to ignore any changes that may occur within a subspace of the form $\{\nu : B\nu = 0\}$. Show that if in 2. (above) we replace the first two equations by

$$\Delta_r^n := (n-r+1)^{-\frac{1}{2}} \sum_{k=r}^{n} X_k$$
$$S_r^n := \Delta_r^{nT} B^T (BRB^T)^{-1} B \Delta_r^n$$

5.8.2 Change Detection in an ARMA Process

The model in question is an ARMA process of the form

$$y_n = \sum_{i=1}^{p} a_i y_{n-i} + \sum_{j=1}^{q} b_j \nu_{n-j} + \nu_n$$

where ν_n is a white noise with variance σ^2. The object is to detect changes using a local method with parameter $\theta^T = (a_1, \ldots, a_p; b_1, \ldots b_q)$.

Part 1. *ELS algorithm.*

This algorithm was described in Paragraph 2.4.3.3. The vector field $H(\theta, X_n)$ associated with this algorithm is given by $\Phi_n e_n$, where

$$\Phi_n^T := (y_{n-1}, \ldots, y_{n-p}; e_{n-1}, \ldots, e_{n-q})$$
$$e_n := y_n - \sum_{i=1}^{p} a_i y_{n-i} - \sum_{j=1}^{q} b_j e_{n-j}$$
$$\theta^T := (a_1, \ldots, a_p; b_1, \ldots, b_q)$$

1. Suppose we have a nominal model θ_0. Show that (notation of Section 5.3):

    $$R(\theta_0) = \sigma_e^2(\theta_0) \Sigma_\phi(\theta_0)$$

    where for $\theta = \theta_0$

    $$\sigma_e^2(\theta_0) = E_{\theta_0}(e_n^2), \quad \Sigma_\phi(\theta_0) = E_{\theta_0}(\Phi_n \Phi_n^T)$$
    $$h_\theta(\theta_0) = -E_{\theta_0}(\Phi_n \tilde\Phi_n)$$

    where, setting $B(Q^{-1}) = 1 - \sum_j b_j q^{-j}$,

    $$\tilde\Phi_n := \frac{1}{B_0(q^{-1})} \cdot \Phi_n$$

2. Using the formulae of Subsection 5.4.1, describe the local test associated with the ELS algorithm.

5.8 Exercises

Part 2. *RML algorithm.*

The associated vector field $H(\theta, X_n)$ is given by $\tilde{\Phi}_n(\theta)e_n$.

1. Show that
$$R(\theta_0) = \sigma_e^2(\theta_0)\Sigma_{\tilde{\Phi}}(\theta_0)$$
where
$$\Sigma_{\tilde{\Phi}}(\theta_0) := E_\theta(\tilde{\Phi}_n \tilde{\Phi}_n^T)$$
and prove that
$$h_\theta(\theta_0) = -\Sigma_{\tilde{\Phi}}(\theta_0)$$

2. Deduce the local test associated with the RML algorithm: it is a local likelihood test: compare this with the local test arising from the ELS algorithm.

Part 3. *ARMAX models.*

Consider the same questions for the ARMAX model below
$$y_n = \sum_{i=1}^{p} a_i y_{n-i} + \sum_{k=1}^{l} c_k u_{n-k} + \sum_{j=1}^{q} b_j \nu_{n-j} + \nu_n$$
where ν_n is an unobserved Gaussian white noise and u_n is a zero-mean, stationary **observed** signal independent of ν_n. Here we are interested in detecting changes in the vector
$$\theta^T := (a_1, \ldots, a_p; c_1, \ldots, c_l; b_1, \ldots, b_q)$$

1. Describe the vector fields of the algorithms corresponding to the ELS and RML methods.
2. Describe the corresponding local tests.

5.8.3 Sensitivity Methods

These methods aim to concentrate attention on a subset of the set of system parameters.

Part 1. *General method.*

The notation is as in Subsection 5.4.1. We suppose that we have a change of coordinates
$$\theta = F(\Psi)$$
where F is a local diffeomorphism in a neighbourhood of the nominal model θ_0, with Jacobian F_Ψ. Given that θ and Ψ have d components, the aim is to detect a change in the first p components ($p \leq d$) $\Psi(1), \ldots, \Psi(p)$ of the vector Ψ, assuming the other components remain fixed.

1. Show that in terms of the statistics $Y_k(\theta_0)$ of Subsection 5.4.1, the change is expressed as

$$Y_k(\theta_0) \approx N(0, R(\theta_0)) \qquad \text{for } k < r$$
$$Y_k(\theta_0) \approx N(-h_\theta(\theta_0) \cdot F_\Psi^{1p}(\Psi_0) \cdot \Psi[p], R(\theta_0)) \qquad \text{for } k \geq r$$

where $\Psi[p] \in \mathbb{R}^p$ is the (unknown) magnitude of the jumps, and F_Ψ^{1p} is the matrix obtained by deleting the columns $p+1, \ldots, d$ of the Jacobian F_Ψ.

2. Using the method of Subsection 5.4.1 and the results of Exercise 5.8.1, give a description of the local test associated with the vector field $H(\theta, X_n)$ for detecting either a change in the components $\Psi(1), \ldots \Psi(p)$ of the system or a change whilst ignoring some of these components.

Part 2. *Application to ARMA processes.*

Here, the aim is to detect a change in the AR part of the ARMA process

$$y_n = \sum_{i=1}^{p} a_i y_{n-i} + \sum_{j=1}^{q} b_j \nu_{n-j} + \nu_n$$

where the MA coefficients are assumed to be unchanged.

1. Apply the method of Part 1 to the local test associated with the ELS algorithm,. Prove that the test is given by the formula of question 2 of Exercise 5.8.1, where

$$X_k := Y_k(\theta_0), \qquad R = R(\theta_0)$$

as given in question 1.1 of Exercise 5.8.2 and where

$$A = \text{columns 1 to } p \text{ of } h_\theta(\theta_0)$$

where $h_\theta(\theta_0)$ is also given in question 1.1 of Exercise 5.8.2.

2. Do the same for the RML algorithm. Show that we have the following simplification

$$S_r^n = \frac{1}{\sigma_e^2(\theta_0)} (\tilde{\Delta}_r^n(AR))^T (\Sigma_{\tilde{\Phi}_{AR}}(\theta_0))^{-1} (\tilde{\Delta}_r^n(AR))$$

where

$$\tilde{\Delta}_r^n(AR) = (n-r+1)^{-\frac{1}{2}} \sum_{k=r}^{n} \tilde{\Phi}_k(AR) e_k$$
$$\tilde{\Phi}_k(AR) := \frac{1}{B_0(q^{-1})} \Phi_k(AR)$$
$$\Phi_k^T(AR) = (y_{k-1}, \ldots, y_{k-p})$$
$$\Sigma_{\tilde{\Phi}_{AR}}(\theta_0) := E_{\theta_0}(\tilde{\Phi}_k(AR) \tilde{\Phi}_k^T(AR))$$

5.8 Exercises

3. Compare with the local test associated with the method of instrumental variables described in Paragraph 5.4.2.2.

Part 3. *Monitoring the poles of an AR process.*

Here we are given an AR process

$$y_n = \sum_{i=1}^{p} a_i y_{n-i} + \nu_n$$

and we wish to monitor the complex conjugate poles separately.

1. Let $p = 2q$, and suppose that for $k = 1, \ldots, q$, $z_k = \rho_k e^{i\omega_k}$ and $\bar{z}_k$ are the roots of the polynomial $1 - \sum_{i=1}^{p} a_i z^{-i}$. Next apply a change of coordinates to the AR process to obtain : $\theta = F(\Psi)$ where

$$\Psi^T = (\rho_1, \omega_1; \ldots; \rho_q, \omega_q)$$

Calculate the Jacobian $F_\Psi(\Psi_0)$ for the nominal model Ψ_0.

2. Using the sensitivity method, describe a local test which allows us to monitor

   - a given pair $(\rho_{k_0}, \omega_{k_0})$
   - a given frequency ω_{k_0}.

Part 4. *Coupling and nuisance parameters.*

We return to the general framework of Part 1. We assume a change of coordinates

$$\theta = F(\Psi)$$

and a decomposition of Ψ into (Ψ', Ψ'').

1. Denote the stopping rule for the local sensitivity test to monitor changes in Ψ' by ν'. Show that ν' is blind to small changes in Ψ'' if and only if the matrix

$$F_\Psi^{-1} \cdot h_\theta^{-1} R h_\theta^{-T} \cdot F_\Psi^{-T}$$

is block diagonal with respect to the partition $\Psi = (\Psi', \Psi'')$ of Ψ. In this case, the monitoring of the variables Ψ' and Ψ'' is said to be free of coupling, and the test ν' to monitor Ψ' is robust with respect to the nuisance parameter Ψ''; otherwise a change in Ψ'' may be wrongly interpreted by the test ν' as a change in Ψ'.

2. Apply this result to the test of question 2.2 and compare, from this new viewpoint, the local RML test for the poles of an ARMA process with the local test derived from the method of instrumental variables.

5.8.4 Monitoring Phase-Locked Loops

This exercise follows on from Exercise 3.4.1.

Denote $r(\phi) = \sum_n \mathrm{cov}(\varepsilon_n(\phi), \varepsilon_0(\phi))$ and refer to Exercise 3.4.1 for results about one or other of the phase-locked loops described there, as required.

1. Prove that the local test associated with the phase-locked loops is given by using
$$S_r^n = \frac{1}{n-r+1} \frac{\sum_{k=r}^n \varepsilon_k^2(\phi_0)}{r(\phi_0)}$$
in (5.4.4). Give explicit formulae for both loops.

2. How can Exercise 5.8.1 be used to develop a more effective test to take advantage of the fact that ϕ is a scalar?

5.8.5 Effect on a Local Test of the Use of an Identified Nominal Model

The notation is as in Subsections 5.3.1 and 5.4.1. As noted in Subsection 5.4.3, in practice, the nominal model θ_0 is often identified by an earlier procedure: θ_0 is then not the so-called "true" model before the change (or before validation), but only an estimator of that model. This exercise will study these points more closely.

1. Mathematical framework. To take into account the fact that the test parameter θ_0 now corresponds to an identified nominal model, we replace the cumulative sums (5.3.5) by
$$\hat{D}_{n,m}(\theta_0, \theta) := n^{-\frac{1}{2}} \sum_{k=1}^m H(\hat{\theta}_0, \theta_0 + \frac{\theta}{\sqrt{n}}; X_k)$$
where
   - $z = \theta_0 + \frac{\theta}{\sqrt{n}}$ is the estimate of the true system value,
   - θ_0 is the true model before change,
   - $\hat{\theta}_0$ is the nominal model actually used; this is given by an estimator of θ_0 derived from previous observations of the system behaviour.

   Set $\tilde{\theta}_0 = \hat{\theta}_0 - \theta_0$, and express $\hat{D}_{n,m}(\theta_0, \theta)$ as a function of $D_{n,m}(\theta_0, \theta)$ and of $\tilde{\theta}_0$.

2. Deduce that if the preliminary estimate $\hat{\theta}_0$ of the true model before change is such that $\mathrm{cov}(\tilde{\theta}_0) \ll n^{-1}$, then it is legitimate to replace θ_0 by $\hat{\theta}_0$.

5.8.6 Validation of a Vector Quantiser

This exercise is a sequel to Exercise 2.7.12, it uses the notation of that exercise.

Suppose we are given a nominal quantiser θ_0. It is desired to construct a local test to validate this quantiser against a new observation of (X_n). For simplicity, we assume in this exercise that (X_n) is a sequence of independent variables.

The test will be based on the adaptive algorithm due to Lloyd which was introduced in Exercise 2.7.12.

1. Calculate $R(\theta_0)$.
2. Describe the local validation test according to the method of Section 5.5

5.9 Comments on the Literature

General Comments.

It should not be forgotten that (Wald 1949) is the bible of sequential analysis, as far as sequential decision methods are concerned. This apart, the description of change detection methods which we give in this chapter is due to (Page 1954), see also (Hinkley 1971), (Lorden 1971) and (Shiryaev 1961).

As far as analysis of the change detection problem for dynamical systems in control science is concerned (Willsky 1976) which is still up to date is noteworthy, see also (Basseville 1988). We mention also (Himmelblau 1978) (more control theory, but does not use statistical techniques) and (Isermann 1984). In parallel, the Soviet school has also produced numerous results, not only from the control science point of view (Mironovski 1980), but also more statistical or signal processing related (Kligene and Tel'ksnis 1983). We have no intention of covering the whole domain, and so we shall simply refer the reader to the book (Basseville and Benveniste 1986) (this seems to be the current reference work for control science and signal processing aspects) and to the review articles (Basseville 1986, 1988).

The origins of the methods we have proposed go back to the notion of *contiguity*, as described by Le Cam (see (Roussas 1972)). The idea of using the contiguity (in other words, the asymptotic local method) not as a method of analysis, but rather as a design method is found in (Roussas 1972) for hypothesis testing (including model validation), and for change detection is due to Nikiforov (Nikiforov 1983, 1986). The extension of this approach to methods other than likelihood methods is due to (Basseville, Benveniste and Moustakides 1986), see also (Moustakides and Benveniste 1986).

Finally, as far as the underlying mathematics is concerned, in addition to the works previously mentioned, the book (Ibragimov and Khas'minskii 1981), and the article (Deshayes and Picard 1986) are well worth noting.

References Used in this Chapter.

Section 5.1 is original. Section 5.2 is largely borrowed from (Basseville 1986), (Benveniste 1986) and (Basseville, Benveniste and Moustakides 1986). Sections 5.3 and 5.4 are largely original, except for the examples: the first example is taken from (Nikiforov 1983) and the second is from (Basseville, Benveniste and Moustakides 1986). Section 5.5 is original. We note also an extension of this method to diagnosis in (Basseville, Benveniste and Moustakides 1986).

Exercise 5.8.2 is taken from (Nikiforov 1983). Exercise 5.8.3 is from (Basseville, Benveniste and Moustakides 1986) and (Basseville, Benveniste, Moustakides and Rougée 1987). The other exercises are original.

Chapter 6
Appendices to Part I

6.1 Rudiments of Systems Theory

6.1.1 Linear Systems, Transfer functions, Filters

A **linear system** or **filter** is a linear, homogeneous function of time which transforms an input signal $(u_n)_{n \in \mathbb{Z}}$ to an output signal $(y_n)_{n \in \mathbb{Z}}$; thus it may be represented by the convolution

$$y_n = \sum_{k \in \mathbb{Z}} s_k u_{n-k} \qquad (6.1.1)$$

where $S := (s_k)_{k \in \mathbb{Z}}$ is the impulse response of the system. To avoid convergence problems, we shall assume in this paragraph that the entries (u_n) are **almost always zero**, which means that they are zero except for a **finite** number of instants n.

The system S is said to be

(i)   causal if $s_k = 0$ for $k < 0$

(ii)   strictly causal if $s_k = 0$ for $k \leq 0$ (6.1.2)

Given a sequence $(u_n)_{n \in \mathbb{Z}}$, we define its **z-transform**[1] to be the formal power series

$$u(z) = \sum_{n \in \mathbb{Z}} u_n z^{-n} \qquad (6.1.3)$$

and we use $Z(u_n) = u(z)$ to denote this transformation. Z is linear, transforms convolution into multiplication, and satisfies

$$Z(u_{n-1}) = z^{-1} u(z) \qquad (6.1.4)$$

The **transfer function** $S(z)$ is the z-transform of the impulse response (s_k) of a linear system.

Amongst causal systems, we distinguish **state variable systems** defined by

$$\begin{aligned} X_n &= A X_{n-1} + B u_n, \quad X_{-\infty = 0} \\ y_n &= C X_{n-1} + D u_n \end{aligned} \qquad (6.1.5)$$

---

[1] In this book we use z or q to denote this variable

where $X_n \in \mathbb{R}^d$ is the **state**, and A, B and C are matrices of appropriate dimensions. The initial condition $X_{-\infty} = 0$ signifies that the state is zero prior to the occurrence of an input $u_n \neq 0$. The impulse response $(s_k)_{k \geq 0}$ of (6.1.5) is

$$s_0 = D, \qquad s_k = CA^{k-1}B \qquad \text{for } k > 1 \qquad (6.1.6)$$

whilst its transfer function $S(z)$ is given by (this follows directly from (6.1.3), (6.1.4) and (6.1.5))

$$S(z) = D + C(zI - A)^{-1}B \qquad (6.1.7)$$

Of all transfer functions, we are primarily interested in **rational** transfer functions

$$S(z) = \frac{B(z^{-1})}{A(z^{-1})} \qquad (6.1.8)$$

where A and B are polynomials:

$$A(z^{-1}) = \sum_{i=0}^{p} a_i z^{-i}$$

$$B(z^{-1}) = \sum_{j=0}^{q} b_j z^{-j}$$

If $a_0 = 1$, the input/output relationship

$$y(z) = S(z)u(z)$$

may be expressed in the time domain as

$$y_n = \sum_{i=1}^{p} a_i y_{n-i} + \sum_{j=0}^{q} b_j u_{n-j}, \qquad y_{-\infty} = 0 \qquad (6.1.9)$$

where the initial condition $y_{-\infty} = 0$ signifies that the output y_n is at the origin prior to the first non-zero input. Formula (6.1.9) shows that a rational filter is **causal**.

6.1 Rudiments of Systems Theory

6.1.2 Equivalence of Rational Filters and State Variable Representations

The following theorem is fundamental.

Theorem A1. *The following three conditions are equivalent:*
(i) the system $S = (s_k)_{k \geq 0}$ *has a state variable form, the minimal dimension of the state being d.*
(ii) rank(S)=d where S is the **Hankel matrix** *of the system, defined by*

$$S = \begin{pmatrix} s_1 & s_2 & s_3 & \cdots \\ s_2 & s_3 & & \\ s_3 & & & \\ & & & \end{pmatrix} \qquad (6.1.10)$$

(iii) the transfer function $S(z)$ is rational of degree d.

The third assertion means that $S(z)$ may be written in the irreducible form (6.1.8), with $\max(p,q) = d$.

Proof. $(i) \Rightarrow (iii)$.
If the matrix $(zI - A)^{-1}$ of (6.1.7) is expressed in terms of the cofactors of A:

$$(zI - A)^{-1} = \frac{A_1 + \ldots + A_d z^{d-1}}{\det(zI - A)} = \frac{A_1 z^{-1} + \ldots + A_d z^{-d}}{1 + a_1 z^{-1} + \ldots + a_d z^{-d}} \qquad (6.1.11)$$

it is immediately clear that $S(z)$ is rational of degree $\leq d$.
$(iii) \Rightarrow (ii)$.
Since $A(z^{-1})S(z) = B(z^{-1})$, we have

$$\sum_{i=0}^{p} a_i s_{k+i} = 0 \text{ for } k > 0 \qquad (6.1.12)$$

from which it follows that the rank of S is $\leq d$.
$(ii) \Rightarrow (i)$.
This is the difficult and interesting part. Define

$$\begin{aligned} U_- &= \text{the space of } (u_n) \text{ such that } u_n = 0 \text{ for } n > 0 \\ Y_+ &= \text{the space of } (y_n)_{n>0} \end{aligned}$$

Then when $n > 0$ and (u_n) belongs to U_-, $y_n = \sum_{k>0} s_k u_{n-k}$ may be written as:

$$\begin{pmatrix} y_n \\ y_{n+1} \\ \vdots \end{pmatrix} = S \begin{pmatrix} u_{n-1} \\ u_{n-2} \\ \vdots \end{pmatrix} \qquad (6.1.13)$$

Since $\mathcal{S}$ has rank d, we have the factorisation

$$\mathcal{S} = \mathcal{O} \cdot \mathcal{C} \tag{6.1.14}$$

where $\mathcal{O}$ and $\mathcal{C}$ are matrices of dimensions (∞, d) and (d, ∞) (respectively), and of rank d. Then we have the diagram

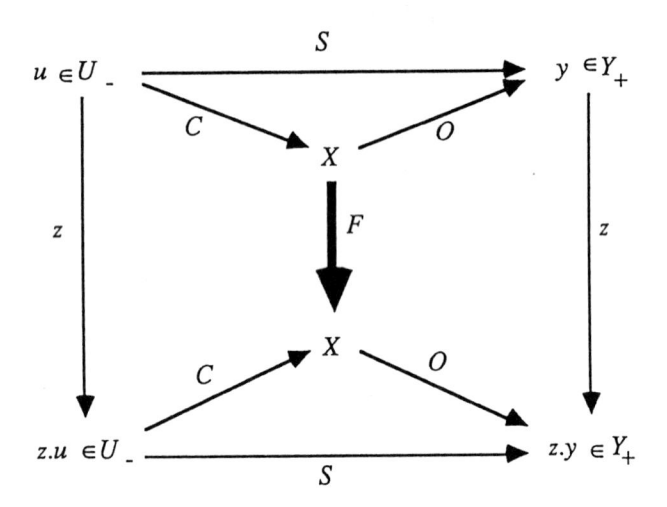

from which we deduce (since $\mathcal{O}$ and $\mathcal{C}$ have full rank) the existence of a unique endomorphism A of the **state space** X which makes the diagram commutative. The mapping

$$z \cdot S : (u_n) \to (y_{n+1})$$

is represented by

$$\mathcal{S}^{\leftarrow} = \mathcal{S}^{\uparrow} = \begin{pmatrix} s_2 & s_3 & s_4 & \cdots \\ s_3 & s_4 & & \\ s_4 & & & \end{pmatrix} \tag{6.1.15}$$

where $M^{\leftarrow}$ (resp. $M^{\uparrow}$) denotes the matrix obtained from M by moving the columns (resp. rows) of the matrix one position to the left (resp. upwards). Using (6.1.15) together with the commutative diagram, we obtain the factorisations

$$\mathcal{O}\mathcal{C}^{\leftarrow} = \mathcal{S}^{\leftarrow} = \mathcal{O}A\mathcal{C} = \mathcal{S}^{\uparrow} = \mathcal{O}^{\uparrow}\mathcal{C} \tag{6.1.16}$$

6.1 Rudiments of Systems Theory

If we now set

$$C := \text{first row of } \mathcal{O}$$
$$B := \text{first row of } \mathcal{C}$$

then from (6.1.16), we have

$$\mathcal{O} = \begin{pmatrix} C \\ CA \\ CA^2 \\ \vdots \end{pmatrix} \qquad \mathcal{C} = (B, AB, A^2B, \ldots) \qquad (6.1.17)$$

and

$$s_k = CA^{k-1}B, \qquad k > 0 \qquad (6.1.18)$$

Thus we have shown that there exists a realisation of state dimension d, using the 4-tuple (A, B, C, D) with $D := s_0$.

To complete the proof, it remains to show that the inequalities we derived on the degree of $S(z)$ and on the rank of $\mathcal{S}$ are in fact equalities, and that d is actually the minimum possible dimension of a state representation of the system. The three partial results derived above give the following inequalities:

minimal dimension of state space
$\geq$ degree of $S(z)$
$\geq$ rank of $\mathcal{S}$
$\geq$ minimal dimension of state space

and thus we have the desired result. This completes the proof of Theorem A1. □

This theorem guarantees the existence of a state variable representation which may be constructed from the impulse response $(s_k)_{k \geq 0}$. The freedom in the choice of the factorisation (6.1.16) corresponds to the arbitrary choice of the basis for the state space. We describe two possible choices of basis which are easy to derive given a transfer function $S(z)$ in the form of a rational function.

6.1.3 Canonical Forms

Set

$$S(z) = \frac{b_0 + b_1 z^{-1} + \ldots + b_d z^{-d}}{1 - (a_1 z^{-1} + \ldots + a_d z^{-d})} \qquad (6.1.19)$$

It is easy to show that the following two state variable representations may be used to represent $S(z)$.

Observability Canonical Form.

This is given by

$$C = (1, 0, \ldots, 0), \qquad A = \begin{pmatrix} 0 & 1 & & & \\ & \ddots & \ddots & & \\ & & \ddots & \ddots & \\ & & & 0 & 1 \\ a_d & \ldots & \ldots & \ldots & a_1 \end{pmatrix}$$

$$B = \begin{pmatrix} s_1 \\ \vdots \\ \vdots \\ \vdots \\ s_d \end{pmatrix}, \qquad D = b_0 \tag{6.1.20}$$

where $(s_k)_{k=1,\ldots,d}$ is the initial impulse response of the system, which is related to equation (6.1.19) by the formulae

$$\begin{aligned} b_1 &= s_1 \\ b_2 &= s_2 - a_1 s_1 \\ &\ldots \\ b_d &= s_d - (a_1 s_{d-1} + \ldots + a_{d-1} s_1) \end{aligned} \tag{6.1.21}$$

This choice of basis corresponds to a factorisation (6.1.16), where the so-called "observability" matrix $\mathcal{O}$ is of the form

$$\mathcal{O} = \begin{bmatrix} I_d \\ \star \\ \star \end{bmatrix}$$

where the starred part depends upon the system under consideration.

Controllability Canonical Form.

This is obtained from the previous form by the substitution

$$C \Rightarrow B^T, \qquad A \Rightarrow A^T, \qquad B \Rightarrow C^T$$

with D unchanged. This corresponds to a particular choice of the factorisation (6.1.16) with

$$\mathcal{C} = (I_d \star \star)$$

6.2 Second Order Stationary Processes

6.2.1 ARMA Processes and Processes with a Gaussian Markov Representation

In what follows, (ν_n) will denote a **white noise**, i.e. a stationary sequence of independent variables with mean zero and finite variance.

An **Auto Regressive Moving Average (ARMA) process** is a sequence of variables $(y_n)_{n\geq 0}$ satisfying an equation of the form

$$y_n = \sum_{i=1}^{p} a_i y_{n-i} + \sum_{j=1}^{q} b_j \nu_{n-j} + \nu_n \qquad (6.2.1)$$

for given initial conditions $y_{-1}, \ldots, y_{-p}$.

Taking $d = \max(p,q)$, it was shown in Section 6.1, that (6.2.1) may be put in the state variable form

$$\begin{aligned} X_n &= AX_{n-1} + B\nu_n, \qquad X_n \in \mathbb{R}^d \\ y_n &= CX_{n-1} + \nu_n \end{aligned} \qquad (6.2.2)$$

If we choose the **observability canonical form** of Section 6.1, we see that in this case the state X_n is given by

$$X_n = \begin{pmatrix} E(y_{n+1} \mid y_n, y_{n-1}, \ldots) \\ \cdots \\ \cdots \\ E(y_{n+d} \mid y_n, y_{n-1}, \ldots) \end{pmatrix} \qquad (6.2.3)$$

where $E(. \mid .)$ is a least squares estimator. Moreover (ν_n) is the **innovation**, characterised by the property

$$\nu_n = y_n - E(y_n \mid y_{n-1}, \ldots) \qquad (6.2.4)$$

Conversely, again using the results of Section 6.1, it can be shown that any process with a Markov representation of the form (6.2.2) defines an ARMA process of order (p,q) with $d \geq \max(p,q)$.

6.2.2 Stability and Stationarity

From the state variable form (6.2.2), it follows directly that an ARMA process is asymptotically stationary if and only if the matrix A has all its eigenvalues strictly inside the unit circle of the complex plane, or equivalently, if and only if the roots of the polynomial $1 - \sum_{i=1}^{p} a_i z^{-i}$ are strictly inside the unit circle.

6.2.3 Spectrum of Stationary Processes

Here we assume we have a stationary process or **signal** $(y_n)_{n \in \mathbb{Z}}$ with zero mean and finite variance. Its **auto-correlation function** is defined by

$$r_k = E(y_{n+k} y_n) \tag{6.2.5}$$

The **spectral measure** $M_y(d\omega)$ is the Fourier transform of $(r_k)_{k \in \mathbb{Z}}$:

$$r_k = \frac{1}{2\pi} \int_{-\pi}^{\pi} e^{ik\omega} M_y(d\omega) \tag{6.2.6}$$

Since the covariance matrix

$$R_y = \begin{pmatrix} r_0 & r_{-1} & r_{-2} \\ r_1 & & \\ r_2 & & \end{pmatrix} \tag{6.2.7}$$

is real, symmetric and positive, we deduce from Bochner's Theorem that $M_y(d\omega)$ is positive and satisfies $M_y(-d\omega) = M_y(d\omega)$. Conversely, given these properties, Bochner's Theorem again applies to show that $M(d\omega)$ is the spectral measure of a stationary process.

Examples.

1. If (ν_n) is a white noise of variance σ^2, we have $M_\nu(d\omega) = \sigma^2 \cdot d\omega$.
2. If

$$y_n = \sum_{k=0}^{q} b_k \nu_{n-k}$$

is a MA process, it follows directly that

$$M_y(d\omega) = \Big| \sum_{k=0}^{q} b_k e^{-ik\omega} \Big|^2 d\omega \tag{6.2.8}$$

6.2.4 Regular or Purely Non-Deterministic Processes

Formula (6.2.8) extends to all processes (y_n) of the form

$$y_n = \sum_{k \geq 0} s_k \nu_{n-k}, \qquad \sum_k s_k^2 < \infty \tag{6.2.9}$$

If $S(z) = \sum_{k \geq 0} s_k z^{-k}$ denotes the transfer function of the system defined by (6.2.9), we have

$$M_y(d\omega) = |S(e^{i\omega})|^2 d\omega \tag{6.2.10}$$

The formula

$$S_y(z) = S(z) S(z^{-1}) \tag{6.2.11}$$

6.2 Second Order Stationary Processes

gives a meromorphic extension to the whole of the complex plane of the **spectral density** $|S(e^{i\omega})|^2$ of the process (y_n) of (6.2.10). The function S_y is usually called the **spectrum** of (y_n). In particular, (6.2.11) shows that if (y_n) is a stationary ARMA process, its spectrum S_y is **rational**.

A difficult theorem of harmonic analysis shows that the process (y_n) may be represented in the form (6.2.9) (then we say that it is **regular** or **purely non-deterministic**) if and only if its spectral measure $M_y(d\omega)$ is of the form

$$M_y(d\omega) = S_y(e^{i\omega})d\omega$$

where the spectrum S_y satisfies

$$\int_{-\pi}^{\pi} \log S_y(e^{i\omega})d\omega > -\infty$$

6.2.5 Representation of Processes with a Rational Spectrum

We have seen that ARMA processes have a rational spectrum. Conversely, let (y_n) be a process with a rational spectrum S_y. As we have seen, S_y has real coefficients, and satisfies the condition

$$S_y(z^{-1}) = S_y(z) \tag{6.2.12}$$

obtained by meromorphic extension of the equality $S_y(e^{-i\omega}) = S_y(e^{i\omega})$. Now from (6.2.12), if α is a pole (resp. zero) of S_y, the same is true of $\alpha^{-1}, \overline{\alpha}$ and $\overline{\alpha}^{-1}$. The poles and the zeroes of S_y are thus distributed as shown in Fig. 24.

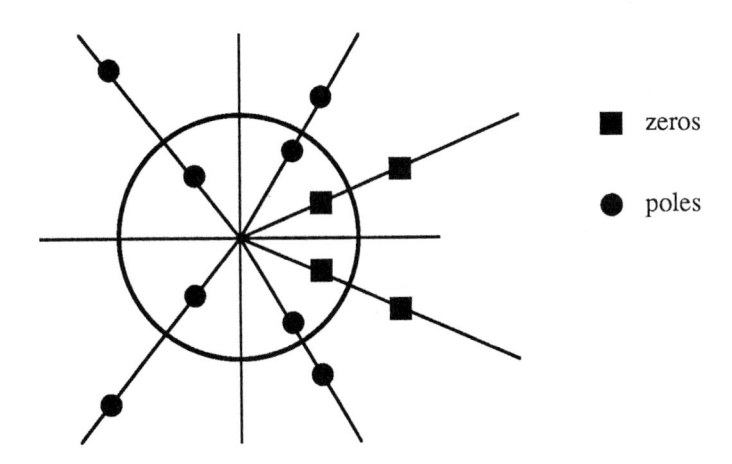

Figure 24. Position of the poles and zeroes of a rational spectrum with respect to the unit circle

But then the unit circle partitions the poles and zeroes, and we have a **spectral factorisation**

$$S_y(z) = S(z)S(z^{-1}) \qquad (6.2.13)$$

where $S(z)$ and $S^{-1}(z)$ are analytic outside the unit circle and define stable and causal filters. If we set

$$S^{-1}(z) = \sum_{k \geq 0} s_k^\times z^{-k}, \qquad \sum s_k^{\times 2} < \infty$$

the equation

$$\nu_n = \sum_{k \geq 0} s_k^\times y_{n-k}$$

then defines a signal with spectrum equal to 1, i.e. a white noise: (ν_n) is the **innovation** of (y_n) and we have the representation of equation (6.2.1). Consequently, and since S as defined by (6.2.13) is a rational filter, it follows that **any process with a rational spectrum is an ARMA process.**

Example.

Suppose (y_n) has a state variable representation

$$\begin{aligned} X_n &= AX_{n-1} + BV_n \quad (A \text{ stable}) \\ y_n &= CX_{n-1} + DV_n \end{aligned} \qquad (6.2.14)$$

where $X_n \in \mathbb{R}^d$, but where now (V_n) is a white noise **vector** of dimension $r > 1$. Since (y_n) is scalar, we cannot apply the previous arguments directly to (6.2.14) to deduce that (y_n) is an ARMA process. However, the spectrum of (y_n) is equal to

$$\begin{aligned} S_y(z) &= S(z)S^T(z^{-1}) \\ S(z) &= D + C(zI - A)^{-1}B \end{aligned}$$

which is rational. Thus the previous results show that (y_n) is an ARMA process; note that its innovation (ν_n) is a scalar white noise not equal to (V_n), which must be obtained by spectral factorisation or by some other method.

6.3 Kalman Filters

6.3.1 Discrete-Time Kalman Filter

Kalman Filter with Finite Time Origin.

For $n \geq 0$, consider the following dynamical system

$$\begin{aligned} X_n &= AX_{n-1} + V_n \\ Y_n &= CX_{n-1} + W_n \end{aligned} \qquad (6.3.1)$$

6.3 Kalman Filters

where X_0 has mean x_0 and covariance Σ_0, and where

$$E\left\{\begin{pmatrix} V_n \\ W_n \end{pmatrix}(V_m^T W_m^T)\right\} = \begin{pmatrix} Q & S \\ S^T & R \end{pmatrix}\delta(n-m) \qquad (6.3.2)$$

where $Q \geq 0$, and $R > 0$. We denote by

$$\widehat{X}_n = E[X_n \mid Y_{n-1}, Y_{n-2}, \ldots] \qquad (6.3.3)$$

the least squares estimate of the state X_n given the observations $Y_1, \ldots, Y_{n-1}$ to time $n-1$. Then $\widehat{X}_n$ satisfies the *Kalman filter equations*

$$
\begin{aligned}
&(i) & \widehat{X}_n &= A\widehat{X}_{n-1} + K_{n-1}[Y_n - C\widehat{X}_{n-1}], \quad \widehat{X}_0 = x_0 \\
&(ii) & K_n &= [A\Sigma_n C^T + S][C\Sigma_n C^T + R]^{-1} \\
&(iii) & \Sigma_n &= A\Sigma_{n-1}A^T - K_{n-1}[C\Sigma_{n-1}C^T + R]K_{n-1}^T + Q
\end{aligned} \qquad (6.3.4)
$$

Stationary Kalman Filter.

Suppose now that we have the same filter problem in a stationary context (this means that the process begins at time $-\infty$, it does not necessarily mean that (6.3.1) defines a stationary second order process), then the *stationary Kalman filter equations* are given by substituting

$$\Sigma_{n-1} \equiv \Sigma \qquad (6.3.5)$$

in (6.3.4-i and ii), where Σ is a *maximal solution of the algebraic Riccati equation*

$$\begin{aligned} \Sigma &= A\Sigma A^T - K[C\Sigma C^T + R]K^T + Q \\ K &= [A\Sigma C^T + S][C\Sigma C^T + R]^{-1} \end{aligned} \qquad (6.3.6)$$

For (6.3.4-i and ii) as modified by (6.3.6) to define a stationary second order process, it is sufficient that

$$|\lambda(A - KC)| < 1 \qquad (6.3.7)$$

where $\lambda(.)$ denotes the eigenvalues of the given matrix. Let

$$\begin{aligned} A' &= A - SR^{-1}C \\ Q' &= Q - SR^{-1}S^T, \quad Q' = DD^T \end{aligned} \qquad (6.3.8)$$

A sufficient condition for (6.3.7) to hold is (cf. (Anderson and Moore 1979))

$$(A', D) \text{ is controllable, and } (C, A') \text{ is observable} \qquad (6.3.9)$$

In this case, Σ as defined in (6.3.6) is positive definite. If further, in (6.3.4), the initial condition Σ_0 is chosen ≥ 0, then the maximal solution of (6.3.6) is the limit as $n \to \infty$ of Σ_n, as defined in (6.3.4).

6.3.2 Continuous-Time Kalman Filter

Kalman Filter with Finite Time Origin.

Consider the continuous-time dynamical system
$$\begin{aligned} dX_t &= AX_t dt + dV_t \\ dY_t &= CX_t dt + dW_t \end{aligned} \qquad (6.3.10)$$

where X_0 has mean x_0 and covariance Σ_0, and where
$$\begin{pmatrix} V_t \\ W_t \end{pmatrix}$$
is a Wiener process with covariance matrix
$$\begin{pmatrix} Q & S \\ S^T & R \end{pmatrix} \qquad (6.3.11)$$

with $Q \geq 0$ and $R > 0$. We denote by
$$\widehat{X}_t = E[X_t \mid Y_s, s \leq t] \qquad (6.3.12)$$

the least squares estimate of the state X_t given the observations Y_s up to time t. Then $\widehat{X}_t$ satisfies the *Kalman filter equations*

$$\begin{aligned} (i) \quad & d\widehat{X}_t = A\widehat{X}_t dt + K_t[dY_t - C\widehat{X}_t dt], \qquad \widehat{X}_0 = x_0 \\ (ii) \quad & K_t = [\Sigma_t C^T + S]R^{-1} \\ (iii) \quad & \frac{d}{dt}\Sigma_t = A\Sigma_t + \Sigma_t^T A^T - K_t R K_t^T + Q \end{aligned} \qquad (6.3.13)$$

Stationary Kalman filter.

Suppose we now have the same filter problem in a stationary context (this means that the process begins at time $-\infty$, it does not necessarily mean that (6.3.10) defines a stationary second order process), the *stationary Kalman filter equations* are given by substituting
$$\Sigma_t \equiv \Sigma \qquad (6.3.14)$$

in (6.3.13-i and ii), where Σ is a *maximal solution of the algebraic Riccati equation*
$$\begin{aligned} 0 &= A\Sigma + \Sigma A^T - KRK^T + Q \\ K &= [\Sigma C^T + S]R^{-1} \end{aligned} \qquad (6.3.15)$$

For (6.3.13-i and ii) as modified by (6.3.15) to define a stationary second order process, it is sufficient that
$$\operatorname{Re} \lambda(A - KC) < 0 \qquad (6.3.16)$$

The conditions which guarantee (6.3.16) are exactly the same as for the discrete-time filter, namely (6.3.8) and (6.3.9).

Part II

Stochastic Approximations: Theory

Chapter 1
O.D.E. and Convergence A.S. for an Algorithm with Locally Bounded Moments

Introduction.
We now begin the mathematical study of the algorithms considered in Part I. In this first chapter, we shall consider their behaviour (approximation by solution of the "mean" differential equation termed ODE in Part I, and asymptotic analysis) in a framework which is sufficiently general to cover most of the cases introduced in Part I. Applications to the previous cases will be discussed in detail in Chapter 2.

We reserve the case of the most general algorithm that it is possible to study for Chapter 3. Consideration of this algorithm imposes real technical complications, although the method of analysis is fundamentally the same.

Section 1.1 establishes the notation and the general assumptions. In Section 1.2 we introduce an assumption particular to this chapter, which we shall call the assumption of "locally bounded" moments (Assumption (A.5)). Sections 1.3 and 1.4 contain preparative "techniques" for the main theorems which are found in Sections 1.5 to 1.10.

1.1 Introduction of the General Algorithm

1.1.1

The general model of the algorithm considered in Part I was of the form

$$\theta_{n+1} = \theta_n + \gamma_{n+1} H(\theta_n, X_{n+1}) + \gamma_{n+1}^2 \rho_{n+1}(\theta_n, X_{n+1}) \qquad (1.1.1)$$

where θ_n evolves in $\mathbb{R}^d$ and the **state vector** X_n lies in $\mathbb{R}^k$ or in a subset of $\mathbb{R}^k$. H and ρ_n are two functions from $\mathbb{R}^d \times \mathbb{R}^k$ to $\mathbb{R}^d$.

We assume that the r.v. $\theta_0, X_0, X_1, \ldots, X_n, \ldots$ are defined on a probability space (Ω, A, P), and we denote the σ-field of events generated by the r.v. $\theta_0, X_0, \ldots, X_n$ by F_n.

In all that follows, we shall make the following assumptions:

(A.1) $(\gamma_n)_{n \in \mathbb{N}}$ *is a decreasing sequence (in the broad sense) of positive real numbers such that* $\sum_n \gamma_n = +\infty$.

(A.2) *There exists a family* $\{\Pi_\theta : \theta \in \mathbb{R}^d\}$ *of transition probabilities* $\Pi_\theta(x, A)$ *on* $\mathbb{R}^k$ *such that, for any Borel subset A of $\mathbb{R}^k$, we have*

$$P[X_{n+1} \in A | F_n] = \Pi_{\theta_n}(X_n, A) \qquad (1.1.2)$$

Note that this implies that

$$E\{g(\theta_n, X_{n+1})|F_n\} = \int g(\theta_n, x)\Pi_{\theta_n}(X_n, dx) \qquad (1.1.3)$$

for any Borel function $g(\theta, x)$ which is positive, or which satisfies $E|g(\theta_n, X_{n+1})| < \infty$. Formula (1.1.3) implies that the random variable $\int g(\theta_n(\omega), x)\Pi_{\theta_n}(X_n(\omega), dx)$ is a version of the conditional expectation of $g(\theta_n, X_{n+1})$ given F_n, i.e. given the values taken by the algorithm and by the state vector up to time n.

Assumption (A.2) says that the 2-tuple $(X_n, \theta_n)_{n\geq 0}$ is a Markov process. Its transition probability depends on n (since γ_n and ρ_n depend on n). It is therefore an inhomogeneous Markov process. If $\gamma_n = \gamma$ (algorithm with constant step size) and $\rho_n = \rho$, it is homogeneous.

Notation.

a. Let $P_{x,a}$ denote the distribution of $(X_n, \theta_n)_{n\geq 0}$ for the initial conditions $X_0 = x$, $\theta_0 = a$.

b. If, more precisely,

$$P_{x,a}^{(\gamma_n, \rho_n; n\geq 0)}$$

denotes the distribution of $(X_n, \theta_n)_{n\geq 0}$ for the given sequence $(\gamma_n, \rho_n)_{n\geq 0}$, with initial conditions $X_0 = x$, $\theta_0 = a$, then the conditional distribution of $(X_{n+k}, \theta_{n+k})_{k\geq 0}$ given F_n is

$$P_{X_n, \theta_n}^{(\gamma_{n+k}, \rho_{n+k}; k\geq 0)}$$

c. In what follows, it will be useful to express the trajectory of the algorithm $n \to \theta_n$ in the form of a continuous-time process. To this end, we set

$$t_0 = 0, t_1 = \gamma_1, \ldots, t_n = \sum_{i=1}^{n} \gamma_i \qquad (1.1.4)$$

$$\theta(t) = \sum_{k\geq 0} I(t_k \leq t < t_{k+1})\theta_k \qquad (1.1.5)$$

where $I(A)$ denotes the characteristic function of the set A (often denoted by 1_A).

The study of the behaviour of $\theta(t)$ between times t_n and $t_n + T$ thus reduces to the study of the behaviour of θ_k for integers k between n and $m(n, T)$, where

$$m(n, T) = \inf\{k : k \geq n, \gamma_{n+1} + \ldots + \gamma_{k+1} \geq T\} \qquad (1.1.6)$$

For simplicity, we shall denote

$$m(T) = m(0, T) \qquad (1.1.7)$$

d. For any function $f(x,\theta)$ on $\mathbb{R}^k \times \mathbb{R}^d$, we shall denote the partial mapping $x \to f(x,\theta)$ by f_θ. In particular, $\Pi_\theta f_\theta$ denotes the function

$$x \to \int f(y,\theta)\Pi_\theta(x,dy)$$

1.1.2 Examples

Example 1. *Algorithms with conditionally linear dynamics.*

In such algorithms, the state $(X_n)_{n\geq 0}$ evolves according to the recurrence

$$X_{n+1} = A(\theta_n)X_n + B(\theta_n)W_{n+1} \tag{1.1.8}$$

where $(W_n)_{n\geq 1}$ is a sequence of independent, identically distributed r.v. and $A(\theta)$ and $B(\theta)$ are matrix valued functions of θ (cf. Part I—Paragraph 1.3.2.4).

Example 2. *Algorithm independent linear dynamics.*

This case arises in particular from Example 1 when A and B are matrices independent of θ. Referring back to the examples of Part I, it is easy to see that the transversal equaliser, derived from the recursive equaliser of Paragraph 1.3.2.1 **by omitting the recursive decision-feedback part**, comes under this heading (note here that $X_n^T = (U_{n+N}^T, y_{n+N}, \ldots, y_{n-N}))$.

Example 3. *Recursive decision-feedback equaliser.*

The case of the recursive decision-feedback equaliser leads to a state vector $\xi_n^T = (Y_n^T, \eta_n^T)$ such that

$$\begin{aligned} Y_{n+1} &= AY_n + BW_{n+1} \\ \eta_{n+1} &= f(\theta_n, Y_{n+1}, \eta_n) \end{aligned} \tag{1.1.9}$$

where A and B are independent of (θ_n), and (W_n) is a sequence of independent, identically distributed r.v. See also Paragraph 1.3.2.1 of Part I.

Example 4. *Robbins–Monro algorithms.*

It is important to make the connection between the general theory presented here and the "classical" theory of stochastic algorithms, as initiated by Robbins–Monro (Robbins and Monro 1951) (cf. also (Wasan 1969), (Hall and Heyde 1980) and (Dvoretsky 1956)). All the algorithms from this theory are of the form (1.1.1) and satisfy Assumptions (A.1) and (A.2), together with the additional "Robbins–Monro" assumption:

(RM). For all x, $\Pi_\theta(x,dz) = \mu_\theta(dz)$, where μ_θ is a probability distribution on $\mathbb{R}^k$.

1.1.3 General Assumptions on H, ρ_n and Π

Notation. We shall frequently denote the function $x \to H(\theta, x)$ by H_θ; unlike the notation of Part I, this does not denote a partial derivative.

Throughout Part II, we shall assume that D is an open subset of $\mathbb{R}^d$. The functions H and ρ_n will be required to satisfy:

(A.3) *For any compact subset Q of D, there exist constants C_1, C_2, q_1, q_2 (depending on Q), such that for all $\theta \in Q$, and all n we have*

(i) $\qquad |H(\theta, x)| \leq C_1(1 + |x|^{q_1})$

(ii) $\qquad |\rho_n(\theta, x)| \leq C_2(1 + |x|^{q_2})$

When we wish to express the dependence of Q explicitly in the above formulae, we shall write $C_i(Q)$ or $q_i(Q)$.

The fundamental assumption which we shall introduce next may seem a little "abstract", but we shall see in what follows, and in the applications, that its verification is central to the study of the algorithm:

(A.4) *There exists a function h on D, and for each $\theta \in D$ a function $\nu_\theta(.)$ on $\mathbb{R}^k$ such that*

(i) h *is locally Lipschitz on* D;

(ii) $(I - \Pi_\theta)\nu_\theta = H_\theta - h(\theta)$ *for all* $\theta \in D$;

(iii) *for all compact subsets Q of D, there exist constants $C_3, C_4, q_3, q_4, \lambda \in [\frac{1}{2}, 1]$, such that for all $\theta, \theta' \in Q$*

$$|\nu_\theta(x)| \leq C_3(1 + |x|^{q_3}) \qquad (1.1.10)$$

$$|\Pi_\theta \nu_\theta(x) - \Pi_{\theta'} \nu_{\theta'}(x)| \leq C_4 |\theta - \theta'|^\lambda (1 + |x|^{q_4}) \qquad (1.1.11)$$

Comments on (A.4).

a. Note that the functions $H_\theta(.)$, $h(\theta)$ and $\nu_\theta(.)$ take their values in $\mathbb{R}^d$. Condition (A.4-ii) implies that for each $i = 1, \ldots, d$

$$(I - \Pi_\theta)\nu_\theta^i = H_\theta^i - h^i(\theta)$$

where the superscript i denotes the i-th coordinate in $\mathbb{R}^d$.

b. Often in examples to verify (A.4), we shall prove the following: there exists a function h, locally Lipschitz on D and, for all $\theta \in D$, a function w_θ on $\mathbb{R}^k$ such that

$$(I - \Pi_\theta)w_\theta = \Pi_\theta H_\theta - h(\theta) \qquad (1.1.12)$$

1.1 Introduction of the General Algorithm

Moreover, for any compact subset Q of D, there exist constants C_3', C_4, q_3', q_4, λ such that

$$|w_\theta(x)| \leq C_3'(1+|x|^{q_3'}) \tag{1.1.13}$$

$$|w_\theta(x) - w_{\theta'}(x)| \leq C_4|\theta - \theta'|^\lambda (1+|x|^{q_4}) \tag{1.1.14}$$

In fact, it is sufficient to set

$$\nu_\theta = w_\theta + H_\theta - h(\theta) \tag{1.1.15}$$

to obtain ν_θ and h satisfying (A.4-i). Then trivially $\Pi_\theta \nu_\theta = w_\theta$, whence (1.1.11) holds by virtue of (1.1.14). Property (1.1.10) results from (A.3-i) and (1.1.13).

c. Concerning the importance of (A.4), note that if for all θ, the Markov chain with transition probability Π_θ is positive, recurrent, with invariant probability Γ_θ, and if we set

$$h(\theta) = \int H_\theta(y) \Gamma_\theta(dy) \tag{1.1.16}$$

(or more concisely $\Gamma_\theta H_\theta$), then the function $H_\theta - h(\theta)$ has a zero integral with respect to Γ_θ, and thus equation (A.4-ii), the so-called **Poisson equation**, has a solution ν_θ. Moreover in most cases, this solution may be expressed in the form

$$\nu_\theta(y) = \sum_{k \geq 0} \Pi_\theta^k (H_\theta - h(\theta))(y) \tag{1.1.17}$$

when the series is convergent (cf. Chapter 2 to follow).
In applications, the existence of $h(\theta)$ results most often from the existence of an invariant Γ_θ, in particular with

$$\Gamma_\theta g = \lim_{n \to \infty} \Pi_\theta^n g$$

for a "sufficiently rich" family of functions g on $\mathbb{R}^k$. It is interesting to note at this point, that even if H_θ is not regular (because for example of a discontinuity in θ), for a sufficiently "regularising" kernel Π_θ, the regularity of the terms $\Pi_\theta^n(H_\theta - h(\theta))$ may, for $n \geq 1$, imply property (1.1.11) via the representation (1.1.17).

d. The main idea behind the study of stochastic algorithms, as presented here in Part II, is to obtain very general results based upon (A.4), and so to reduce the study of individual algorithms to the verification of this condition.

1.1.4 Examples (Continued)

We review the algorithms given in 1.1.2.

Example 1.

This is the algorithm (1.1.8). Suppose that

$$E|W_n|^q = \bar{\nu}_q < \infty \qquad (1.1.18)$$

If we denote $\prod_{i=l}^{m} A_i = A_m \cdot A_{m-1} \ldots A_l$, then formula (1.1.8) gives

$$X_{n+1} = \prod_{i=0}^{n} A(\theta_i) X_0 + \sum_{k=0}^{n-1} \left\{ \prod_{i=k+1}^{n} A(\theta_i) \right\} B(\theta_k) W_{k+1} + B(\theta_n) W_{n+1} \qquad (1.1.19)$$

Let us assume that:

$$\sup_{\theta \in Q} |B(\theta)| \leq M \qquad (1.1.20)$$

and for the moment that

$$\sup_{\theta \in Q} |A(\theta)| \leq \rho < 1 \qquad (1.1.21)$$

Now we see that for any function g on $\mathbb{R}^k$,

$$\Pi_\theta g(x) = E\{g(A(\theta)x + B(\theta)W_1)\} \qquad (1.1.22)$$

whenever this expectation exists; that is, for any function g such that $|g(x)| \leq \text{Constant}\,(1 + |x|^q)$. Similarly, for all n

$$\Pi_\theta^n g(x) = E\{g(A^n(\theta)x + \sum_{k=0}^{n-1} A^{n-k}(\theta) B(\theta) W_{k+1})\}$$

and, using the fact that the $(W_k)_{k=1,\ldots n}$ are identically distributed, we may write

$$\Pi_\theta^n g(x) = E\{g(A^n(\theta)x + \sum_{k=1}^{n} A^k(\theta) B(\theta) W_k)\} \qquad (1.1.23)$$

Assumptions (1.1.18), (1.1.20) and (1.1.21) show that the sequence

$$\left(A^n(\theta)x + \sum_{k=1}^{n} A^k(\theta) B(\theta) W_k \right)_{n \geq 0}$$

of random variables converges in L^q to

$$U_\infty(\theta) = \sum_{k=1}^{\infty} A^k(\theta) B(\theta) W_k$$

Thus

$$\lim_{n \to \infty} \Pi_\theta^n g(x) = E\{g(U_\infty)\}$$

for any Borel function g such that $|g(x)| \leq C(1 + |x|^q)$. This shows the existence of the invariant measure Γ_θ. Then we have (if $q_1 \leq q$)

$$h(\theta) = \int H_\theta(y)\Gamma_\theta(dy) = E(H_\theta(U_\infty(\theta)))$$

and the existence of a solution ν_θ of the Poisson equation (A.4-ii) can be proved by considering the series (1.1.17) together with (1.1.23) (see also Chapter 2 to follow).

Thus we see that if $A(\theta)$ and $B(\theta)$ are locally Lipschitz functions of θ, then for any Lipschitz function g, the expression $\Pi_\theta^n g(x)$ given by (1.1.23) shows that for all x, $\Pi_\theta^n g(x)$ is a Lipschitz function of θ. If the function $H_\theta(y)$ is itself regular in θ and y, then, as can be seen, it is not difficult to verify all of Assumption (A.4). We shall return to this more fully in Chapter 2. We shall show there that (A.4-ii) holds for less rigorous conditions on $A(\theta)$ ((1.1.21) will be replaced by the condition that the eigenvalues of $A(\theta)$ are uniformly strictly less than 1), and on H (which may have discontinuities if the densities of the W_k are sufficiently regular).

Example 2. *Robbins–Monro algorithms (cf. 1.1.2 Example 4).*

Since $\Pi_\theta(x, A) = \mu_\theta(A)$, condition (A.4) may be simplified considerably. Let us assume that for all θ the integral:

$$h(\theta) = \int H(\theta, x)\mu_\theta(dx) \tag{1.1.24}$$

exists. Then $\nu(\theta, x) = H(\theta, x)$ satisfies (A.4-ii) and $\Pi_\theta \nu(\theta, x) = h(\theta)$ is independent of x. Thus the whole of (A.4) can be deduced from the single assumption that h as defined by (1.1.24) is locally Lipschitz. In fact, the Robbins–Monro algorithm lends itself to direct study without the use of (A.4) (cf. Subsection 1.4.7 and Chapter 5 to follow).

1.2 Assumptions Peculiar to Chapter 1

In this chapter, we shall place ourselves in the situation in which the moments of a given order q of the state vector X_n are bounded in n, at least whilst θ_n remains in some compact set. We shall use the boundedness of the moments to derive L^2 upper bounds and to obtain directly results covering most of the applications considered in this book.

In Chapter 3 we shall examine the case in which the moments of the (X_n) may become infinite with n, and we shall replace the L^2 estimates of this chapter with L^p estimates which will enable us to carry out asymptotic analysis for $\gamma_n = 1/n^\alpha$.

1.2.1 Assumptions on the Moments

Firstly we have the "local boundedness" assumption:

(A.5) *For any compact subset Q of D and any $q > 0$, there exists $\mu_q(Q) < \infty$ such that for all $n, x \in \mathbb{R}^k, a \in \mathbb{R}^d$*

$$E_{x,a}\{I(\theta_k \in Q, k \leq n)(1 + |X_{n+1}|^q)\} \leq \mu_q(Q)(1 + |x|^q) \quad (1.2.1)$$

Remarks.

1. If (1.2.1) is true for q, then it is true for $q' < q$ with

$$\mu_{q'}(Q) \leq 2^{(q-q')q'/q} \mu_q(Q)$$

2. In the definition of (A.5), an inequality (1.2.1) is assumed for all $q > 0$. In fact, in the proofs in this chapter, we shall always use a weaker assumption, namely that (1.2.1) is valid for a sufficiently large q, i.e. larger than a well-defined function of the exponents q_i in (A.3) and (A.4).

1.2.2 Examples

Example 1.

For algorithm (1.1.8), formula (1.19) and assumptions (1.1.18), (1.1.20) and (1.1.21) imply that

$$\begin{aligned} E\{|X_{n+1}|^q I(\theta_k \in Q, k \leq n)\} &\leq E\{[\rho^n |x| + \sum_{k=0}^{n} \rho^{n-k} M |W_{k+1}|^q]\} \\ &\leq C_q \{|x|^q + \sum_{k=0}^{n} (\rho^{n-k} M)^q \bar{\nu}_q\} \end{aligned}$$

for some suitable constant C_q.

Condition (A.5) follows.

Example 2.

If in the above example A and B are also given to be independent of θ, then we have an inequality stronger than (1.2.1):

$$E(|X_{n+1}|^q) \leq \mu_q(1 + |x|^q)$$

1.3 Decomposition of the General Algorithm

1.3.1

In Part I, we saw heuristically that if γ_1 tends to zero, the algorithm $\theta(t)$ has a tendency to follow the solution of the differential equation (deterministic) with initial condition $a = \theta_0$

$$\bar{\theta}' = h(\bar{\theta}(t)) \quad (1.3.1)$$

1.3 Decomposition of the General Algorithm

This is because $\bar{\theta}(t_n)$ is close (Euler's approximation) to the solution $\bar{\theta}_n$ of

$$\begin{aligned}\bar{\theta}_{n+1} &= \bar{\theta}_n + \gamma_{n+1} h(\bar{\theta}_n) \\ \bar{\theta}_0 &= a\end{aligned} \qquad (1.3.2)$$

and because algorithm (1.1.1) may be written in the form

$$\theta_{n+1} = \theta_n + \gamma_{n+1} h(\theta_n) + \varepsilon_n \qquad (1.3.3)$$

where

$$\begin{aligned}\varepsilon_n &= \theta_{n+1} - \theta_n - \gamma_{n+1} h(\theta_n) \\ &= \gamma_{n+1}[H(\theta_n, X_{n+1}) - h(\theta_n) + \gamma_{n+1}\rho_{n+1}(\theta_n, X_{n+1})]\end{aligned} \qquad (1.3.4)$$

is a "small" fluctuation for small γ_1.

Comparison of the behaviour of the solution (1.3.1) with the behaviour of the algorithm thus depends on obtaining upper bounds on the fluctuation ε_n. More generally, in the sequel, we shall require upper bounds for the expressions

$$\varepsilon_n(\phi) = \phi(\theta_{n+1}) - \phi(\theta_n) - \gamma_{n+1}\phi'(\theta_n) \cdot h(\theta_n) \qquad (1.3.5)$$

Let ϕ be a C^2 function from $\mathbb{R}^d$ to $\mathbb{R}$ with **bounded second derivatives**. For the compact subset Q of D we denote

$$\begin{aligned}M_0(Q) &= \sup_{\theta \in Q} |\phi(\theta)| \\ M_1(Q) &= \sup_{\theta \in Q} |\phi'(\theta)| \\ M_2(Q) &= \sup_{\theta \in Q} |\phi''(\theta)| \\ M_2 &= \sup_{\theta \in \mathbb{R}^d} |\phi''(\theta)|\end{aligned} \qquad (1.3.6)$$

Then there exists a matrix $R(\phi, \theta, \theta')$ such that

$$\phi(\theta') - \phi(\theta) - (\theta' - \theta) \cdot \phi'(\theta) = R(\phi, \theta, \theta') \qquad (1.3.7)$$

with, for all $\theta, \theta' \in \mathbb{R}^d$

$$|R(\phi, \theta, \theta')| \leq M_2 |\theta' - \theta|^2 \qquad (1.3.8)$$

Thus for all k

$$\begin{aligned}\varepsilon_k(\phi) &= \phi'(\theta_k) \cdot [(\theta_{k+1} - \theta_k) - \gamma_{k+1} h(\theta_k)] + R(\phi, \theta_k, \theta_{k+1}) \\ &= \gamma_{k+1}\phi'(\theta_k) \cdot [H(\theta_k, X_{k+1}) - h(\theta_k)] + \gamma_{k+1}^2 \phi'(\theta_k) \cdot \rho_{k+1}(\theta_k, X_{k+1}) \\ &\quad + R(\phi, \theta_k, \theta_{k+1})\end{aligned} \qquad (1.3.9)$$

with

$$|R(\phi, \theta_k, \theta_{k+1})| \leq \gamma_{k+1}^2 M_2 |H(\theta_k, X_{k+1}) + \gamma_{k+1}\rho_{k+1}(\theta_k, X_{k+1})|^2 \qquad (1.3.10)$$

1.3.2 Decomposition of $\varepsilon_n(\phi)$

If (1.3.9) is written in the form
$$\varepsilon_k(\phi) = \gamma_{k+1}\phi'(\theta_k) \cdot [H(\theta_k, X_{k+1}) - h(\theta_k)] + A_k^1$$
then using (A.4-ii)
$$\begin{aligned}
\phi(\theta_{k+1}) &- \phi(\theta_k) - \gamma_{k+1}\phi'(\theta_k) \cdot h(\theta_k) \\
&= \gamma_{k+1}\phi'(\theta_k)[H(\theta_k, X_{k+1}) - h(\theta_k)] + A_k^1 \\
&= \gamma_{k+1}\phi'(\theta_k)[\nu_{\theta_k}(X_{k+1}) - \Pi_{\theta_k}\nu_{\theta_k}(X_{k+1})] + A_k^1 \\
&= \gamma_{k+1}\phi'(\theta_k) \cdot [\nu_{\theta_k} - \Pi_{\theta_k}\nu_{\theta_k}(X_k)] \\
&\quad + \gamma_{k+1}\phi'(\theta_k)[\Pi_{\theta_k}\nu_{\theta_k}(X_k) - \Pi_{\theta_k}\nu_{\theta_k}(X_{k+1})] + A_k^1 \\
&= A_k^2 + A_k^3 + A_k^1
\end{aligned}$$

Actually this calculation only makes sense if $\theta_k \in D$, since h is only defined on D; this is why we introduce for a fixed compact subset Q of D
$$\tau = \tau(Q) = \inf(n; \theta_n \notin Q) \tag{1.3.11}$$
Let
$$\psi_\theta(x) = \phi'(\theta) \cdot \Pi_\theta \nu_\theta(x) \tag{1.3.12}$$
Then in $\{\tau \geq n\}$ and for $r < n$ we have:
$$\begin{aligned}
\sum_{k=r}^{n-1} \varepsilon_k &= \sum_{k=r}^{n-1}(A_k^1 + A_k^2) + \sum_{k=r}^{n-1} \gamma_{k+1}(\psi_{\theta_k}(X_k) - \psi_{\theta_k}(X_{k+1})) \\
&= \sum_{k=r}^{n-1}(A_k^1 + A_k^2) + \gamma_{r+1}\psi_{\theta_r}(X_r) \\
&\quad + \sum_{k=r+1}^{n-1} \gamma_{k+1}(\psi_{\theta_k}(X_k) - \psi_{\theta_{k-1}}(X_k)) \\
&\quad + \sum_{k=r+1}^{n-1}(\gamma_{k+1} - \gamma_k)\psi_{\theta_{k-1}}(X_k) - \gamma_n \psi_{\theta_{n-1}}(X_n)
\end{aligned}$$
whence we have:

Lemma 1. *For $r < n$ in $\{n \leq r\}$ we have*
$$\sum_{k=r}^{n-1} \varepsilon_k(\phi) = \sum_{k=r}^{n-1} \varepsilon_k^{(1)} + \sum_{k=r+1}^{n-1} \varepsilon_k^{(2)} + \sum_{k=r+1}^{n-1} \varepsilon_k^{(3)} + \sum_{k=r}^{n-1} \varepsilon_k^{(4)} + \eta_{n;r}$$
where
$$\begin{aligned}
\varepsilon_k^{(1)} &= \gamma_{k+1}\phi'(\theta_k) \cdot (\nu_{\theta_k}(X_{k+1}) - \Pi_{\theta_k}\nu_{\theta_k}(X_k)) \\
\varepsilon_k^{(2)} &= \gamma_{k+1}(\psi_{\theta_k}(X_k) - \psi_{\theta_{k-1}}(X_k)), \quad \psi_\theta(x) = \phi'(\theta) \cdot \Pi_\theta \nu_\theta(x) \\
\varepsilon_k^{(3)} &= (\gamma_{k+1} - \gamma_k)\psi_{\theta_{k-1}}(X_k) \\
\varepsilon_k^{(4)} &= \gamma_{k+1}^2 \phi'(\theta_k) \cdot \rho_{k+1} + R(\phi, \theta_k, \theta_{k+1}) \\
\eta_{n;r} &= \gamma_{r+1}\psi_{\theta_r}(X_r) - \gamma_n \psi_{\theta_{n-1}}(X_n)
\end{aligned}$$

Remark 1. Using (A.1) and (A.4-iii) we have

$$\Pi_\theta \nu_\theta(x) = E_{x;\theta}(\nu_\theta(X_1)) \leq C_3 E_{x;\theta}(1+|X_1|^{q_3}) \leq C_3 \mu_{q_3}(1+|x|^{q_3})$$

for all $\theta \in Q$, i.e.

$$\sup_{\theta \in Q} |\Pi_\theta \nu_\theta(x)| \leq C_3 \mu_{q_3}(1+|x|^{q_3}) \qquad (1.3.13)$$

Remark 2. From (1.36), (1.1.10) and (1.1.11) we have:

$$\sup_{\theta \in Q} |\psi_\theta(x)| \leq M_1 C_3 \mu_{q_3}(1+|x|^{q_3}) \qquad (1.3.14)$$

$$\sup_{\theta,\theta' \in Q} |\psi_\theta(x) - \psi_{\theta'}(x)| \leq M_1 C_4(1+|x|^{q_4})|\theta-\theta'|^\lambda + M_2 C_3 \mu_{q_3}(1+|x|^{q_3})|\theta-\theta'|$$

$$(1.3.15)$$

1.4 L^2 Estimates

The aim of this paragraph is to prove Proposition 7 (below), which gives a mean squares upper bound for the "fluctuation"

$$\sup_{n \leq m \wedge \tau} |\sum_{k=0}^{n-1} \varepsilon_k(\phi)|$$

(cf. the introduction of the $\varepsilon_k(\phi)$ in the previous paragraph) where τ is the time at which the process θ_n leaves the compact subset Q.

In this paragraph, Q is a fixed compact set. The "constants" which appear in the results may depend upon Q just as they depend upon the parameters C_i, μ_q and λ of the assumptions and upon the numbers $M_i(\phi)$ associated with the given function ϕ (cf. (1.3.6)). On the other hand, they are valid for all sequences $(\gamma_n)_{n\geq 0}$ such that $\gamma_1 \leq \gamma$, for fixed γ. When they do not depend on Q, we shall say so explicitly.

1.4.1

Lemma 2. *There exists a constant A_1 such that:*

$$E_{x,a}\{\sup_{n\leq m} I(n \leq \tau)|\sum_{k=0}^{n-1} \varepsilon_k^{(1)}|^2\} \leq A_1(1 + |x|^{2q_3}) \sum_{k=0}^{m-1} \gamma_{k+1}^2$$

where, using the constants of Assumptions (A.3) and (A.4)

$$A_1 \leq \tilde{A}_1 \mu_{2q_3}(Q) M_1^2(Q) C_3^2(Q)$$

the constant $\tilde{A}_1$ being independent of Q. Moreover, on $\{\tau = +\infty\}$, $\sum_{k=0}^{n-1} \varepsilon_k^{(1)}$ converges a.s. and in L^2 if $\sum \gamma_{k+1}^2 < \infty$.

Proof. If we set

$$Z_n = \sum_{k=0}^{n-1} \gamma_{k+1} I(k+1 \leq \tau) \phi'(\theta_k) \cdot [\Pi_{\theta_k} \nu_{\theta_k}(X_k) - \nu_{\theta_k}(X_{k+1})]$$

we note that $I(n \leq \tau) \cdot |\sum_{k=0}^{n-1} \varepsilon_k^{(1)}| \leq |Z_n|$. Since

$$E\{I(k+1 \leq \tau) \nu_{\theta_k}(X_{k+1})|F_k\} = I(k+1 \leq \tau) \Pi_{\theta_k} \nu_{\theta_k}(X_k)$$

Z_n is a martingale and (the conditional expectation is a contraction in L^2)

$$E\{I(k+1 \leq \tau)|\phi'(\theta_k) \cdot \Pi_{\theta_k} \nu_{\theta_k}(X_k)|^2\} \leq E\{I(k+1 \leq \tau)|\phi'(\theta_k) \cdot \nu_{\theta_k}(X_{k+1})|^2\}$$

Then

$$\begin{aligned}E|Z_n|^2 &= \sum_{k=0}^{n-1} \gamma_{k+1}^2 E\{I(k+1 \leq \tau)|\phi'(\theta_k) \cdot (\Pi_{\theta_k} \nu_{\theta_k}(X_k) - \nu_{\theta_k}(X_{k+1}))|^2\} \\ &\leq 2M_1^2 \sum_{k=0}^{n-1} \gamma_{k+1}^2 E\{I(k+1 \leq \tau)|\nu_{\theta_k}(X_{k+1})|^2\} \\ &\leq K \sum_{k=0}^{n-1} \gamma_{k+1}^2 E\{I(k+1 \leq \tau)(1 + |X_{k+1}|^{q_3})^2\} \\ &\leq K \sum_{k=0}^{n-1} \gamma_{k+1}^2 E\{I(k+1 \leq \tau)(1 + |X_{k+1}|^{2q_3})\} \\ &\leq K \mu_{2q_3}(1 + |x|^{2q_3}) \sum_{k=0}^{n-1} \gamma_{k+1}^2\end{aligned}$$

with $K \leq \overline{K} C_3^2 M_1^2 \mu_{2q_3}$, $\overline{K}$ being independent of Q.

The first part of the lemma then results from (Doob's inequality)

$$E\{\sup_{n\leq m} |Z_n|^2\} \leq 4 \sup_{n\leq m} E|Z_n|^2 \leq K(1 + |x|^{2q_3}) \sum_{k=0}^{m-1} \gamma_{k+1}^2$$

1.4 L^2 Estimates

The second part is proved since on $\{\tau = \infty\}$

$$|\sum_{k=r}^{n-1} \varepsilon_k^{(1)}| \leq |Z_n - Z_r|$$

and since if $\sum \gamma_{k+1}^2 < \infty$, the martingale Z_n converges a.s. and also in L^2 since it is bounded in L^2. (Martingale convergence theorems. See for example (Neveu 1972).) $\square$

1.4.2

In considering the following terms we note that for $i = 2, 3, 4$

$$E\{\sup_{n \leq m} I(n \leq \tau)|\sum_{k=0}^{n-1} \varepsilon_k^{(i)}|^2\} \leq E\left(\sum_{k=0}^{m\wedge\tau-1} |\varepsilon_k^{(i)}|\right)^2$$

$$= E\left(\sum_{k=0}^{m-1} |\varepsilon_k^{(i)}| I(k+1 \leq \tau)\right)^2$$

with the convention: $\varepsilon_0^{(2)} = \varepsilon_0^{(3)} = 0$.

Lemma 3. *There exists a constant A_2 such that for all m:*

$$E_{x,a}\left\{\sum_{k=1}^{m\wedge\tau-1} |\varepsilon_k^{(2)}|\right\}^2 \leq A_2(1+|x|^{s_1})(\sum_{k=1}^{m-1} \gamma_{k+1}^{1+\lambda})^2$$

with $s_1 = \max(2q_4 + 2\lambda(q_1 \vee q_2), 2q_3 + 2(q_1 \vee q_2))$, and using the constants of (A.3), (A.4) and (A.5) and denoting $C_1(Q) + C_2(Q)\gamma_1$ by $\overline{C}(Q)$:

$$A_2 \leq \tilde{A}_2 \mu_{s_1}(Q)[\overline{C}^{2\lambda}(Q)M_1^2(Q)C_4^2(Q) + \overline{C}^2(Q)M_2^2(Q)C_3^2(Q)]$$

$\tilde{A}_2$ being a constant independent of Q.

Proof. We use (1.3.15) and note that on $\{k+1 \leq \tau\}$:

$$|\theta_k - \theta_{k-1}| \leq [C_1(Q) + C_2(Q)\gamma_1]\gamma_k(1+|X_k|^{q_1 \vee q_2})$$

Then

$$E\{\sum_{k=1}^{m\wedge\tau-1} |\varepsilon_k^{(2)}|\}^2 \leq K_1 E\left\{\sum_{k=1}^{m-1} \gamma_{k+1}^{1+\lambda}(1+|X_k|^{q_4+\lambda(q_1 \vee q_2)})I(k+1 \leq \tau)\right\}^2$$

$$+ K_2 E\left\{\sum_{k=1}^{m-1} \gamma_{k+1}^2(1+|X_k|^{q_3+(q_1 \vee q_2)})I(k+1 \leq \tau)\right\}^2$$

with (denoting $\overline{C}(Q) = C_1(Q) + C_2(Q)\gamma_1$)

$$K_1 \leq A\overline{C}^{2\lambda}(Q)M_1^2(Q)C_4^2(Q) \text{ and } K_2 \leq A\overline{C}^2(Q)M_2^2(Q)C_3^2(Q)\mu_{q_3}^2$$

where A depends only on q_1, q_2, q_3 and q_4. Using Schwartz's inequality, and since $1 + \lambda \leq 2$, we have

$$E\left\{\sum_{k=1}^{m\wedge\tau-1} |\varepsilon_k^{(2)}|\right\}^2 \leq \left(\sum_{k=1}^{m-1} \gamma_{k+1}^{1+\lambda}\right) \cdot$$

$$E\{\sum_{k=1}^{m-1} \gamma_{k+1}^{1+\lambda} I(k+1 \leq \tau)[K_1(1 + |X_k|^{2q_4 + 2\lambda(q_1 \vee q_2)})$$

$$+ K_2(1 + |X_k|^{2q_3 + 2(q_1 \vee q_2)})]\}$$

The lemma is finally proved using (A.5). □

1.4.3

Lemma 4. *There exists a constant A_3, such that for all n*

$$E_{x,a}\left\{\sum_{k=1}^{m\wedge\tau-1} |\varepsilon_k^{(3)}|\right\}^2 \leq A_3(1 + |x|^{2q_3})\gamma_1^2$$

with

$$A_3 \leq \tilde{A}_3 M_1^2(Q) C_3^2(Q) \mu_{2q_3}^3(Q)$$

A_3 being a constant independent of Q.

Proof. Using (1.3.14) we obtain

$$E\left\{\sum_{k=1}^{m\wedge\tau-1} |\varepsilon_k^{(3)}|\right\}^2$$

$$\leq KE\left\{\sum_{k=1}^{m-1} (\gamma_k - \gamma_{k+1})(1 + |X_k|^{q_3}) I(k+1 \leq \tau)\right\}^2$$

$$\leq K \sum_{k=1}^{m-1} (\gamma_k - \gamma_{k+1}) \sum_{k=1}^{m-1} (\gamma_k - \gamma_{k+1}) E\{(1 + |X_k|^{q_3})^2 I(k+1) \leq \tau\}$$

with $K \leq M_1^2(Q) C_3^2(Q) \mu_{q_3}^2(Q)$. Thus from (A.5)

$$E\left\{\sum_{k=1}^{m\wedge\tau-1} |\varepsilon_k^{(3)}|\right\}^2 \leq K(1 + |x|^{2q_3})\gamma_1^2$$

with $K \leq \tilde{A}_3 M_1^2(Q) C_3^2(Q) \mu_{2q_3}^3(Q)$. □

1.4 L^2 Estimates

1.4.4

Lemma 5. *Denote* $s_2 = \sup(4q_1, 4q_2)$. *There exists a constant* A_4 *such that for all* m

$$E_{x,a}\left\{\sum_{k=0}^{m\wedge\tau-1}|\varepsilon_k^{(4)}|\right\}^2 \leq A_4(1+|x|^{s_2})(\sum_{k=0}^{m-1}\gamma_{k+1}^2)^2$$

with

$$A_4 \leq \tilde{A}_4\mu_{s_2}(Q)[C_2^2(Q)M_1^2(Q) + C_1^4(Q) + \gamma_1^4 C_2^4(Q)]$$

$\tilde{A}_4$ *being a constant independent of* Q.

Proof. We have

$$|\varepsilon_k^{(4)}| \leq K\gamma_{k+1}^2(1+|X_{k+1}|^{s_1})$$

with, following (1.3.10) and (A.3):

$$s_1 = \sup(2q_1, 2q_2) \text{ and } K \leq C_2(Q)M_1(Q) + M_2[C_1(Q) + \gamma_1 C_2(Q)]^2$$

whence

$$E\left\{\sum_{k=0}^{m\wedge\tau-1}|\varepsilon_k^{(4)}|\right\}^2$$

$$\leq K^2 E\left\{\sum_{k=0}^{m-1}\gamma_{k+1}^2(1+|X_{k+1}|^{s_1})I(k+1\leq\tau)\right\}^2$$

$$\leq K^2 E\left\{(\sum_{k=0}^{m-1}\gamma_{k+1}^2)(\sum_{k=0}^{m-1}\gamma_{k+1}^2(1+|X_{k+1}|^{s_1})^2 I(k+1\leq\tau))\right\}$$

The lemma is proved by applying Assumption (A.5). □

1.4.5

Lemma 6. *There exists a constant* A_5 *such that*

$$E_{x,a}\{\sup_{1\leq n\leq m} I(n\leq\tau)|\eta_{n,0}|^2\} \leq A_5(1+|x|^{2q_3})\sum_{k=0}^{m-1}\gamma_{k+1}^2$$

with

$$A_5 \leq \tilde{A}_5 M_1^2(Q)C_3^2(Q)\mu_{2q_3}^2(Q)$$

$\tilde{A}_5$ *being independent of* Q. *Moreover* $\eta_{n;0}$ *converges a.s. and in* L^2 *on* $\{\tau = +\infty\}$ *when* $\sum \gamma_{k+1}^2 < \infty$.

Proof.

$$|\gamma_1\psi_{\theta_0}(X_0)|^2 \leq K\gamma_1^2(1+|x|^{2q_3})$$

and

$$E\{\sup_{1\leq n\leq m} I(n\leq\tau)|\gamma_n\psi_{\theta_{n-1}}(X_n)|^2\} \leq KE\{\sup_{n\leq m}\gamma_n^2 I(n\leq\tau)(1+|X_n|^{2q_3})\}$$

with $K \leq \overline{K}M_1^2(Q)C_3^2(Q)\mu_{2q_3}(Q)$, $\overline{K}$ being a constant independent of Q. Whence

$$E\{\sup_{1\leq n\leq m} I(n\leq\tau)|\eta_{n;0}|^2\} \leq KE\sum_{k=0}^{m-1}\gamma_{k+1}^2 I(k\leq\tau)(1+|X_k|^{2q_3})$$
$$\leq K\mu_{2q_3}(1+|x|^{2q_3})\sum_{k=0}^{m-1}\gamma_{k+1}^2$$

This proves the first part of the lemma. Finally, since

$$E\{I(\tau=\infty)|\gamma_n\psi_{\theta_{n-1}}(X_n)|^2\} \leq K\gamma_n^2(1+|x|^{2q_3})$$

we have

$$E\{I(\tau=\infty)\sum_n|\gamma_n\psi_{\theta_{n-1}}(X_n)|^2\} < \infty$$

which implies that $\lim_{n\to\infty}\gamma_n\psi_{\theta_{n-1}}(X_n) = 0$ a.s. $\square$

1.4.6

Proposition 7. *For any compact subset Q of D, and for any C^2 function ϕ on $\mathbb{R}^d$ with bounded second derivatives, there exist constants B_1, B_2 and s such that for all m:*

1. *We have*

$$E_{x,a}\left\{\sup_{n\leq m} I(n\leq\tau(Q))\left|\sum_{k=0}^{n-1}\varepsilon_k(\phi)\right|^2\right\} \leq B_1(1+|x|^s)(1+\sum_{k=0}^{m-1}\gamma_{k+1}^{2\lambda})\sum_{k=0}^{m-1}\gamma_{k+1}^2 \quad (1.4.1)$$

where λ is the constant $\in [1/2,1]$ of (A.5); and similarly making explicit the constants of Assumptions (A.3), (A.4) and (A.5)

$$B_1 \leq \tilde{B}_1(1+\mu_s^3(Q))$$
$$[M_1^4(Q)+C_1^4(Q)+C_2^4(Q)+C_3^4(Q)+\overline{C}^{4\lambda}(Q)C_4^4(Q)]$$

where $\overline{C}(Q) = C_1(Q)+\gamma_1 C_2(Q)$, $\tilde{B}_1$ being independent of Q. Lastly we may take $s = \max(2q_4+2\lambda(q_1\vee q_2), 2q_3+2(q_1\vee q_2), 4q_1, 4q_2)$.

1.4 L^2 Estimates

2. If $\sum \gamma_k^{1+\lambda} \leq 1$

(i) $E_{x,a}\left\{\sup_n I(n \leq \tau(Q))|\sum_{k=0}^{n-1}\varepsilon_k(\phi)|\right\}^2 \leq B_2(1+|x|^s)\sum_{k\geq 1}\gamma_k^{1+\lambda}$

(1.4.2)

where $B_2 \leq CB_1$ for some constant C independent of Q.

(ii) On $\{\tau(Q) = \infty\}$ the series $\sum_k \varepsilon_k(\phi)$ converges a.s. and in L^2.

Proof. Lemmas 2 to 6 tell us that the first term of (1.4.1) is bounded above by a sum of terms each of which is itself bounded above by expressions of the form

$$B(1+|x|^s)\sum_{k=0}^{m-1}\gamma_{k+1}^2 \quad \text{or} \quad B(1+|x|^s)(\sum_{k=0}^{m-1}\gamma_{k+1}^{1+\lambda})^2$$

Point 1. of the proposition is true since

$$(\sum_{k=0}^{m-1}\gamma_{k+1}^{1+\lambda})^2 \leq (\sum_{k=0}^{m-1}\gamma_{k+1}^2)(\sum_{k=0}^{m-1}\gamma_{k+1}^{2\lambda})$$

Point 2. is also a trivial consequence of these upper bounds. □

We also have:

Corollary. For all $T > 0$

$$E_{x,a}\left\{\sup_{n\leq m(T)} I(n \leq \tau(Q))|\sum_{k=0}^{n-1}\varepsilon_k(\phi)|\right\}^2$$

$$\leq B_1(1+|x|^s)(1+T\gamma_1^{2\lambda-1})\sum_{k=1}^{m(T)}\gamma_k^2 \qquad (1.4.3)$$

1.4.7 Remark: the Case of a Robbins–Monro Algorithm

For such an algorithm, the decomposition of the "fluctuation" $\varepsilon_{n+1}(\phi)$ may be greatly simplified since

$$\varepsilon_k(\phi) = \gamma_{k+1}\phi'(\theta_k)\cdot(H(\theta_k, X_{k+1}) - h(\theta_k)) + \varepsilon_k^{(4)}(\phi)$$

and the process

$$Z_n = \sum_{k=0}^{n-1}\gamma_{k+1}\phi'(\theta_k)\cdot(H(\theta_k, X_{k+1}) - h(\theta_k))$$

is a martingale which can be bounded above using Doob's inequality as in Lemma 2. In this case there is no need to use Assumption (A.4), which in any case is almost trivial (cf. 1.1.4).

1.5 Approximation of the Algorithm by the Solution of the O.D.E.

In this section we prove one of the theorems announced in Chapter 2 of Part I of this book, with the assumptions of this chapter.

1.5.1

For $t \geq t_0$, $\bar{\theta}(t; t_0, a_0)$ denotes the solution of:

$$\bar{\theta}'(t) = h(\bar{\theta}(t)), \quad t \geq t_0, \quad \bar{\theta}(t_0) = a_0 \qquad (1.5.1)$$

We choose $T > 0$ and two compact subsets Q_1, Q_2 of D, with $Q_1 \subset Q_2$, which we assume to satisfy:

$$\text{for all } a \in Q_1, \text{ all } t \leq T, \ d(\bar{\theta}(t; 0, a), Q_2^c) \geq \delta_0 > 0 \qquad (1.5.2)$$

This condition only applies to the solutions of the differential equation (1.5.1). It implies that any tube of radius $\delta < \delta_0$ around the solution $\bar{\theta}(t; 0, a)$, $0 \leq t \leq T$ is contained in Q_2. Following (A.4), there exist constants $L_1 = L_1(Q_2)$, $L_2 = L_2(Q_2)$, such that:

$$|h(\theta)| \leq L_1, \quad |h(\theta) - h(\theta')| \leq L_2|\theta - \theta'| \text{ for all } \theta, \theta' \in Q_2 \qquad (1.5.3)$$

Then, for the t_n defined by (1.1.4), $t_{n+1} \leq T$,

$$\begin{aligned}\bar{\theta}(t_{n+1}) - \bar{\theta}(t_n) &= \int_{t_n}^{t_{n+1}} h(\bar{\theta}(s))ds \\ &= \gamma_{n+1} h(\bar{\theta}(t_n)) + \alpha_n\end{aligned}$$

where

$$|\alpha_n| \leq \gamma_{n+1}^2 \cdot L_2 \qquad (1.5.4)$$

We wish to compare $\theta(t)$ and $\bar{\theta}(t; 0, a)$ on $[0, T]$, or more precisely θ_n and $\bar{\theta}(t_n; 0, a)$ for $n \leq m(T)$ as given by (1.1.7).

1.5.2

We fix a, $\theta_0 = a$ and write $\bar{\theta}(t)$ for $\bar{\theta}(t; 0, a)$. Consider for $\delta < \delta_0$ the set

$$\{ \sup_{n \leq m(T)} |\theta_n - \bar{\theta}(t_n)| \geq \delta \}$$

If we set (cf. (1.3.11) for the definition of $\tau(Q)$)

$$\nu = m(T) \wedge \tau(Q_2) \qquad (1.5.5)$$

this set is equal to $\{\sup(|\theta_n - \bar{\theta}(t_n)|; n \leq \nu) \geq \delta\}$, since at time $n = \tau(Q_2)$, from (1.5.2), we must have $|\theta_n - \bar{\theta}(t_n)| \geq \delta$. If as before, we denote

$$\varepsilon_k = \theta_{k+1} - \theta_k - \gamma_{k+1} h(\theta_k) \qquad (1.5.6)$$

1.5 Approximation of the Algorithm by the Solution of the O.D.E.

then, applying (1.4.3) to the coordinate functions $\phi_i(\theta) = \theta^i$, we have

$$E_{x,a}\left\{\sup_{n\leq m(T)} I(n \leq \tau(Q_2))|\sum_{k=0}^{n-1}\varepsilon_k|\right\}^2 \leq B(1+|x|^s(1+T\gamma_1^{2\lambda-1})\sum_{k=1}^{m(T)}\gamma_k^2 \tag{1.5.7}$$

for some constants B and s.

Since

$$\theta_r - \bar{\theta}(t_r) = \theta_{r-1} - \bar{\theta}(t_{r-1}) + \gamma_r(h(\theta_{r-1}) - h(\bar{\theta}(t_{r-1}))) + \varepsilon_{r-1} + \alpha_{r-1}$$

we have

$$\theta_r - \bar{\theta}(t_r) = \sum_{k=0}^{r-1}\gamma_{k+1}(h(\theta_k) - h(\bar{\theta}(t_k))) + \sum_{k=0}^{r-1}\varepsilon_k + \sum_{k=0}^{r-1}\alpha_k$$

$$|\theta_r - \bar{\theta}(t_r)| \leq L_2\sum_{k=0}^{r-1}\gamma_{k+1}|\theta_k - \bar{\theta}(t_k)| + |\sum_{k=0}^{r-1}\varepsilon_k| + L_2\sum_{k=0}^{r-1}\gamma_{k+1}^2$$

On $\{n \leq \nu\}$, for $r = 0, 1, \ldots, n$ we have:

$$|\theta_r - \bar{\theta}(t_r)| \leq L_2\sum_{k=0}^{r-1}\gamma_{k+1}|\theta_k - \bar{\theta}(t_k)| + \sup_{m\leq m(T)}\{I(m \leq \nu)|\sum_{k=0}^{m-1}\varepsilon_k|\} + L_2\sum_{k=1}^{m(T)}\gamma_k^2$$

$$\leq L_2\sum_{k=1}^{r-1}\gamma_{k+1}|\theta_k - \bar{\theta}(t_k)| + U_1 + U_2$$

Lemma 8. *If $\nu_r \leq r_1\sum_{i=1}^{r}\gamma_i\nu_{i-1} + r_2$, $\nu_0 = 0$ for $r = 0, 1, \ldots, n$, with r_1, r_2, γ_i positive, then $\nu_n \leq r_2 \cdot \exp(r_1\sum_{i=1}^{n}\gamma_i)$.*

Proof. We may suppose that $r_1 = 1$. We let $P(r)$ denote the property:

$$1 + \sum_{i=1}^{r}\gamma_i\exp(\sum_{j=1}^{i-1}\gamma_j) \leq \exp(\sum_{i=1}^{r}\gamma_i), \quad \nu_r \leq r_2 \cdot \exp(\sum_{i=1}^{r}\gamma_i) \tag{1.5.8}$$

Then $P(1)$ reduces to $1 + \gamma_1 \leq \exp(\gamma_1)$ and $\nu_1 \leq r_2$, which is clearly true.

Let us suppose that $P(r)$ is true. Then on the one hand

$$\exp(\sum_{i=1}^{r+1}\gamma_i) = \exp(\sum_{1}^{r}\gamma_i)\exp(\gamma_{r+1})$$

$$\geq \exp(\sum_{1}^{r}\gamma_i) + \gamma_{r+1}\exp(\sum_{1}^{r}\gamma_i)$$

$$\geq 1 + \sum_{1}^{r}\gamma_i\exp(\sum_{j=1}^{i-1}\gamma_j) + \gamma_{r+1}\exp(\sum_{1}^{r}\gamma_i)$$

$$= 1 + \sum_{i=1}^{r+1}\gamma_i\exp(\sum_{j=1}^{i-1}\gamma_j)$$

On the other hand

$$\nu_{r+1} \leq r_2\{1 + \sum_{i=1}^{r+1} \gamma_i \exp(\sum_{j=1}^{i-1} \gamma_j)\} \leq r_2 \exp(\sum_{i=1}^{r+1} \gamma_i)$$

□

Applying the lemma, on $\{n \leq \nu\}$ we have

$$|\theta_n - \bar{\theta}(t_n)| \leq \exp(L_2 \sum_{k=1}^{r} \gamma_k)\{U_1 + U_2\}$$

$$\sup_{n \leq \nu} |\theta_n - \bar{\theta}(t_n)|^2 \leq \exp(2L_2 T)(2U_1^2 + 2U_2^2)$$

$$E\{\sup_{n \leq \nu} |\theta_n - \bar{\theta}(t_n)|^2\} \leq \exp(2L_2 T)\{2E(U_1^2) + 2U_2^2\}$$

Noting that

$$U_2^2 \leq L_2^2 \sum_{k=1}^{m(T)} \gamma_k \cdot \sum_{k=1}^{m(T)} \gamma_k^3 \leq L_2^2 \cdot T \sum_{k=1}^{m(T)} \gamma_k^3$$

and using (1.5.7), we obtain:

$$\{\sup_{n \leq \nu} |\theta_n - \bar{\theta}(t_n)|^2\} \leq B(1 + |x|^s)(\gamma_1^{2\lambda-1} T + 1)(1 + L_2^2) \exp(2L_2 T) \sum_{k=1}^{m(T)} \gamma_k^2$$

(1.5.9)

Theorem 9. *We assume that (A.1) to (A.5) hold and that $\gamma_1 \leq 1$. Let $Q_1 \subset Q_2$ be two compact subsets of D. Then there exist constants B_3, L_2, s such that for all $T > 0$ satisfying (1.5.2), all $\delta < \delta_0$, all $a \in Q$, and all x, we have:*

$$P_{x,a}\{\sup_{n \leq m(T)} |\theta_n - \bar{\theta}(t_n; 0, a)| \geq \delta\} \leq \frac{B_3}{\delta^2}(1 + |x|^s)(1 + T) \exp(2L_2 T) \sum_{k=1}^{m(T)} \gamma_k^2$$

In particular if $\gamma_k = \gamma \leq 1$ for all k, then

$$P_{x,a}\{\sup_{n \leq [T/\gamma]} |\theta_n - \bar{\theta}(n\gamma; 0, a)| \geq \delta\} \leq \frac{B_3}{\delta^2}(1 + |x|^s)(T + T^2) \exp(2L_2 T) \cdot \gamma$$

Remark 1. L_2 is the Lipschitz constant of h on Q_2.

Remark 2. The assumption $\gamma_1 \leq 1$ is introduced into the theorem in order to simplify the expression of the constants. It is unimportant, since it can always be obtained by modifying H and ρ_n.

1.5.3

If $P_{n;x,a}$ denotes the distribution of (X_{n+k}, θ_{n+k}) with $X_n = x, \theta_n = a$, then as we have already seen, $P_{n;x,a}$ is equal to $\widetilde{P}_{x,a}$, where $\widetilde{P}_{x,a}$ is the distribution of $(\widetilde{X}_k, \tilde{\theta}_k)$ given by (1.1.1), where γ_k is replaced by γ_{n+k} and ρ_n is replaced by ρ_{n+k}. Thus, with the same assumptions as in Theorem 9, in view of the uniformity in n of (A.3-ii), for all n we have:

$$P_{n;x,a}\{\sup_{n \leq r \leq m(n,T)} |\theta_r - \bar{\theta}(t_r; t_n, a)| \geq \delta\}$$
$$\leq \frac{B_3}{\delta^2}(1 + |x|^s)(1 + T)\exp(2L_2 T) \cdot \sum_{k \geq n} \gamma_k^2$$

This inequality shows that if the series $\sum \gamma_n^2$ converges, the trajectories of the algorithm have an increasing tendency to follow the solutions of the D.E. (1.5.1). This implies that if the D.E. (1.5.1) has an attractor θ_*, the trajectories of the algorithm tend to converge to this attractor. We shall now formulate this more precisely.

1.6 Asymptotic Analysis of the Algorithm

1.6.1

Let us assume that there exists a point θ_* in D which is a point of asymptotic stability for the D.E. (1.5.1) with domain of attraction D; this means that any solution of (1.5.1) for, $a \in D$ remains indefinitely in D and converges to θ_* as t tends to $+\infty$. It can then be shown ((Krasovskii 1963) Th. 5.3, p. 31) that there exists a function U on D of class C^2 such that

(i) $U(\theta_*) = 0$; $U(\theta) > 0$ for all $\theta \in D, \theta \neq \theta_*$
(ii) $U'(\theta) \cdot h(\theta) < 0$ for all $\theta \in D, \theta \neq \theta_*$
(iii) $U(\theta) \to +\infty$ if $\theta \to \partial D$ or $|\theta| \to +\infty$

We shall study this type of situation later with slightly more general assumptions, in that the attractor may be a compact subset of D. Thus we introduce the following assumptions:

(A.6). $\sum \gamma_n^{1+\lambda} < +\infty$, where λ is given by (A.4-iii).

(A.7). There exists a positive function U of class C^2 on D such that $U(\theta) \to C \leq +\infty$ if $\theta \to \partial D$ or $|\theta| \to +\infty$ and $U(\theta) < C$ for $\theta \in D$ satisfying:

$$U'(\theta) \cdot h(\theta) \leq 0 \text{ for all } \theta \in D \qquad (1.6.1)$$

Let us introduce some notation:

$$K(c) = \{\theta; U(\theta) \leq c\} \qquad (1.6.2)$$
$$\tau(c) = \inf(n; \theta_n \notin K(c)) \qquad (1.6.3)$$
$$\nu(c) = \inf(n; \theta_n \in K(c)) \qquad (1.6.4)$$

Proposition 10. *Suppose $c_1 < c_2 < C$. There exist constants B_3 and s such that for all $a \in K(c_1)$, all x,*

$$P_{x,a}\{\tau(c_2) < +\infty\} \leq B_3(1+|x|^s)\sum_{k=1}^{+\infty}\gamma_k^{1+\lambda}$$

Proof. Let ϕ be a C^2 function on $\mathbb{R}^d$ which coincides with U on $K(c_2)$, and which satisfies $\inf\{\phi(\theta); \theta \notin K(c_2)\} = c_2$. Since from (1.3.5) we have $\phi(\theta_{k+1}) - \phi(\theta_k) = \gamma_{k+1}\phi'(\theta_k) \cdot h(\theta_k) + \varepsilon_k(\phi)$, then for all n:

$$\phi(\theta_n) - \phi(\theta_0) = \sum_{k=0}^{n-1}\gamma_{k+1}\cdot\phi'(\theta_k)\cdot h(\theta_k) + \sum_{k=0}^{n-1}\varepsilon_k(\phi)$$

Thus on $\{\tau(c_2) < +\infty\}$

$$\phi(\theta_{\tau(c_2)}) - \phi(\theta_0) = \sum_{k=0}^{\tau(c_2)-1}\gamma_{k+1}\phi'(\theta_k)\cdot h(\theta_k) + \sum_{k=0}^{\tau(c_2)-1}\varepsilon_k(\phi)$$

But on the one hand, if $a \in K(c_1)$, then $\phi(\theta_{\tau(c_2)}) - \phi(\theta_0) \geq c_2 - c_1$, whilst on the other hand $\phi'(\theta_k) \cdot h(\theta_k) = U'(\theta_k) \cdot h(\theta_k) \leq 0$ for $k < \tau(c_2)$. Thus

$$(c_2 - c_1)I(\tau(c_2) < +\infty) \leq I(\tau(c_2) < +\infty)|\sum_{k=0}^{\tau(c_2)-1}\varepsilon_k(\phi)|$$

$$\leq \sup_n I(n \leq \tau(c_2))\cdot|\sum_{k=0}^{n-1}\varepsilon_k(\phi)|$$

whence

$$P(\tau(c_2) < +\infty) \leq (c_2 - c_1)^{-2}E\left\{\sup_n I(n \leq \tau(c_2))|\sum_{k=0}^{n-1}\varepsilon_k(\phi)|\right\}^2$$

and Proposition 10 is proved using Proposition 7, (1.4.2). □

1.6.2

Let us now consider a compact subset F of D satisfying:

$$F = \{\theta; U(\theta) \leq c_0\} \supset \{\theta; U'(\theta)\cdot h(\theta) = 0\} \qquad (1.6.5)$$

1.6 Asymptotic Analysis of the Algorithm

This condition is satisfied by $F = \{\theta_*\}$, $c_0 = 0$ if θ_* has domain of attraction D.

Proposition 11. *Suppose $c < c_2 < C$. For all $a \in K(c)$, all x, θ_n converges to F, $P_{x,a}$ a.s. on $\{\tau(c_2) = +\infty\}$.*

To prove Proposition 11, we begin with a lemma:

Lemma 12. *Suppose $c_0 < c_1 < c_2 < C$. For all $a \in K(c_2)$, all x, $\nu(c_1) < +\infty$ $P_{x,a}$ a.s. on $\{\tau(c_2) = +\infty\}$.*

Proof of Lemma 12. From (1.6.1) we have $-U'(\theta_n) \cdot h(\theta_n) \geq \alpha > 0$ for all θ such that $c_1 \leq U(\theta) \leq c_2$. If we choose the same function ϕ as in Proposition 10 and $T > 0$, then on $\{\nu(c_1) = \tau(c_2) = +\infty\}$ we have:

$$\phi(\theta_{m(n,T)}) - \phi(\theta_n) = \sum_{k=n}^{m(n,T)-1} \gamma_{k+1} \phi'(\theta_k) \cdot h(\theta_k) + \sum_{k=n}^{m(n,T)-1} \varepsilon_k(\phi)$$

But

$$-\sum_{k=n}^{m(n,T)-1} \gamma_{k+1} \phi'(\theta_k) \cdot h(\theta_k) \geq \alpha \sum_{k=n}^{m(n,T)-1} \gamma_{k+1} \geq \alpha(T-1)$$

and on the other hand

$$\phi(\theta_{m(n,T)}) - \phi(\theta_n) \geq -(c_2 - c_1)$$

Thus

$$\sum_{k=n}^{m(n,T)-1} \varepsilon_k(\phi) \geq \alpha(T-1) - (c_2 - c_1) \geq 1$$

for T sufficiently large. This contradicts Proposition 7, Point 2-ii. The proof of Lemma 12 is complete. $\square$

Proof of Proposition 11. To prove the proposition, we shall show that, for all $c > c_0$, $\limsup U(\theta_n) \leq c$.

Suppose $c_0 \leq c_1 < c < c_2 < C$ and consider $\{\tau(c_2) = +\infty, \limsup U(\theta_n) > c\}$. We define

$$\nu_1 = \inf(n; \theta_n \in K(c_1)), \qquad \tau_1 = \inf(n > \nu_1; \theta_n \notin K(c))$$
$$\nu_k = \inf(n > \tau_{k-1}; \theta_n \in K(c_1)), \qquad \tau_k = \inf(n > \nu_k; \theta_n \notin K(c))$$

All these values are finite and clearly $\nu_k \geq k$. Thus for the same function ϕ we have:

$$\phi(\theta_{\tau_n}) - \phi(\theta_{\nu_n}) = \sum_{k=\nu_n}^{\tau_n-1} \gamma_{k+1} \phi'(\theta_k) \cdot h(\theta_k) + \sum_{k=\nu_n}^{\tau_n-1} \varepsilon_k(\phi)$$

$$\leq \sum_{k=\nu_n}^{\tau_n-1} \varepsilon_k(\phi)$$

But
$$\phi(\theta_{\tau_n}) - \phi(\theta_{\nu_n}) \geq c - c_1 > 0$$

Again this contradicts Proposition 7, Point 2-ii. □

1.6.3

We recall that the distributions $P_{n,x,a}$ were defined in 1.5.3.

Theorem 13. *We assume that (A.1) to (A.7) hold and that F is a compact set satisfying (1.6.5). Then for any compact $Q \subset D$, there exist constants B_4 and s such that for all $n \geq 0$, all $a \in Q$, all x,*

$$P_{n,x,a}\{\theta_k \text{ converges to } F\} \geq 1 - B_4(1+|x|^s) \sum_{k=n+1}^{+\infty} \gamma_k^{1+\lambda}$$

Proof. There exists c_2 such that $Q \subset K(c_2)$, then, following Proposition 11, on $\{\tau(c_2) = +\infty\}$, θ_k converges to F a.s. We obtain the lower bound for $P(\tau(c_2) = +\infty)$ using Proposition 10 applied to the algorithm

$$\theta_{n+k+1} = \theta_{n+k} + \gamma_{n+k+1} H(\theta_{n+k}, X_{n+k+1}) + \rho_{n+k+1}(\theta_{n+k}, X_{n+k+1}), \quad k \geq 0$$

and taking into account that the constants of Lemma 8 and Proposition 11 are valid for all sequences $(\gamma_{n+k})_{k\geq 0}$ (cf. the remarks at the beginning of Section 1.4). □

1.7 An Extension of the Previous Results

The aim of this section is to give a mild generalisation of Theorem 13. This will only be used in the analysis of the example in Section 2.5 of Chapter 2; thus this section may be omitted on a first reading.

1.7.1 Weakening of Assumption (A.4-iii)

Assumption (A.4-iii) requires that for all points x we have:

$$|\Pi_\theta \nu_\theta(x) - \Pi_{\theta'}\nu_{\theta'}(x)| \leq C_4(1+|x|^{q_4})|\theta - \theta'|^\lambda$$

In fact, in certain cases this does not hold for each x, but is only true on average on the trajectories of the process (X_n, θ_n). This is why we introduce the following:

(A.4-iii)'. *For any compact subset Q of D, there exist constants C_5, q_5, $\mu \in]0, 2]$, such that, for all $a \in Q$, all x, all k:*

$$E_{x,a}\{I(k+1 < \tau(Q))|\Pi_{\theta_{k+1}}\nu_{\theta_{k+1}}(X_{k+1}) - \Pi_{\theta_k}\nu_{\theta_k}(X_{k+1})|^2\}$$
$$\leq C_5(1+|x|^{q_5})\gamma_{k+1}^\mu \tag{1.7.1}$$

1.7.2

Assumption (A.4-iii) was only used in Lemma 3, so we shall repeat the proof of that lemma using (A.4-iii)'.

$$\begin{aligned}\varepsilon_{k+1}^{(2)} &= \gamma_{k+2}\{\phi'(\theta_{k+1}) \cdot \Pi_{\theta_{k+1}}\nu_{\theta_{k+1}}(X_{k+1}) - \phi'(\theta_k)\Pi_{\theta_k}\nu_{\theta_k}(X_{k+1})\} \\ &= \gamma_{k+2}(\phi'(\theta_{k+1}) - \phi'(\theta_k))\Pi_{\theta_{k+1}}\nu_{\theta_{k+1}}(X_{k+1}) \\ &\quad + \gamma_{k+2}\phi'(\theta_k)(\Pi_{\theta_{k+1}}\nu_{\theta_{k+1}}(X_{k+1}) - \Pi_{\theta_k}\nu_{\theta_k}(X_{k+1})) \\ &= V_{k+1} + W_{k+1}\end{aligned}$$

The same upper bounds as those used in the proof of Lemma 3 lead to

$$E\left\{\sum_{k=1}^{n-1}|V_k|I(k+1 \leq \tau)\right\}^2 \leq K_2 E\left\{\sum_{k=1}^{n-1}\gamma_{k+1}^2(1+|X_{k+1}|^{q_3+q_1})I(k+1\leq \tau)\right\}^2$$

$$E\{\phi'(\theta_k)[\Pi_{\theta_k}\nu_{\theta_k}(X_k) - \Pi_{\theta_{k-1}}\nu_{\theta_{k-1}}(X_k)]^2 I(k+1 \leq \tau)\}$$
$$\leq K_1(1+|x|^{q_5})[\sum_{k=1}^{n-1}\gamma_k^{1+\mu/2}]^2$$

from which, as in Lemma 3 we deduce that

$$E\left\{I(n \leq \tau)\sum_{k=0}^{n-1}|W_k|\right\}^2 \leq A_2(1+|x|^s)(\sum_{k=0}^{n-1}\gamma_{k+1}^{1+\mu/2})^2$$

with moreover

$$A_2 \leq \tilde{A}_2 \mu_s(Q)[C_1^{2\mu}(Q)M_1^2(Q)C_5(Q) + C_1^2(Q)M_2^2(Q)C_3^2(Q)]$$

Thus we need change nothing in Sections 1.5 and 1.6 to obtain the following theorem:

Theorem 14. *We assume (A.1) to (A.5) with (A.4-iii) replaced by (A.4-iii)'. Then:*

1. *For any compact $Q \subset D$ there exist constants B_5, B_6 and s, such that for all $n \geq 0$ such that $\gamma_n \leq 1$, all $a \in Q$, all x and all $\delta < \delta_0$:*

$$P_{n,x,a}\{\sup_{n \leq r \leq m(n,T)} |\theta_r - \bar{\theta}(t_r; t_n, a)| \geq \delta\}$$
$$\leq \frac{B_5}{\delta}(1+|x|^s)(1+T)\exp(2L_2 T) \cdot \sum_{k \geq n}\gamma_k^{2\wedge(1+\mu/2)}$$

2. *If we now assume (A.6) and (A.7), then if F is a compact set satisfying (1.6.5) and $\sum \gamma_n^{1+\mu/2} \leq 1$ for some suitable constant B_6 we have*

$$P_{n,x,a}\{\theta_k \text{ converges to } F\} \geq 1 - B_6(1+|x|^s)\sum_{k \geq n+1}\gamma_k^{1+\mu/2}$$

1.8 Alternative Formulation of the Convergence Theorem

Theorems 13 and 14 may be presented in a different way.

Theorem 15. *Suppose that (A.1) to (A.5) hold for an open subset D of $\mathbb{R}^d$. Let θ_* be a point of asymptotic stability of the D.E. (1.5.1). Suppose that S is a compact subset of D, and that Q is a compact subset of the domain of attraction of θ_*. Let*

$$\Omega(S, Q) = \{\theta_n \in S \text{ for all } n, \theta_n \in Q \text{ for infinitely many } n\}$$

Then for all x, a, θ_n converges to θ_*, $P_{x,a}$ a.s. on $\Omega(S, Q)$. The same is true if (A.4) is modified by replacing (A.4-iii) with (A.4-iii)' as introduced in 1.7.1

Corollary 16. *Suppose that $D = \mathbb{R}^d$; if*

1. *θ_n is bounded a.s.;*
2. *θ_n visits a.s. a compact subset of the domain of attraction of θ_* infinitely often;*

then θ_n converges to θ_* a.s.

Proof. Theorem 15 may be deduced from Proposition 7 using the Kushner–Clark lemma (Kushner and Clark 1978). However, it also follows easily from Theorem 13. Let Δ be an open set containing Q, and let U be a Lyapunov function relative to θ_*. Then on Δ, U satisfies (A.6) and (A.7), and if we set $F = \{\theta_*\}$, then F satisfies (1.6.5) with $c_0 = 0$. Again denoting $K(c) = \{U \leq c\}$, we note that on $\{\tau(S) = +\infty\}$,

$$\limsup\{\theta_n \in K(c_1)\} \subset \liminf\{\theta_n \in K(c_2)\}$$

for $c_1 < c_2 < C$. For if not, then if ϕ denotes a suitable extension of U, for successive visits ν_n, τ_n to $K(c_1)$ and $K(c_2)^c$, and we would have

$$c_2 - c_1 \leq \phi(\theta_{\tau_n}) - \phi(\theta_{\nu_n}) \to 0$$

Then choosing c_1 and c_2 such that $Q \subset K(c_1)$, we have

$$P(\Omega(S,Q) \cap \{\theta_k \to \theta_*\}^c)$$
$$= P((\tau(S) = +\infty) \cap \limsup_n\{\theta_n \in Q\} \cap \{\theta_k \to \theta_*\}^c)$$
$$\leq P((\tau(S) = +\infty) \cap \liminf\{\theta_n \in K(c_2)\} \cap \{\theta_k \to \theta_*\}^c)$$
$$\leq \liminf_n P((\tau(S) = +\infty) \cap (\theta_n \in K(c_2)) \cap \{\theta_k \to \theta_*\}^c)$$
$$\leq \liminf_n E(I(\tau(S) > n; \theta_n \in K(c_2)) P_{n, X_n, \theta_n}\{\theta_k \to \theta_*\}^c)$$
$$\leq B_4 \liminf_n E\{I(\tau(S) > n)(1 + |X_n|^s)\} \sum_{k \geq n} \gamma_k^{1+\lambda}$$
$$\leq B_4 \mu_s (1 + |x|^s) \liminf_n \sum_{k \geq n} \gamma_k^{1+\lambda} = 0$$

□

1.9 A Global Convergence Theorem

1.9.1 Assumptions

We shall suppose that the constants $C_i(Q)$ of Assumptions (A.3) and (A.4) grow at most linearly with the diameter of Q, the constant C_4 being independent of Q if $\lambda = 1$ and of the order of $(\text{diam}(Q))^{1-\lambda}$ if $\lambda < 1$. Similarly, in (A.5), we suppose that the constants μ_q are independent of Q. Thus we suppose that there exist constants $\overline{C}_i, q_i$, $i = 1,\ldots,4$ and μ_q $(q > 0)$, such that for all $\theta \in \mathbb{R}^d, a \in \mathbb{R}^d, n \geq 0, R > 0$ we have:

$$|H(\theta, x)| \leq \overline{C}_1(1 + |\theta|)(1 + |x|^{q_1}) \tag{1.9.1}$$

$$|\rho_n(\theta, x)| \leq \overline{C}_2(1 + |\theta|)(1 + |x|^{q_2}) \tag{1.9.2}$$

$$E_{x,a}\{1 + |X_{n+1}|^q\} \leq \mu_q(1 + |x|^q) \tag{1.9.3}$$

$$|\nu_\theta(x)| < \overline{C}_3(1 + |\theta|)(1 + |x|^{q_3}) \tag{1.9.4}$$

with, for all θ, θ' such that $|\theta| \leq R, |\theta'| \leq R$ and some $\lambda \in [1/2, 1]$,

$$|\Pi_\theta \nu_\theta(x) - \Pi_{\theta'} \nu_{\theta'}(x)|$$
$$\leq \overline{C}_4(1 + R^{1-\lambda})|\theta - \theta'|^\lambda (1 + |x|^{q_4}) \tag{1.9.5}$$

where ν_θ satisfies (A.4-ii). We further assume that

$$\sum_k \gamma_k^{1+\lambda} < +\infty \tag{1.9.6}$$

1.9.2

Theorem 17. *We suppose that assumptions (A.1), (A.2) and (1.9.1) to (1.9.6) are satisfied. Then we have the following properties:*

a. *if there exists a positive function U on $\mathbb{R}^d$ of class C^2 with bounded second derivatives such that for all $\theta, |\theta| \geq \rho_0$*
  (i) $U'(\theta) \cdot h(\theta) \leq 0$
  (ii) $U(\theta) \geq \alpha |\theta|^2$, $\quad \alpha > 0$
  then for all $a \in \mathbb{R}^d, x \in \mathbb{R}^k$, the sequence (θ_n) is $P_{x,a}$ a.s. bounded;

b. *if further there exists $\theta_* \in \mathbb{R}^d$ such that*
  (i)' $U'(\theta) \cdot h(\theta) < 0$ *for all $\theta \neq \theta_*$*
  (iii) $U(\theta) = 0$ *iff $\theta = \theta_*$*
  then the sequence (θ_n) converges $P_{x,a}$ a.s. to θ_*.

Remark. If $V(\theta)$ is a positive regular function on $\mathbb{R}^d$ which tends to infinity with $|\theta|$ and which satisfies (i) or (i)', and (ii), then for any regular function

$$\psi : \mathbb{R}_+ \to \mathbb{R}_+, \quad \psi(0) = 0, \quad \psi'(t) > 0, \quad \lim_{t \to +\infty} \psi(t) = +\infty$$

$U = \psi \circ V$ also satisfies (i) or (i)' and (ii) and tends to infinity with $|\theta|$. Taking a function ψ whose derivatives tend to zero sufficiently rapidly at infinity, we can always return to a function U with bounded second derivatives. On the other hand, in order to obtain (ii), we must start with a function V which is uniformly increasing on the sets $\{|\theta| = R\}$.

1.9.3

Proof of Theorem 17. Since the second derivatives of U are bounded, we have:

$$|U'(\theta)| \le \overline{K}_1(1+|\theta|), \qquad |U(\theta)| \le \overline{K}_2(1+|\theta|)^2 \qquad (1.9.7)$$

We may assume that $\alpha \le 1$. Set

$$A = (1+\overline{K}_2)(1+\rho_0^2) \qquad (1.9.8)$$
$$\sigma_n = \inf(k; U(\theta_k) > A2^n) \qquad (1.9.9)$$
$$\tau_n = 1 + \sup(k < \sigma_{n+1}, U(\theta_k) \le A2^n) \qquad (1.9.10)$$

Note that we have $\sigma_n \le \tau_n \le \sigma_{n+1}$, $U(\theta_{\tau_n-1}) \le A2^n$ and $|\theta_n| \ge \rho_0$ if $k \in [\tau_n, \sigma_{n+1}]$.

We shall construct a sequence of sets B_n such that $\lim_n P_{x,a}(B_n) = 1$, with for all $n \ge 2$

$$U(\theta_{\sigma_{n+1}}) - U(\theta_{\tau_n}) \ge \frac{A}{4} \cdot 2^n \text{ on } B_n \cap \{\sigma_{n+1} < +\infty\} \qquad (1.9.11)$$

$$\lim_n 2^{-2n} E\{|U(\theta_{\sigma_{n+1}}) - U(\theta_{\tau_n})|^2 \cdot I(B_n \cap (\sigma_{n+1} < +\infty))\} = 0 \qquad (1.9.12)$$

We deduce immediately from (1.9.11) and (1.9.12) that

$$2^{-2n} E\{|U(\theta_{\sigma_{n+1}}) - U(\theta_{\tau_n})|^2 I(B_n \cap (\sigma_{n+1} < +\infty))\}$$
$$\ge \frac{A^2}{16} P\{B_n \cap (\sigma_{n+1} < +\infty)\} \text{ whence } \lim_n P(\sigma_{n+1} < +\infty) = 0$$

This proves that $P_{x,a}(\sup_n |\theta_n| < +\infty) = 1$

We shall construct B_n and establish (1.9.11) and (1.9.12) in a sequence of lemmas. Note firstly that for all $q \ge 0$ we have:

$$E\{\sum_k \gamma_k^{1+\lambda}(1+|X_k|^{q(1+\lambda)})\} \le \mu_q(1+|x|^{q(1+\lambda)}) \sum_k \gamma_k^{1+\lambda} < +\infty$$

and so

$$\lim_k \gamma_k |X_k|^q = 0 \qquad P_{x,a} \text{ a.s.} \qquad (1.9.13)$$

1.9 A Global Convergence Theorem

Lemma 18. *Let*

$$C_n = \{\text{for all } k \geq n, U(\theta_{k+1}) - U(\theta_k) \leq \frac{1}{2}(U(\theta_k) + A)\}$$

Then $P(C_n)$ tends to 1.

Proof of Lemma 18. On the one hand we have

$$U(\theta_{k+1}) - U(\theta_k) \leq |U'(\theta_k)||\theta_{k+1} - \theta_k| + \frac{1}{2}\sup_{\theta}|U''(\theta)| \cdot |\theta_{k+1} - \theta_k|^2$$

and on the other hand, by assumption,

$$|\theta_{k+1} - \theta_k| \leq \gamma_{k+1}\overline{C}_1(1+|\theta_k|)(1+|X_{k+1}|^{q_1}) + \gamma_{k+1}^2\overline{C}^2(1+|\theta_k|)(1+|X_{k+1}|^{q_2})$$
$$|U'(\theta_k)| \leq \overline{K}_1(1+|\theta_k|), \qquad |U''(\theta)| \leq M_2$$

whence

$$U(\theta_{k+1}) - U(\theta_k) \leq (1+|\theta_k|^2) \cdot Z_k$$

where $Z_k \to 0$ a.s. following (1.9.13).
But $1+|\theta_k|^2 \leq 1+\rho_0^2$ if $|\theta_k| \leq \rho_0$ and $1+|\theta_k|^2 \leq 1+1/\alpha U(\theta_k)$ if $|\theta_k| \geq \rho_0$.
In all cases $1+|\theta_k|^2 \leq \alpha^{-1}(U(\theta_k) + A)$, thus

$$U(\theta_{k+1}) - U(\theta_k) \leq \alpha^{-1}Z_k(U(\theta_k) + A)$$

and Lemma 18 is proved. □

Lemma 19. $\lim_n P(n < \sigma_n) = 1$

Proof of Lemma 19. Following Lemma 18 it is sufficient to show that for each r,

$$\lim_n P[C_r \cap (\sigma_n \leq n)] = 0$$

If $V(\theta) = U(\theta) + A$, then on C_r for $k \geq r$ we have $V(\theta_{k+1}) \leq 3/2 \cdot V(\theta_k)$, whence $V(\theta_k) \leq (3/2)^{k-r}V(\theta_r)$ and so

$$\text{on } C_r \quad U(\theta_k) \leq A + (3/2)^{k-r}(U(\theta_r) + A) \qquad (1.9.14)$$

Since

$$\{\sigma_n \leq n\} = \{\sup_{r \leq k \leq n} U(\theta_k) \geq A2^n\} \cup \bigcup_{k=1}^r \{U(\theta_k) \geq A \cdot 2^n\}$$

and since clearly $P(U(\theta_k) \geq A2^n)$ tends to 0 with n, in order to prove the lemma, it is sufficient to note that

$$P[C_r \cap (\sup_{r<k\leq n} U(\theta_k) \geq A2^n)] \leq P[U(\theta_r) + A \geq A2^n(3/2)^{r-n}] \to 0$$

□

Then we set
$$B_n = C_n \cap (n < \sigma_n) \qquad (1.9.15)$$

Lemma 20. *On $B_n \cap (\sigma_{n+1} < +\infty)$ we have*
$$U(\theta_{\sigma_{n+1}}) - U(\theta_{\tau_n}) \geq \frac{A}{4} 2^n, \quad n \geq 2$$

Proof of Lemma 20. On $\{\sigma_{n+1} < +\infty\}$, $U(\theta_{\sigma_{n+1}}) \geq A 2^{n+1}$. Moreover on B_n, $n < \sigma_n \leq \tau_n$, and so $\tau_n - 1 \geq n$, whence:
$$U(\theta_{\tau_n}) \leq \frac{3}{2} U(\theta_{\tau_n}) + \frac{A}{2} \leq \frac{3}{2} A 2^n + \frac{A}{2}$$

and
$$U(\theta_{\sigma_{n+1}}) - U(\theta_{\tau_n}) \geq A[2^{n+1} - \frac{3}{2} 2^n - \frac{1}{2}] \geq \frac{A}{4} 2^n \text{ if } n \geq 2$$
□

Lemma 21. *We have:*
$$E\{|U(\theta_{\sigma_{n+1}}) - U(\theta_{\tau_n})|^2 I(B_n \cap (\sigma_{n+1} < +\infty))\} \leq 2^{2n} \cdot \alpha_n$$
where α_n is a sequence of real variables which tends to zero.

Proof of Lemma 21. On $B_n \cap \{\sigma_{n+1} < +\infty\}$,
$$U(\theta_{\sigma_{n+1}}) - U(\theta_{\tau_n}) = \sum_{k=\tau_n}^{\sigma_{n+1}-1} \gamma_{k+1} U'(\theta_k) \cdot h(\theta_k) + \sum_{k=\tau_n}^{\sigma_{n+1}-1} \varepsilon_k(U)$$
$$\leq \sum_{k=\tau_n}^{\sigma_{n+1}-1} \varepsilon_k(U)$$

since for $k \in [\tau_n, \sigma_{n+1}]$, $|\theta_k| \geq \rho_0$, by virtue of the choice of A. Then
$$I(B_n \cap (\sigma_{n+1} < +\infty)) \cdot \{U(\theta_{\sigma_{n+1}}) - U(\theta_{\tau_n})\}$$
$$\leq I(B_n \cap (\sigma_{n+1} < +\infty)) \sup_{n < k < \sigma_{n+1}} |\sum_{i=k}^{\sigma_{n+1}-1} \varepsilon_i(U)|$$
$$\leq 2 I(B_n \cap (\sigma_{n+1} < +\infty)) \sup_{n < k \leq \sigma_{n+1}} |\sum_{i=n}^{k-1} \varepsilon_i(U)|$$
$$\leq 2 I(n < \sigma_n) \sup_{n < k} I(k \leq \sigma_{n+1}) |\sum_{i=n}^{k-1} \varepsilon_i(U)|$$

But following Proposition 7, on $\{n < \sigma_n\}$,
$$E\{\sup_{k > n} I(k \leq \sigma_{n+1}) |\sum_{i=n}^{k-1} \varepsilon_i(U)|^2 | F_n\} \leq \overline{K}_4 B_1(Q_n)(1 + |X_n|^s) \cdot \sum_{k \geq n} \gamma_{k+1}^{1+\lambda}$$

1.9 A Global Convergence Theorem

where
$$Q_n = \{\theta; U(\theta) \leq A 2^{n+1}\} \tag{1.9.16}$$

Thus

$$E\{I(B_n \cap (\sigma_{n+1} < +\infty)).|U(\theta_{\sigma_{n+1}}) - U(\theta_{\tau_n})|^2\} \leq \overline{K}_5 B_1(Q_n)(1 + |x|^s) \cdot \alpha_n$$

It remains to evaluate $B_1(Q_n)$. For $\theta \in Q_n$ we have

$$|\theta|^2 \leq \frac{1}{\alpha} U(\theta) + A \leq \frac{1}{\alpha}(2^{n+1} + A)$$

Thus the constants $C_i(Q_n), (i = 1,2,3)$ and $M_1(Q_n)$ of Proposition 7 increase as const.$2^{n/2}$, and $C_4(Q_n)$ grows as const.$2^{n(1-\lambda)/2}$, consequently $B_2(Q_n) \leq \overline{K}_6 2^{2n}$, which implies Lemma 21. □

Proof of Theorem 17 (Continued). The second part of Theorem 17 now follows from Theorem 15. □

1.9.4 Remarks

It is clear that in the statement of Theorem 17, as in Theorems 13 and 14, the assumption (1.9.5) may be replaced by the weaker condition: there exist a constant $\overline{C}_4$, and $\mu \in]0,2]$, such that for all $R > 0$, all a, $|a| \leq R$, all k

$$\begin{aligned}
&E_{x,a}\{I(k+1 \leq \tau_R)|\Pi_{\theta_{k+1}}\nu_{\theta_{k+1}}(X_{k+1}) - \Pi_{\theta_k}\nu_{\theta_k}(X_{k+1})|^2\} \\
&\leq \overline{C}_4(1+R)^2 \gamma_{k+1}^\mu (1 + |x|^{q_4})
\end{aligned} \tag{1.9.17}$$

1.10 Rate of L^2 Convergence of Some Algorithms

In this section, we discuss the problem of obtaining an upper bound of the form $E(|\theta_n - \theta_*|^2) \leq \lambda \gamma_n^\beta$ for a suitable constant λ and exponent β. Such an upper bound is easily derived in the case of Robbins–Monro algorithms. It is harder to derive a bound of this type for more general algorithms, and indeed it appears difficult to obtain a general theorem with practical assumptions. In the case of the equaliser in the learning phase, (Eweda and Macchi 1983) gives a result of this type. For general algorithms we shall simply prove a theorem with a "local L^2 upper bound", that is to say with an upper bound of the form:

$$E(|\theta_n - \theta_*|^2 I(n \leq \tau(Q))) \leq \lambda \gamma_n$$

where, as usual $\tau(Q)$, is the time at which θ_n leaves the compact set Q.

1.10.1 Robbins–Monro Algorithm

We shall consider the special class of so-called Robbins–Monro algorithms, as given in Example 4 of Subsection 1.1.2 (cf. also 1.1.4). The algorithm is given by:

$$\theta_{n+1} = \theta_n + \gamma_{n+1} H(\theta_n, X_{n+1}) \tag{1.10.1}$$

where

$$E[H(\theta_n, X_{n+1}) - h(\theta_n)|F_n] = 0 \tag{1.10.2}$$

with

$$h(\theta) = \int H(\theta, x) \mu_\theta(dx) \tag{1.10.3}$$

Note that $E_{x,a}$ is independent of x.; we shall denote it by E_a. The useful assumptions here amount to the following:

For all $a \in \mathbb{R}^k$

$$E_a[|H(\theta_n, X_{n+1})|^2 | F_n] \leq C_1(1 + |\theta_n|^2) \tag{1.10.4}$$

for some suitable constant C_1. We suppose that there exists a constant $\delta > 0$ such that for all θ

$$(\theta - \theta_*) \cdot h(\theta) \leq -\delta |\theta - \theta_*|^2 \tag{1.10.5}$$

with, for some $\beta \leq 1$

$$\liminf_{n \to \infty} 2\delta \frac{\gamma_n^\beta}{\gamma_{n+1}} + \frac{\gamma_{n+1}^\beta - \gamma_n^\beta}{\gamma_{n+1}^2} > 0 \tag{1.10.6}$$

Note regarding 1.10.6. If

$$\gamma_n = \frac{A}{n^\alpha + B} \quad 0 \leq \alpha \leq 1$$

then condition (1.10.6) is true for all $\beta < 1$. It is true for $\beta = 1$ if $2\delta > \frac{\alpha}{A}$.

Theorem 22. *Under the assumptions (1.10.2) to (1.10.6) algorithm (1.10.1) has the following property*

$$E_a(|\theta_n - \theta_*|^2) \leq \lambda(a) \gamma_n^\beta$$

for some suitable constant $\lambda(a)$.

Proof. It is sufficient to show that for some suitable n_0, there exists $\lambda(a, n_0)$, such that for all $n \geq n_0$

$$E_a(|\theta_n - \theta_*|^2) \leq \lambda(a, n_0) \gamma_n^\beta \tag{1.10.7}$$

1.10 Rate of L^2 Convergence of Some Algorithms

Set $T_n := \theta_n - \theta_*$ and write

$$E(|T_{n+1}|^2|F_n) = |T_n|^2 + 2\gamma_{n+1}T_n \cdot h(\theta_n)$$
$$+2\gamma_{n+1}T_n \cdot E[H(\theta_n, X_{n+1}) - h(\theta_n)|F_n] + \gamma_{n+1}^2 E[|H(\theta_n, X_{n+1})|^2|F_n]$$

Suppose that n is sufficiently large that $1 \geq 2\gamma_{n+1}\delta$. From (1.10.2) and (1.10.3), and taking expectations

$$E(|T_{n+1}|^2) \leq (1 - 2\gamma_{n+1}\delta + \overline{C}_1 \gamma_{n+1}^2) E|T_n|^2 + \overline{C}_1 \gamma_{n+1}^2 \qquad (1.10.8)$$

where $\overline{C}_1$ is a constant such that

$$C_1(1 + |\theta|^2) \leq \overline{C}_1(1 + |\theta - \theta_*|^2)$$

We shall use the following lemma:

Lemma 23. *There exist λ_0 and n_0 such that for all $\lambda > \lambda_0$ and $n \geq n_0$, the sequence $u_n = \lambda \gamma_n^\beta$ satisfies*

$$u_{n+1} \geq (1 - 2\gamma_{n+1}\delta + \gamma_{n+1}^2 \overline{C}_1) u_n + \gamma_{n+1}^2 \overline{C}_1$$

Proof of Lemma 23. In fact the condition

$$\lambda \gamma_{n+1}^\beta \geq (1 - 2\gamma_{n+1}\delta + \gamma_{n+1}^2 \overline{C}_1)\lambda \gamma_n^\beta + \gamma_{n+1}^2 \overline{C}_1 \qquad (1.10.9)$$

may be rewritten as

$$\lambda(\gamma_{n+1}^\beta - \gamma_n^\beta + \gamma_{n+1}\gamma_n^\beta(2\delta - \gamma_{n+1}\overline{C}_1)) \geq \gamma_{n+1}^2 \overline{C}_1$$

The existence of λ_0 and n_0 such that (1.10.9) is true for all $n \geq n_0$ and $\lambda \geq \lambda_0$ follows directly from condition (1.10.6). □

Proof of Theorem 22 (continued). Suppose n_0 and λ_0 are as in Lemma 23. Choose $\lambda(n_0, a) \geq \lambda_0$ such that

$$E|T_{n_0}|^2 \leq \lambda(n_0, a)\gamma_{n_0}^\beta$$

It follows immediately by induction on n that the sequence $u_n = \lambda(n_0, a)\gamma_n^\beta$, $n \geq n_0$, satisfies $E|T_n|^2 \leq u_n$.

1.10.2 Local L^2 Upper Bounds for General Algorithms

We return to the case of the algorithm

$$\theta_{n+1} = \theta_n + \gamma_{n+1}H(\theta_n, X_{n+1}) + \gamma_{n+1}^2 \rho_{n+1}(\theta_n, X_{n+1}) \qquad (1.10.10)$$

which satisfies the general Assumptions (A.1) to (A.5) of this chapter together with the following additional conditions: there exist $\delta > 0$ and θ_* such that for all $\theta \in \mathbb{R}^d$

$$(\theta - \theta_*) \cdot h(\theta) \leq -\delta |\theta - \theta_*|^2 \qquad (1.10.11)$$

and

$$\liminf_{n \to \infty} 2\delta \frac{\gamma_n}{\gamma_n + 1} + \frac{\gamma_{n+1} - \gamma_n}{\gamma_{n+1}^2} > 0 \qquad (1.10.12)$$

Theorem 24. *Under Assumptions (A.1) to (A.5) with $\lambda = 1$, and conditions (1.10.11) and (1.10.12), algorithm (1.10.10) has the following property: for any compact subset $Q \subset \mathbb{R}^d$, for all $x \in \mathbb{R}^k, a \in Q$, there exists a constant $\lambda(x, a, Q)$ such that*

$$E_{x,a}(|\theta_n - \theta_*|^2 I(n \leq \tau(Q))) \leq \lambda(x, a, Q) \gamma_n$$

where

$$\tau(Q) = \inf\{n : \theta_n \notin Q\}$$

Proof. Writing $T_n = \theta_n - \theta_*$, and with the decomposition

$$H(\theta, x) = h(\theta) + \nu_\theta(x) - \Pi_\theta \nu_\theta(x)$$

we obtain

$$\begin{aligned} |T_{n+1}|^2 &= |T_n|^2 + 2\gamma_{n+1} h(\theta_n) + 2\gamma_{n+1} T_n \cdot [\nu_{\theta_n}(X_{n+1}) - \Pi_{\theta_n} \nu_{\theta_n}(X_{n+1})] \\ &\quad + 2\gamma_{n+1}^2 T_n \cdot \rho_{n+1}(\theta_n, X_{n+1}) \\ &\quad + \gamma_{n+1}^2 |H(\theta_n, X_{n+1}) + \gamma_{n+1} \rho_{n+1}(\theta_n, X_{n+1})|^2 \end{aligned} \qquad (1.10.13)$$

Then, decomposing as in Subsection 1.3.2:

$$\begin{aligned} &\nu_{\theta_n}(X_{n+1}) - \Pi_{\theta_n} \nu_{\theta_n}(X_{n+1}) \\ &= \nu_{\theta_n}(X_{n+1}) - \Pi_{\theta_n} \nu_{\theta_n}(X_n) + \Pi_{\theta_{n-1}} \nu_{\theta_{n-1}}(X_n) \\ &\quad - \Pi_{\theta_n} \nu_{\theta_n}(X_{n+1}) + \Pi_{\theta_n} \nu_{\theta_n}(X_n) - \Pi_{\theta_{n-1}} \nu_{\theta_{n-1}}(X_n) \end{aligned}$$

Note that

$$E\{T_n \cdot [\nu_{\theta_n}(X_{n+1}) - \Pi_{\theta_n} \nu_{\theta_n}(X_n)]\} = 0 \qquad (1.10.14)$$

and that

$$\begin{aligned} &T_n \cdot [\Pi_{\theta_{n-1}} \nu_{\theta_{n-1}}(X_n) - \Pi_{\theta_n} \nu_{\theta_n}(X_{n+1})] \\ &= z_n - z_{n-1} + (T_{n+1} - T_n) \cdot \Pi_{\theta_n} \nu_{\theta_n}(X_{n+1}) \end{aligned}$$

where

$$z_n = T_n \cdot \Pi_{\theta_{n-1}} \nu_{\theta_{n-1}}(X_n) \qquad (1.10.15)$$

1.10 Rate of L^2 Convergence of Some Algorithms

with

$$|(T_{n+1} - T_n) \cdot \Pi_{\theta_n}\nu_{\theta_n}(X_{n+1})|$$
$$\leq \gamma_{n+1}|H(\theta_n, X_{n+1}) \cdot \Pi_{\theta_n}\nu_{\theta_n}(X_{n+1})|$$
$$+\gamma_{n+1}^2|\rho_{n+1}(\theta_n, X_{n+1})|\|\Pi_{\theta_n}\nu_{\theta_n}(X_{n+1})\|$$

Thus from (1.10.13), (1,10.14), (A.3) and (A.4), we deduce that

$$E_{x,a}\{|T_{n+1}|^2 I(n+1 \leq \tau(Q))\}$$
$$\leq (1 - 2\gamma_{n+1}\delta + \gamma_{n+1}^2 \overline{C}_1(Q, x))E_{x,a}\{|T_n|^2 I(n \leq \tau(Q))\}$$
$$+\gamma_{n+1}^2 \overline{C}_2(Q, x) + 2\gamma_{n+1}E(z_n - z_{n+1}) \quad (1.10.16)$$

for suitable constants $\overline{C}_1(Q, x)$ and $\overline{C}_2(Q, x)$. Note that because of (A.4)

$$E(|z_n|) \leq \overline{C}_3(a, x, Q) \quad (1.10.17)$$

□

We now have the following lemmas.

Lemma 25. *Suppose n_0 is such that $1 - 2\gamma_{n+1}\delta - \gamma_{n+1}^2 \overline{C}_1(a, x, Q) \geq 0$ for all $n \geq n_0$ and*

$$\inf_{n \geq n_0} \frac{\gamma_{n+1} - \gamma_n}{\gamma_n \gamma_{n+1}} + 2\delta - \gamma_{n+1}\overline{C}_1 > 0$$

Consider for $n > n_0$ the finite sequence $(\Lambda_k^n)_{k=n_0,\ldots,n}$

$$\Lambda_k^n = \begin{cases} 2\gamma_k \prod_{j=k}^{n-1}(1 - 2\gamma_{j+1}\delta + \gamma_{j+1}^2 \overline{C}_1) & \text{if } k \leq n-1 \\ 2\gamma_n & \text{if } k = n \end{cases}$$

Then the sequence $(\Lambda_k^n)_{k=n_0,\ldots,n}$ is increasing.

Proof of Lemma 25. If $k+1 < n$

$$\Lambda_{k+1}^n - \Lambda_k^n = (\prod_{j=k+1}^{n-1}(1 - 2\gamma_{j+1}\delta + \gamma_{j+1}^2 \overline{C}_1))$$
$$2(\gamma_{k+1} - \gamma_k + 2\gamma_k\gamma_{k+1}\delta - \gamma_k\gamma_{k+1}^2 \overline{C}_1)$$

If $k+1 = n$

$$\Lambda_n^n - \Lambda_{n-1}^n = 2(\gamma_n - \gamma_{n-1} + 2\gamma_n\gamma_{n-1}\delta - \gamma_{n-1}\gamma_n^2 \overline{C}_1)$$

Thus if $n > n_0$ we have $\Lambda_{k+1}^n - \Lambda_k^n \geq 0$ for all $n_0 \leq k < n$. □

Lemma 26. Let $(u_n)_{n \geq n_0}$ be a sequence of real numbers such that for all $n \geq n_0$
$$u_{n+1} \geq u_n(1 - 2\gamma_{n+1}\delta + \gamma_{n+1}^2 \overline{C}_1) + \gamma_{n+1}^2 \overline{C}_2 \qquad (1.10.18)$$
with additionally
$$E\{|T_{n_0}|^2 I(n_0 \leq \tau(Q))\} \leq u_{n_0} \qquad (1.10.19)$$
Then for all $n \geq n_0 + 1$.
$$E\{|T_n|^2 I(n \leq \tau(Q))\} \leq u_n + \sum_{k=n_0+1}^{n} \Lambda_k^n(z_{k-1} - z_k) \qquad (1.10.20)$$

Proof of Lemma 26. If (1.10.20) is true for n, then following (1.10.16) and (1.10.18)

$$\begin{aligned}
E(|T_{n+1}|^2) &\leq u_{n+1} + (1 - 2\gamma_{n+1}\delta + \gamma_{n+1}^2 \overline{C}_1)\left(\sum_{k=n_0+1}^{n} \Lambda_k^n(z_{k-1} - z_k)\right) \\
&\quad + 2\gamma_{n+1}(z_n - z_{n+1}) \\
&\leq u_{n+1} + \sum_{k=n_0+1}^{n} \Lambda_k^{n+1}(z_{k-1} - z_k) + \Lambda_{n+1}^{n+1}(z_n - z_{n+1}) \\
&\leq u_{n+1} + \sum_{k=n_0+1}^{n+1} \Lambda_k^{n+1}(z_{k-1} - z_k)
\end{aligned}$$

□

Proof of Theorem 24 (Continued). It is clearly sufficient to show that for some suitable N, there exists $\lambda(a, x, N, Q)$, such that for all $n \geq N$
$$E_{x,a}\{|\theta_n - \theta_*|^2 I(n \leq \tau(Q))\} \leq \lambda(a, x, N, Q)\gamma_n$$

We take $N \geq n_0$ where n_0 is as defined in Lemma 25. Having fixed N, we choose λ such that
$$E(|T_N|^2) \leq \lambda \gamma_N$$

For $n \geq N$ we have
$$\sum_{k=N+1}^{n} \Lambda_k^n(z_{k-1} - z_k) = \sum_{k=N+1}^{n-1}(\Lambda_{k+1}^n - \Lambda_k^n)z_k - 2\gamma_n z_n + \Lambda_{N+1}^n z_N$$

Following (1.10.20) and (1.10.17), for any sequence $(u_n)_{n \geq N}$ satisfying (1.10.18) and (1.10.19), we have
$$E\{|T_n|^2 I(n \leq \tau(Q))\} \leq u_n + 3\overline{C}_3(a, x, Q)\Lambda_n^n \qquad (1.10.21)$$

As in Lemma 23, we see that the sequence $u_n = \lambda \gamma_n$ satisfies property (1.10.18) for $n > N$. The theorem now follows from (1.10.21) and since $\Lambda_n^n = 2\gamma_n$. □

1.11 Comments on the Literature

The "classical" theory of stochastic algorithms was born with the work of Robbins and Monro (Robbins and Monro 1951). As previously explained in the bibliographic comments at the end of Chapter 1 of Part I, this classical theory related in the main to algorithms having the so-called "Robbins–Monro" property as described in this chapter (1.1.2 Example 4). We mentioned previously several references which take a systematic approach to this theory. The idea of introducing martingales to study algorithm convergence was presented in (Blum 1954), see also the article (Gladyshev 1965). This point of view is the basis of the work of (Hall and Heyde 1980).

The first major systematic investigation of algorithms which deviate from these assumptions, and which may be applied to control identification problems (cf. the end of Chapter 1 of Part I) is the book (Kushner and Clark 1978) in which the stochastic algorithm is viewed as a perturbation of a deterministic algorithm; this perturbation has two parts, the first by assumption satisfies an inequality which is formally the same as Doob's inequality for martingales, and the second is a process which (it is assumed) is bounded and tends to zero at infinity. Under appropriate conditions on the moments, these general assumptions serve to derive, a convergence theorem which (as far as the conclusion is concerned) is analogous to that given here in Theorem 15. Kushner and Clark also consider the convergence of the algorithm in distribution (and thus in probability) to the solution of the ODE. They clearly include many variations on these results.

The basic article of Ljung (Ljung 1977a,b) describes in full the use of dynamical Markov systems (in Ljung's case the linear dynamical Markov system given in 1.1.2 Example 1) to model specific situations. It can be shown (cf. (Métivier and Priouret 1984)) that the general assumptions of Kushner and Clark may be derived from the Markov situation at the expense of additional assumptions (for example the boundedness of $h(\theta)$).

The Markov situation is a natural setting for a large number of applications which we shall review in Chapter 2. In the form presented in this Chapter, it is first found in (Métivier and Priouret 1984) and independently in (Kushner and Shwartz 1984). However the results of these two papers are quite different: the first shows that this setting is suitable for asymptotic trajectorial analysis, whilst the second derives weak convergence results.

However we must also note that there is a wealth of literature relating to convergence analysis when the Markov situation is replaced by stationary ergodic and mixing assumptions on the process (X_n). For example one such case is given in (Kushner and Clark 1978), see also (Ruppert 1982).

Furthermore numerous works have considered the important case of algorithms with constraints: see for example (Kushner and Clark 1978),

(Kushner and Sanvicente 1975), (Hiriart-Urruty 1977), (Schmetterer 1980), (Monnez 1982), (Pflug 1981a, 1986, 1987), (Walk 1983-4), (Ermoliev 1983) and (Marti 1979).

We have not broached the subject of infinite dimensional algorithms such as may arise in function identification problems; for details on this subject see for example (Walk 1977, 1980, 1985), (Walk and Zsido 1989) and (Györfi 1980).

The presentation of this chapter follows that of (Métivier and Priouret 1987) (with simplifications including assumptions on the moments which are sufficient for the applications of the following chapter). The results may be used to consider problems such as ODE approximations for algorithms in which $H(\theta, x)$ has discontinuities similar to those considered in (Benveniste, Goursat and Ruget 1980b).

Chapter 2
Application to the Examples of Part I

Introduction

The aim of this chapter is to provide criteria under which Assumption (A.4) of the previous chapter is valid. We shall see that the algorithms given in Part I satisfy the general assumptions of Chapter 1; thus the conclusions of the theorems involving approximation by solution of the ODE (Theorems 9 and 14) and of the theorems involving asymptotic approximation (Theorems 13, 14, 17) may be applied to these algorithms. As previously noted in Chapter 1, Subsection 1.1.3, comment c., in verifying that Assumption (A.4) is satisfied, we shall require that the Markov chain with transition probability Π_θ is ergodic, and the functions $\Pi_\theta H_\theta$ are regular in θ.

In Section 2.1 we consider the geometric ergodicity of Markov chains in a setting appropriate to our algorithm analysis. In Section 2.2 we give sufficient conditions under which (A.4) is satisfied. In Section 2.3 we study the transitions Π_θ associated with linear dynamics (Chapter 1, Subsection 1.1.3, Example 1). Section 2.4 provides a more exact coverage of three examples given in Part I, namely the transversal equaliser, the least squares algorithm and the phase-locked loop. In Section 2.5, a more refined analysis of the decision-feedback recursive equaliser is given. The first four sections are relatively elementary (except 2.1.4). Section 2.5 and Subsection 2.1.4 are more difficult and may be omitted by readers who are not particularly interested in algorithms with quantisation or decision feedback.

2.1 Geometric Ergodicity of Certain Markov Chains

2.1.1 Preliminary Lemma

We shall consider Markov chains with values in $\mathbb{R}^k$ or in $\mathbb{R}^k \times E$ where E is a finite subset of $\mathbb{R}^r$. In fact we shall consider the latter case, of which the first case is a special instance.

Thus let Π be the transition function of a Markov chain on $\mathbb{R}^k \times E$. We shall denote the points of $\mathbb{R}^k \times E$ by $z, z_i, \ldots$ with

$$z_i = (x_i, e_i), \qquad x_i \in \mathbb{R}^k \ \ e_i \in E \tag{2.1.1}$$

$$\mu_p(x_1) = \sup_{e_1} \int \Pi(x_1, e_1; dx_2 de_2)|x_2|^p \tag{2.1.2}$$

Lemma 1. *Let g be a function on $\mathbb{R}^k \times E$. Suppose that there exist $K_1 \geq 0, q \geq 0$ and $\rho \in]0,1[$, such that for all z_1, z_2, n:*

$$|\Pi^n g(z_1) - \Pi^n g(z_2)| \leq K_1 \rho^n (1 + |x_1|^q + |x_2|^q) \tag{2.1.3}$$

and that $\mu_q(x)$ is finite for all x. Then there exists a constant γ such that for all z, n

$$|\Pi^n g(z) - \gamma| \leq K_1 \frac{\rho^n}{1-\rho}(1 + |x|^q + \mu_q(x)) \tag{2.1.4}$$

Moreover, if for all z_1, $1 + |x_2| + \mu_q(x_2)$ is $\Pi(z_1, .)$ integrable then $u = \sum_{n \geq 0}(\Pi^n g - \gamma)$ is a solution of $(I - \Pi)u = g - \gamma$.

Proof. We have

$$\begin{aligned}|\Pi^{n+1}g(z) - \Pi^n g(z)| &= |\int \Pi(z, dz_1)[\Pi^n g(z_1) - \Pi^n g(z)]| \\ &\leq K_1 \rho^n \int \Pi(z, dx_1, de_1)(1 + |x|^q + |x_1|^q) \\ &\leq K_1 \rho^n (1 + |x|^q + \mu_q(x))\end{aligned}$$

Whence for all $k \geq 0$,

$$|\Pi^{n+k}g(z) - \Pi^n g(z)| \leq K_1 \frac{\rho^n}{1-\rho}(1 + |x| + \mu_q(x)) \tag{2.1.5}$$

This implies that $\Pi^n g(z)$ converges to some value γ, but (2.1.3) implies that $\Pi^n g(z')$ converges to the same limit γ whatever the value of z'. Letting k tend to $+\infty$ in (2.1.5), we obtain (2.1.4).

Thus the series $\sum_{n \geq 0}(\Pi^n g(z_2) - \gamma)$ is convergent, and since

$$\sum |\Pi^n g(z_2) - \gamma| \leq K(1-\rho)^{-2}(1 + |x_2| + \mu_q(x_2))$$

which is $\Pi(z_1, .)$ integrable, we have, following Lebesgue's theorem

$$\Pi u = \sum_{n \geq 0} \Pi(\Pi^n g - \gamma) = \sum_{n \geq 1}(\Pi^n g - \gamma) = u - (g - \gamma)$$

□

2.1.2 Invariant Probability and Solution of Poisson's Equation

Given a function g on $\mathbb{R}^k \times E$, for $p \geq 0$ we set

$$\|g\|_{\infty,p} = \sup_{x,e} \frac{|g(x,e)|}{1 + |x|^p} \tag{2.1.6}$$

$$[g]_p = \sup_{x_1 \neq x_2, e \in E} \frac{|g(x_1, e) - g(x_2, e)|}{|x_1 - x_2|(1 + |x_1|^p + |x_2|^p)} \tag{2.1.7}$$

$$Li(p) = \{g; [g]_p < +\infty\}$$

2.1 Geometric Ergodicity of Certain Markov Chains

Observe that if $[g]_p < +\infty$, then we have $\|g\|_{\infty,p+1} < +\infty$; this leads us to introduce

$$N_p(g) = \sup\{\|g\|_{\infty,p+1}, [g]_p\} \tag{2.1.8}$$

Now fix $p \geq 0$.

Proposition 2. *Suppose that there exist positive constants $K_1, q, \mu, \rho < 1$ such that for all $g \in Li(p), x \in \mathbb{R}^k, z_1, z_2$:*

$$|\Pi^n g(z_1) - \Pi^n g(z_2)| \leq K_1 \rho^n N_p(g)(1 + |x_1|^q + |x_2|^q) \tag{2.1.9}$$
$$\mu_q(x) \leq \mu(1 + |x|^q) \tag{2.1.10}$$

Then there exists a constant K_2 depending only on K_1, μ, ρ, and for all $g \in Li(p)$ a number Γg such that for all z

(i) $\quad |\Pi^n g(z) - \Gamma g| \leq K_2 N_p(g) \rho^n (1 + |x|^q)$

(ii) $\quad u = \sum_{n \geq 0}(\Pi^n g - \Gamma g)$ satisfies $(I - \Pi)u = g - \Gamma g \tag{2.1.11}$

Moreover, if for any C^∞ function ϕ having a compact support on $\mathbb{R}^k \times E$, $\Pi\phi$ is continuous, the chain with transition function Π has a unique invariant probability m which satisfies

$$\int |x|^{p+1} m(dx.de) < +\infty$$

and for all $g \in Li(p)$, $\Gamma g = \int g dm$.

Proof. (i) and (ii) result from Lemma 1 for some constant Γg. We let $(Li)_c$ denote the space of Lipschitz functions with compact support on $\mathbb{R}^k \times E$. Then $(Li)_c \subset Li(p)$ and $g \to \Gamma g$ defines a positive linear form on $(Li)_c$. Thus there exists a positive Radon measure m on $\mathbb{R}^k \times E$, such that for all $g \in (Li)_c$, $\Gamma g = \int g dm$.

For $s \in \mathbb{N}^*$, we define $\phi_s(t)$ (for $t \in \mathbb{R}_+$) by $\phi_s(t) = 1$ if $0 \leq t \leq \frac{s}{2}$, $\phi_s(t) = 0$ if $t \geq s$, and ϕ_s is linear between $\frac{s}{2}$ and s. If we set $f_s(x,e) = \phi_s(|x|)$, then f_s increases towards 1 and satisfies $|f_s(x_1, e_1) - f_s(x_2, e_2)| \leq \frac{2}{s}|x_1 - x_2|$.

Let $g \in Li^+(p)$, and evaluate $\sup_e |(gf_s)(x_1, e) - (gf_s)(x_2, e)|$. If $|x_1| \geq s$ and $|x_2| \geq s$, it is zero. Thus let us suppose that $|x_1| \leq s$, then,

$$|(gf_s)(x_2, e) - (gf_s)(x_1, e)|$$
$$\leq |g(x_1, e)||f_s(x_2, e) - f_s(x_1, e)| + |f_s(x_2, e)||g(x_1, e) - g(x_2, e)|$$
$$\leq \frac{2}{s} N_p(g)|x_1 - x_2|(1 + |x_1|^{p+1}) + |g(x_1, e) - g(x_2, e)|$$
$$\leq 2 N_p(g)|x_1 - x_2|(1 + |x_1|^p) + N_p(g)|x_1 - x_2|(1 + |x_1|^p + |x_2|^p)$$

Whence $N_p(gf_s) \leq 3 N_p(g)$.

Since $gf_s \in (Li)_c$, following (2.1.11-i), we have for all s,

$$|\Pi^n(gf_s)(x,e) - \int gf_s dm| \leq 3K_2 N_p(g)\rho^n(1+|x|^q)$$

Thus, by passing to the increasing limit as s tends to $+\infty$,

$$|\Pi^n(g)(x,e) - \int gdm| \leq 3K_2 N_p(g)\rho^n(1+|x|^q)$$

This proves that for $g \in Li^+(p)$, $\int gdm < +\infty$ and has value Γg. In particular, taking $g(x,e) = |x|^{p+1}$, we have $\int |x|^{p+1} m(dx,de) < +\infty$ and taking $g=1$ we have $m(\mathbb{R}^k \times E) = 1$. Since $g \in Li(p)$ implies that $g^+ \in Li(p)$ and $g^- \in Li(p)$, the extension to $g \in Li(p)$ is immediate. Now if C_p denotes the space of continuous functions g on $\mathbb{R}^k \times E$ such that $\|g\|_{\infty,p} < +\infty$, then for $g \in C_{p+1}$, $\lim_n \Pi^n g(z) = \int gdm$. In fact, if we denote $\phi_p(x,e) = |x|^p$, then **for fixed** z, the sequence $(\Pi^n \phi_{p+1}(z); n \geq 0)$ is bounded and converges to $\int \phi_{p+1} dm$; thus the sequence of positive linear forms $g \to \int \Pi^n(z,dz_1)g(z_1)$ is defined on C_{p+1}, it is also equicontinuous, since

$$\left|\int \Pi^n(z,dz_1)g(z_1)\right| \leq |g|_{p+1}(1 + \Pi^n \phi_{p+1}(z)) \leq c_p \|g\|_{\infty,p+1}$$

Since this sequence converges on the dense subset $Li(p)$, it converges for all $g \in C_{p+1}$, to some limit which can only be $\int gdm$ since $g \to \int gdm$ is continuous on C_{p+1}.

Suppose $g \in C_k^\infty$ and that Π_g is continuous (thus it belongs to C_{p+1}), then $\int gdm = \lim_n \Pi^n g(z) = \lim_n \Pi^n(\Pi g) = \int \Pi g dm$. This proves that m is an invariant probability for the Markov chain with transition function Π. This invariant probability is unique, since if m' is another such, then for all $\phi \in (Li)_c$, and for all n, $\int \phi dm' = \int \Pi^n \phi dm'$ and so

$$\int \phi dm' = \lim_n \int \Pi^n \phi dm' = \int \Gamma \phi dm = \Gamma \phi = \int \phi dm$$

□

2.1.3 Case of a Continuous Transition Function Π from $Li(p)$ to $Li(p)$

As we shall see, Proposition 2 is well suited for the study of certain stochastic algorithms. However, since it involves all the Π^n, the result given below is more satisfactory. We shall restrict our attention to a Markov chain with values in $\mathbb{R}^k$, in order to apply a result due to Sunyach (1975). Extension to the case of a Markov chain with values in $\mathbb{R}^k \times E$ is straightforward, but complicates the assumptions. The integer p is still fixed.

2.1 Geometric Ergodicity of Certain Markov Chains

Proposition 3. *Suppose that there exist constants $r \in \mathbb{N}^*, K \in \mathbb{R}_+, \rho \in]0,1[$ such that, for all g in $Li(p)$, all $x_1, x_2 \in \mathbb{R}^k$,*

(i) $\quad \mu_{p+1}(x_0) < +\infty$ for some $x_0 \in \mathbb{R}^k$
(ii) $\quad |\Pi g(x_1) - \Pi g(x_2)| \leq K[g]_p |x_1 - x_2|(1 + |x_1|^p + |x_2|^p)$
(iii) $\quad |\Pi^r g(x_1) - \Pi^r g(x_2)| \leq \rho[g]_p |x_1 - x_2|(1 + |x_1|^p + |x_2|^p)$ $\quad$ (2.1.12)

Then the Markov chain with transition function Π has a unique invariant probability m and $\int |x|^{p+1} dm(x) < +\infty$. Furthermore
(j) there exist $K_1 > 0, \rho_1 \in]0,1[$ such that for all $g \in Li(p), x, n$:

$$|\Pi^n g(x) - \int g \, dm| \leq K_1 \rho_1^n [g]_p (1 + |x|^{p+1})$$

(jj) for $g \in Li(p), u = \sum_{n \geq 0}(\Pi^n g - \int g \, dm)$ is the unique solution in $Li(p)$ of $(I - \Pi)u = g - \int g \, dm$ such that $\int u \, dm = 0$.

Proof. Since the function $\phi_p(x) = |x|^{p+1}$ satisfies $[\phi]_p < +\infty$, we have

$$\begin{aligned} |\mu_{p+1}(x) - \mu_{p+1}(x_0)| &= |\Pi\phi_p(x) - \Pi\phi_p(x_0)| \\ &\leq K[\phi]_p |x - x_0|(1 + |x|^p + |x_0|^p) \end{aligned}$$

and so $\mu_{p+1}(x) \leq K_2(1 + |x|^{p+1})$. Moreover (2.1.12) may be written as $[\Pi g]_p \leq K[g]_p$, and so $[\Pi^n g]_p \leq K^n [g]_p$; thus there exists K_3 such that for $k = 0, 1, \ldots, r-1$,

$$[\Pi^k g]_p \leq K_3 [g]_p \qquad (2.1.13)$$

Similarly, (2.1.13) may be written in the form $[\Pi^r g] \leq \rho[g]_p$, whence $[\Pi^{nr} g]_p \leq \rho^n [g]_p$. Then, writing $n = qr + k, 0 \leq k \leq r$, we have

$$[\Pi^n g]_p = [\Pi^k \Pi^{qr} g]_p \leq K_3 [\Pi^{qr} g]_p \leq K_3 \rho^q [g]_p$$

which, setting $K_4 = K_3 \rho^{-1}$ and $\rho_1 = \rho^{1/r}$, may be rewritten as

$$|\Pi^n g(x_1) - \Pi^n g(x_2)| \leq K_4 \rho_1^n [g]_p |x_1 - x_2|(1 + |x_1|^p + |x_2|^p) \qquad (2.1.14)$$

Conditions (2.1.9) and (2.1.10) of Proposition 2 are both true for $q = p + 1$; moreover, from (2.1.12), if ϕ is C_∞ bounded, $\Pi\phi$ is locally Lipschitz and therefore continuous. Thus we have proved the first part of the proposition. Moreover $u = \sum_{n \geq 0}(\Pi^n g - \int g \, dm)$ is a solution of $(I - \Pi)u = g - \int g \, dm$, satisfies $\int u \, dm = 0$ and is in $Li(p)$ (following (2.1.14)). Finally, if $u' \in Li(p)$ satisfies $(I - \Pi)u' = g - \int f \, dm$, the function $\nu = u - u'$ satisfies $\nu \in Li(p)$, and $(I - \Pi)\nu = 0$, and consequently $\nu = \Pi\nu = \ldots = \Pi^n \nu$ which converges to $\int \nu \, dm$; thus $\nu = \int \nu \, dm$ and ν is constant. $\square$

2.1.4 Sufficient Conditions for the Validity of the Assumptions of Proposition 2

This subsection is fairly technical. The results will only be used in our investigation of the example in Section 2.5.

In applications, the Markov chains considered are often of the form

$$Z_n = \psi_n(Z_0, \nu_1, \ldots, \nu_n) = \psi_n(Z_0, V)$$

where $V = (\nu_n)$ is a sequence of independent r.v. In other words, if Π denotes the transition function of the chain, then for any bounded Borel function g:

$$\Pi^n g(z) = E\{\psi_n(z, \nu_1, \ldots, \nu_n)\}$$

To evaluate expressions such as (2.1.9), it is natural to consider the following process for any 2-tuple (z_1, z_2):

$$(Z_n, Z'_n) = (\psi_n(z_1, V), \psi_n(z_2, V))$$

This process is itself a Markov chain. If its distribution for the initial conditions (z_1, z_2) is denoted by P_{z_1, z_2}, then we have:

$$\Pi^n g(z_1) - \Pi^n g(z_2) = E_{z_1, z_2}(g(Z_n) - g(Z'_n))$$

Thus let $(\zeta_n)_{n \geq 0} = (X_n, \eta_n, X'_n, \eta'_n)$ be a homogeneous Markov chain with values in $(\mathbb{R}^k \times E)^2$, which is canonical, with distribution P_{z_1, z_2}. We assume that the conditions (X) (below) are satisfied (here $z = (x, e)$ denotes a generic point of $\mathbb{R}^k \times E$):

(X.1). *For all $p \in \mathbb{N}^*$, there exists a constant $C > 0$ such that:*

(i) $$\sup_{x', e, e'} E_{x, e, x', e'}(|X_n|^p) \leq C \cdot (1 + |x|^p)$$

(ii) $$\sup_{x, e, e'} E_{x, e, x', e'}(|X'_n|^p) \leq C \cdot (1 + |x'|^p)$$

(X.2). *There exist positive constants $K_1, \rho_1 < 1$ such that*

$$\sup_{e, e'} E_{x, e, x', e'}(|X_n - X'_n|^2) \leq K_1 \rho_1^n (1 + |x|^2 + |x'|^2)$$

(X.3). *There exist positive constants $K_2, s_1, \rho_2 < 1$ such that for all n, z, z':*

$$P_{z, z'}(\eta_k \neq \eta'_k, k = 0, \ldots, n) \leq K_2 \rho_2^n (1 + |x|^{s_1} + |x'|^{s_1})$$

(X.4). *There exist $r \in \mathbb{N}^*, K_3 \in \mathbb{R}_+, s_2 \in \mathbb{R}_+, \rho_3 < 1$ such that for all $n \geq r, z, z'$:*

$$P_{z, z'}(\eta_{n-1} = \eta'_{n-1}, \eta_n \neq \eta'_n) \leq K_3 \rho_3^n (1 + |x|^{s_2} + |x'|^{s_2})$$

2.1 Geometric Ergodicity of Certain Markov Chains

Then we have:

Proposition 4. *If the process* $(\zeta_n)_{n\geq 0} = (Z_n, Z'_n)_{n\geq 0}$ *satisfies (X.1) to (X.4); then for all $p > 0$, there exist positive constants $K, q, \rho < 1$, depending only on the K_i, s_i, ρ_i, such that for all $g \in Li(p), n, z, z'$:*

$$|E_{z,z'}(g(Z_n) - g(Z'_n))| \leq K\rho^n N_p(g)(1 + |x|^q + |x'|^q) \qquad (2.1.15)$$

In particular, if each of the processes (Z_n) and (Z'_n) is a Markov chain with transition function Π, then for all $g \in Li(p)$ we have

$$|\Pi^n g(x_1, e_1) - \Pi^n g(x_2, e_2)| \leq K\rho^n N_p(g)(1 + |x|^q + |x'|^q)$$

Proof. Since following (X.1)

$$|E_{z,z'}(g(Z_n))| \leq C \cdot N_p(g)(1 + |x|^{p+1}) \qquad (2.1.16)$$

it is sufficient to establish (2.1.15) for $n \geq 2r + 1$ where r is given by (X.3). For $n \geq 2r + 1$, set

$$n_1 = [\tfrac{n}{2}] + 1, \qquad n_2 = n_1 + r \qquad (2.1.17)$$
$$\tau = \inf(k \geq n_2, \eta_k = \eta'_k) \qquad (2.1.18)$$

where $[.]$ denotes the integer part. Note that then $n_2 \leq n$, and $n_1 \geq r$. We have

$$a = E_{z,z'}(g(X_n, \eta_n) - g(X'_n, \eta'_n)) = \sum_{k=n_2}^{n-1} a_1(k) + a_2$$

where

$$a_1(k) = E_{z,z'}\{I(\tau = k)(g(Z_n) - g(Z'_n))\}$$
$$a_2 = E_{z,z'}\{I(\tau \geq n)(g(Z_n) - g(Z'_n))\}$$

We evaluate $a_1(k)$

$$|a_1(k)| \leq E_{z,z'}\{I(\tau = k)I(\eta_n \neq \eta'_n)(|g(Z_n)| + |g(Z'_n)|\}$$
$$+ \sum_{e \in E} E_{z,z'}\{I(\tau = k)I(\eta_n \neq \eta'_n = e)|g(X_n, e) - g(X'_n, e)|\}$$
$$\leq a'_1(k) + a''_1(k)$$

Then following (X.4), since $k \geq n_1 \geq r$

$$P_{z,z'}[\eta_k = \eta'_k, \eta_n \neq \eta'_n] = \sum_{j=k}^{n-1} P_{z,z'}[\eta_j = \eta'_j, \eta_{j+1} \neq \eta'_{j+1}]$$
$$\leq \frac{K_3}{1-\rho_3}\rho_3^{n_1}(1 + |x|^{s_2} + |x'|^{s_2})$$

Then using (X.1) and (2.1.15) we have:

$$[a'_1(k)]^2 \leq K_4 \rho_3^{n_1}(1 + |x|^{s_2} + |x'|^{s_2}) \cdot E_{z,z'}\{|g(Z_n)|^2 + |g(Z'_n)|^2\}$$
$$\leq K_5 \rho_3^{n_1}(N_p(g))^2(1 + |x|^{s_2} + |x'|^{s_2})(1 + |x|^{2p+2} + |x'|^{2p+2})$$

Finally, since $n_1 \geq \frac{n}{2}$

$$a'_1(k) \leq K_6 \rho_3^{n/4} N_p(g)(1 + |x|^{p_1} + |x'|^{p_1}) \qquad (2.1.19)$$

Using (X.1), (X.2) and the Schwarz inequality

$$E_{z,z'}(|g(X_n, e) - g(X'_n, e)|) \leq N_p(g) E_{z,z'}\{|X_n - X'_n|(1 + |X_n|^p + |X'_n|^p)\}$$
$$\leq K_7 N_p(g) \rho_1^{n/2}(1 + |x|^{p+1} + |x'|^{p+1})$$

whence

$$a''_1(k) \leq K_8 N_p(g) \rho_1^{n/2}(1 + |x|^{p+1} + |x'|^{p+1}) \qquad (2.1.20)$$

From (2.1.18) and (2.1.19), we deduce that there exist $K', q_1, \rho_4 \in]0, 1[$ such that:

$$|\sum_{k=n_2}^{n-1} a_1(k)| \leq K' N_p(g) \rho_4^n (1 + |x|^{q_1} + |x'|^{q_1}) \qquad (2.1.21)$$

It remains to evaluate a_2. But

$$|a_2|^2 \leq P_{z,z'}(\tau \geq n) E_{z,z'}(|g(Z_n) + g(Z'_n)|^2)$$

We now note that

$$\{\tau \geq n\} \subset \{\eta_{n_2+k} \neq \eta'_{n_2+k}, k = 1, \ldots, n - n_2 - 1\}$$

and using (X.2) and the Markov property, we obtain

$$P_{z,z'}\{\tau \geq n\} \leq K_2 \rho_2^{-1} \rho^{n-n_2}(1 + |x|^{s_1} + |x'|^{s_1})$$

Making the observation that $n - n_2 \geq \frac{n}{2} - r - 1$ and that $E_{z,z'}(|g(Z_n)|^2) \leq (N_p(g))^2(1 + |x|^{2p+2})$ we deduce that there exist K'', q_2 such that

$$|a_2| \leq K'' N_p(g) \rho_2^{n/2}(1 + |x|^{q_2} + |x'|^{q_2}) \qquad (2.1.22)$$

(2.1.15) now follows from (2.1.20) and (2.1.21). The second part of the proposition is obtained by simple reformulation of (2.1.15). □

2.2 Markov Chains Dependent on a Parameter θ

We consider a family $(\Pi_\theta, \theta \in Q)$ of transition probabilities on $\mathbb{R}^k \times E$, where E is a finite subset of $\mathbb{R}^m$ and Q is a subset of $\mathbb{R}^d$. Given a function $f(\theta, x, e)$ on $\mathbb{R}^d \times \mathbb{R}^k \times E$, we shall study the regularity in θ of the solutions ν_θ of the Poisson equation $(I - \Pi_\theta)\nu_\theta = f_\theta - \Gamma_\theta f_\theta$. We recall that f_θ denotes the mapping $(x, e) \to f(\theta, x, e)$ and that $z_i = (x_i, e_i)$ denotes a generic point of $\mathbb{R}^k \times E$.

2.2.1 Holder Regularity in θ of ν_θ

A function $f(\theta, x, e)$ will be said to be of class $Li(Q, L_1, L_2, p_1, p_2)$ if

(i)   for all $\theta \in Q$, $N_{p_1}(f_\theta) \leq L_1$

(ii)   for all $\theta_1, \theta_2 \in Q$, all $(x, e) \in \mathbb{R}^k \times E$,
$$|f(\theta_1, x, e) - f(\theta_2, x, e)| \leq L_2 |\theta_1 - \theta_2|(1 + |x|^{p_2}) \tag{2.2.1}$$

$Li(Q)$ will denote the set of f which belong to $Li(Q, L_j, p_j)$ for some values of L_j, p_j.

Theorem 5. *Given $p_1 \geq 0, p_2 \geq 0$, we assume that there exist positive constants $K_1, K_2, K_3, q_1, q_2, \rho < 1$ such that:*
(i) for all $g \in Li(p_1), \theta \in Q, n \geq 0, z_1, z_2$:

$$|\Pi_\theta^n g(z_1) - \Pi_\theta^n g(z_2)| \leq K_1 \rho^n N_{p_1}(g)(1 + |x_1|^{q_1} + |x_2|^{q_2})$$

(ii) for all $\theta \in Q, n \geq 0, z$ and all $m \leq q_1 \vee q_2$,

$$\sup_e \int \Pi_\theta^n(x, e; dx_1 de_1)(1 + |x_1|^m) \leq K_2(1 + |x|^m)$$

(iii) for all $g \in Li(p_1), \theta, \theta' \in Q, n \geq 0, z$,

$$|\Pi_\theta^n g(z) - \Pi_{\theta'}^n g(z)| \leq K_3 N_{p_1}(g)|\theta - \theta'|(1 + |x|^{q_2})$$

Then for any function $f(\theta, z)$ of class $Li(Q, L_1, L_2, p_1, p_2)$, there exist functions $h(\theta), \nu_\theta(.)$ and constants $C_1, C_2, C(\lambda), 0 < \lambda < 1$ depending only on the L_j, p_j, such that:
(j) for all $\theta, \theta' \in Q$, $|h(\theta) - h(\theta')| \leq C_1 |\theta - \theta'|$
(jj) for all $\theta \in Q$, $|\nu_\theta(x, e)| \leq C_2(1 + |x|^{q_1})$
(jjj) for all $\theta, \theta' \in Q$ all $\lambda \in]0, 1[$ and for $s = \max(p_2, q_1, q_2)$

$$|\nu_\theta(x, e) - \nu_{\theta'}(x, e)| \leq C(\lambda)|\theta - \theta'|^\lambda (1 + |x|^s)$$
$$|\Pi_\theta \nu_\theta(x, e) - \Pi_{\theta'} \nu_{\theta'}(x, e)| \leq C(\lambda)|\theta - \theta'|^\lambda (1 + |x|^s)$$

(jv) $(I - \Pi_\theta)\nu_\theta = f_\theta - h(\theta)$

2.2.2 Consequences of Theorem 5

Let D be an open subset of $\mathbb{R}^d$ and suppose that for every compact subset Q of D, the family of transition functions $(\Pi_\theta, \theta \in D)$ satisfies Assumptions (i), (ii), (iii) above. Then Assumption (A.4) of Chapter 1 is satisfied by every function $H(\theta, z)$ of class $Li(Q)$. The same is true if $\Pi_\theta H_\theta$ is of class $Li(Q)$. In fact applying Theorem 5 to $f_\theta = \Pi_\theta H_\theta$ we obtain a solution w_θ which satisfies (j), (jj), (jjj) and it is sufficient to take $\nu_\theta = w_\theta + H_\theta - h(\theta)$ (see Comment b. of Chapter 1, Subsection 1.1.3). It is this which allows us to cover the case in which H_θ is discontinuous but the transition function Π_θ has a regularising effect. We shall see a number examples of this.

2.2.3 Proof of Theorem 5

Suppose $f(\theta, z) \in Li(Q, L_1, L_2, p_1, p_2)$

1. First fix θ. Following (ii) for $n = 1$ we have $\mu_{q_1}(z) \leq K_2(1 + |x|^{q_1})$, where μ_q is as defined in (2.1.2). Following (i), since $N_{q_1}(f_\theta) \leq L_1$,

$$|\Pi_\theta^n f_\theta(z_1) - \Pi_\theta^n f_\theta(z_2)| \leq L_1 K_1 \rho^n (1 + |x_1|^{q_1} + |x_2|^{q_1})$$

Assumptions (2.1.9) and (2.1.10) of Proposition 2 are thus satisfied, and so there exists a function $h(\theta)$—namely $\Gamma_\theta f_\theta$—and a constant K, independent of $\theta \in Q$, such that for $\theta \in Q$

$$|\Pi_\theta^n f_\theta(z) - h(\theta)| \leq K L_1 \rho^n (1 + |x|^{q_1}) \qquad (2.2.2)$$

Moreover $\nu_\theta(z) = \sum_{n \geq 0} (\Pi_\theta^n f_\theta(z) - h(\theta))$ satisfies $(I - \Pi_\theta)\nu_\theta = f_\theta - h(\theta)$, and following (2.2.2)

$$|\nu_\theta(z)| \leq \frac{K L_1}{1 - \rho}(1 + |x|^{q_1}) \qquad (2.2.3)$$

2. For $\theta_1, \theta_2 \in Q$, from (iii) we have

$$|\Pi_{\theta_1}^n f_{\theta_1}(z) - \Pi_{\theta_2}^n f_{\theta_1}(z)| \leq K_3 L_1 |\theta_1 - \theta_2|(1 + |x|^{q_2})$$

and from (ii) and (2.2.1)

$$\begin{aligned} |\Pi_{\theta_2}^n f_{\theta_1}(z) - \Pi_{\theta_2}^n f_{\theta_2}(z)| &= |\Pi_{\theta_2}^n (f_{\theta_1} - f_{\theta_2})(z)| \\ &\leq L_2 |\theta_1 - \theta_2| \int \Pi_{\theta_2}^n (z, dx_1, de_1)(1 + |x_1|^{p_2}) \\ &\leq K_2 L_2 |\theta_1 - \theta_2|(1 + |x|^{q_2}) \end{aligned}$$

Whence for $L = \max(K_3 L_1, K_2 L_2)$, $q = \max(p_2, q_2)$

$$|\Pi_{\theta_1}^n f_{\theta_1}(z) - \Pi_{\theta_2}^n f_{\theta_2}(z)| \leq L |\theta_1 - \theta_2|(1 + |x|^q) \qquad (2.2.4)$$

2.2 Markov Chains Dependent on a Parameter θ

If we consider the case where $z = (0, e)$, then letting n tend towards $+\infty$, we obtain

$$|h(\theta_1) - h(\theta_2)| \leq L|\theta_1 - \theta_2| \quad (2.2.5)$$

Then for $\theta_1, \theta_2 \in Q$ and for any integer N, following (2.2.2), (2.2.4) and (2.2.5) we have

$$\begin{aligned}
|\nu_{\theta_1}(z) - \nu_{\theta_2}(z)| &= |\sum_{n=0}^{\infty}(\Pi_{\theta_1}^n f_{\theta_1}(z) - h(\theta_1) - \Pi_{\theta_2}^n f_{\theta_2}(z) + h(\theta_2))| \\
&\leq \sum_{n=0}^{N-1}(|\Pi_{\theta_1}^n f_{\theta_1}(z) - \Pi_{\theta_2}^n f_{\theta_2}(z)| + |h(\theta_1) - h(\theta_2)|) \\
&\quad + 2\sup_{\theta \in Q}\sum_{n=N}^{+\infty}|\Pi_\theta^n f_\theta(z) - h(\theta)| \\
&\leq 2[NL|\theta_1 - \theta_2| + \frac{KL_1}{1-\rho}\rho^N](1 + |x|^s)
\end{aligned}$$

where $s = \max(q, q_1)$.

If $|\theta_1 - \theta_2| \geq 1$, (jjj) is a trivial consequence of (jj). Suppose therefore that $|\theta_1 - \theta_2| \leq 1$, and since the previous inequality is true for all N, choose $N = \frac{\log|\theta_1-\theta_2|}{\log \rho} + u$ where $0 \leq u < 1$. Then if we set $L' = \frac{KL_1}{1-\rho}$, we have:

$$\begin{aligned}
& NL|\theta_1 - \theta_2| + L'\rho^N \\
&\leq \frac{\log|\theta_1-\theta_2|}{\log \rho}L|\theta_1-\theta_2| + L|\theta_1-\theta_2| + L'\rho^{\log|\theta_1-\theta_2|/\log \rho} \\
&\leq K(\lambda)|\theta_1-\theta_2|^\lambda + C|\theta_1-\theta_2| \leq C(\lambda)|\theta_1-\theta_2|^\lambda
\end{aligned}$$

since for $0 \leq t \leq 1$, $|t \log t| \leq A(\lambda)t^\lambda$. For the second inequality of (jjj), it is sufficient to note that

$$\Pi_\theta \nu_\theta(z) = \sum_{n \geq 1}(\Pi_\theta^n f_\theta(z) - h(\theta))$$

$\square$

2.2.4 Case where ν_θ is Lipschitz in θ

We shall provide conditions under which the solution ν_θ of Poisson's equation is Lipschitz in θ. We shall restrict out attention to Markov chains with values in $\mathbb{R}^k$. Let $f(\theta, x)$ be a function on $\mathbb{R}^d \times \mathbb{R}^k$; note that definitions (2.1.6), (2.1.7), (2.1.8) and (2.2.1) are valid with an evident modification, namely the suppression of e. If $f(\theta, x)$ is differentiable with respect to x, then $f'(\theta, x)$ will denote the gradient with respect to x. A function $f(\theta, x)$ which is differentiable with respect to x will be said to be of class $Li^1(Q, L_1, L_2, p_1, p_2)$ if:

(i) for all $\theta \in Q$, $|f(\theta, 0) + N_{p_1}(f'_\theta)| \leq L_1$ \hfill (2.2.6)

(ii) for all $\theta, \theta' \in Q$, $|f(\theta, 0) - f(\theta', 0)| \leq L_2|\theta - \theta'|$ \hfill (2.2.7)

(iii) for all $\theta, \theta' \in Q, x \in \mathbb{R}^k$:
$$|f'(\theta, x) - f'(\theta', x)| \leq L_2|\theta - \theta'|(1 + |x|^{p_2}) \qquad (2.2.8)$$

It is easy to see that this implies

$$[f_\theta]_{p_1+1} \leq L_1 \qquad (2.2.9)$$

and for an L'_2 function of L_2, p_2

$$|f(\theta, x) - f(\theta', x)| \leq L'_2|\theta - \theta'|(1 + |x|^{p_2+1}) \qquad (2.2.10)$$

We shall say that f is of class $Li^1(Q)$ if it is of class $Li^1(Q, L_1, L_2, p_1, p_2)$ for some values of L_j, p_j. Clearly $Li^1(Q) \subset Li(Q)$.

Theorem 6. *Suppose that for all $p \geq 0$, there exist positive constants $K_j, q_j, \rho_j < 1$ such that for any differentiable function $g(x)$, all $n \geq 0$, $\theta, \theta' \in Q, x_1, x_2 \in \mathbb{R}^k$,*

(i) $\displaystyle\int \Pi^n(x_1, dx_3)(1 + |x_3|^p) \leq K_1(1 + |x_1|^p)$

(ii) $|\Pi_\theta^n g(x_1) - \Pi_\theta^n g(x_2)| \leq K_2[g]_p \rho_1^n |x_1 - x_2|(1 + |x_1|^p + |x_2|^p)$

(iii) $|\Pi_\theta^n g(x_1) - \Pi_{\theta'}^n g(x_1)| \leq K_3[g]_p |\theta - \theta'|(1 + |x_1|^{q_1})$

(iv) $|\Pi_\theta^n g(x_1) - \Pi_\theta^n g(x_2) - \Pi_{\theta'}^n g(x_1) + \Pi_{\theta'}^n g(x_2)|$
$\leq K_4 N_p(g') \rho_2^n |\theta - \theta'|(1 + |x_1|^{q_2} + |x_2|^{q_2})$

Then for all $f(\theta, x)$ of class $Li^1(Q, L_1, L_2, p_1, p_2)$, there exist functions $h(\theta), \nu_\theta(x)$, and constants C_i, s depending only on the L_j, p_j, such that $(I - \Pi_\theta)\nu_\theta = f_\theta - h(\theta)$, with for all $\theta, \theta' \in Q, x$,

(j) $|h(\theta) - h(\theta')| \leq C_1|\theta - \theta'|$

(jj) $|\nu_\theta(x)| \leq C_2|\theta - \theta'|$

(jjj) $|\nu_\theta(x) - \nu_{\theta'}(x)| \leq C_3|\theta - \theta'|(1 + |x|^s)$

(jv) $|\Pi_\theta \nu_\theta(x) - \Pi_{\theta'} \nu_{\theta'}(x)| \leq C_3|\theta - \theta'|(1 + |x|^s)$

Remark. The comments of Subsection 2.2.2 apply word for word to Theorem 6.

2.2 Markov Chains Dependent on a Parameter θ

2.2.5 Proof of Theorem 6

We may apply Theorem 5. Note in particular that for $g \in Li(p)$, Πg is continuous, and so the Markov chain with transition function Π_θ has a unique invariant probability m_θ. Moreover $h(\theta) = \Gamma_\theta f_\theta = \int f(\theta, x) m_\theta(dx)$ is Lipschitz and

$$\nu_\theta = \sum_{n \geq 0}(\Pi_\theta^n f_\theta - h(\theta)), \qquad \Pi_\theta \nu_\theta = \sum_{n \geq 1}(\Pi_\theta^n f_\theta - h(\theta))$$

We must show that $\nu_\theta(x)$ and $\Pi_\theta \nu_\theta(x)$ are Lipschitz in θ.

Note that $\nu_\theta(x) - \nu_{\theta'}(x) = u_1(\theta, \theta', x) + u_2(\theta, \theta', x)$ where

$$u_1(\theta, \theta', x) = \sum_{n \geq 0} \{\Pi_\theta^n f_\theta(x) - \Gamma_\theta f_\theta - \Pi_{\theta'}^n f_\theta(x) + \Gamma_{\theta'} f_\theta\} \quad (2.2.11)$$

$$u_2(\theta, \theta', x) = \sum_{n \geq 0} \{\Pi_{\theta'}^n (f_\theta - f_{\theta'})(x) - \Gamma_{\theta'}(f_\theta - f_{\theta'})\} \quad (2.2.12)$$

For $f \in Li^1(Q, L_1, L_2, p_1, p_2)$, we shall consider u_1 and u_2 separately. Let B_j, s_j denote constants which only depend on f through L_j, p_j.

1. Consider

$$\begin{aligned}
A_n &= \Pi_\theta^{n+1} f_\theta(x) - \Pi_\theta^n f_\theta(x) - \Pi_{\theta'}^{n+1} f_\theta(x) + \Pi_{\theta'}^n f_\theta(x) \\
&= \int \Pi_\theta(x, dy) \{\Pi_\theta^n f_{\theta'}(y) - \Pi_\theta^n f_\theta(x) - \Pi_{\theta'}^n f_\theta(y) + \Pi_{\theta'}^n f_\theta(x)\} \\
&\quad + \int \Pi_\theta(x, dy) \Pi_{\theta'}^n f_\theta(y) - \int \Pi_{\theta'}(x, dy) \Pi_{\theta'}^n f_\theta(y) \\
&= A_n^1 + A_n^2
\end{aligned}$$

Using (iv) then (i) we have

$$|A_n^1| \leq K_4 N_{p_1}(f_\theta') \rho_2^n |\theta - \theta'| \int \Pi(x, dy)(1 + |x|^{s_1} + |y|^{s_1})$$
$$\leq B_1 \rho_2^n |\theta - \theta'|(1 + |x|^{s_1})$$

Note that since $[f_\theta]_{p_1+1} \leq L_1$ (from (2.2.9)), we have, following (ii), for all $\theta' \in Q$

$$[\Pi_{\theta'} f_\theta]_{p_1+1} \leq K_2 \rho_1^n [f_\theta]_{p_1+1} \leq B_2 \rho_1^n$$

Then using (iii) we have:

$$|A_n^2| \leq K_3 [\Pi_{\theta'} f_\theta]_{p_1+1} |\theta - \theta'|(1 + |x|^{s_2}) \leq B_3 \rho_1^n |\theta - \theta'|(1 + |x|^{s_2})$$

Finally, we obtain for $\rho = \max(\rho_1, \rho_2)$,

$$|A_n| \leq B_4 \rho^n |\theta - \theta'|(1 + |x|^{s_3})$$

Thus we deduce that

$$|\Pi_\theta^{n+k} f_\theta(x) - \Pi_\theta^n f_\theta(x) - \Pi_{\theta'}^{n+k} f_\theta(x) + \Pi_{\theta'}^n f_\theta(x)|$$
$$\leq \frac{B_4}{1-\rho} \rho^n |\theta - \theta'|(1 + |x|^{s_3})$$

and, letting k tend to $+\infty$

$$|\Pi_\theta^n f_\theta(x) - \Gamma_\theta f_\theta(x) - \Pi_{\theta'}^n f_\theta(x) + \Gamma_{\theta'} f_\theta|$$
$$\leq \frac{B_4}{1-\rho} \rho^n |\theta - \theta'|(1 + |x|^{s_3})$$

Thus we have

$$|u_1(\theta, \theta', x)| \leq B_5 |\theta - \theta'|(1 + |x|^{s_3}) \qquad (2.2.13)$$

2. Next we evaluate $[f_\theta - f_{\theta'}]_{p_2}$ for $\theta, \theta' \in Q$

$$|f(\theta', x_1) - f(\theta, x_1) - f(\theta', x_2) + f(\theta, x_2)|$$
$$= |(x_1 - x_2) \int_0^1 \{f'(\theta', x_1 + t(x_2 - x_1)) - f'(\theta, x_1 + t(x_2 - x_1))\} dt|$$
$$\leq B_6 |x_2 - x_1| \cdot |\theta - \theta'|(1 + |x_1|^{p_2} + |x_2|^{p_2})$$

whence $[f_\theta - f_{\theta'}]_{p_2} \leq B_6 |\theta - \theta'|$. Following (ii),

$$|\Pi_{\theta'}^n (f_\theta - f_{\theta'})(x_1) - \Pi_{\theta'}^n (f_\theta - f_{\theta'})(x_2)|$$
$$\leq B_7 \rho_1^n |x_1 - x_2|(1 + |x_1|^{p_2+1} + |x_2|^{p_2+1})$$

and thus, using Lemma 1 and (i)

$$|\Pi_{\theta'}^n (f_\theta - f_{\theta'})(x) - \Gamma_{\theta'}(f_\theta - f_{\theta'})| \leq B_8 \rho_1^n |\theta - \theta'|(1 + |x|^{p_2+1})$$

We deduce immediately that

$$|u_2(\theta, \theta', x)| \leq B_9 |\theta - \theta'|(1 + |x|^{p_2+1}) \qquad (2.2.14)$$

Inequalities (2.2.13) and (2.2.14) imply (jjj).

2.2.6 Case in which the Transition Probability Π_θ is Independent of θ

In many algorithms (see the examples of Part I, which are also investigated below), the family of transition probabilities Π_θ is independent of θ, and the parameter θ only appears in $H(\theta, x)$. Then, Theorems 5 and 6 may be considerably simplified; in each case, it is sufficient that the transition probability Π satisfies conditions (i) and (ii). It is also much easier to obtain global results which allow us to apply Theorem 17 of Chapter 1 (see (1.9.4) and (1.9.5)). In what follows, we shall use the theorem given below.

2.2 Markov Chains Dependent on a Parameter θ

A function $f(\theta, x)$ which is differentiable with respect to x will be said to be **of class** $\overline{Li}(\mathbb{R}^d, L_1, L_2, p_1, p_2)$ if for all $\theta, \theta' \in \mathbb{R}^d$, all $x, x_1, x_2 \in \mathbb{R}^k$ it satisfies:

$$|f(\theta, x_1) - f(\theta, x_2)| \leq L_1(1 + |\theta|)|x_1 - x_2|(1 + |x_1|^{p_1} + |x_2|^{p_1}) \tag{2.2.15}$$

$$|f(\theta, 0) - f(\theta', 0)| \leq L_2|\theta - \theta'| \tag{2.2.16}$$

$$|f'(\theta, x) - f'(\theta', x)| \leq L_2|\theta - \theta'|(1 + |x|^{p_2}) \tag{2.2.17}$$

It will be said to be of class $\overline{Li}(\mathbb{R}^d)$ if it is of class $\overline{Li}(\mathbb{R}^d, L_1, L_2, p_1, p_2)$ for some values of L_j, p_j.

Theorem 7. *Let Π be a transition probability on $\mathbb{R}^k$. Suppose that for all $p \geq 0$, there exist constants $K_1, K_2, \rho < 1$ such that for all $g \in Li(p)$, $n \geq 0, x_1, x_2 \in \mathbb{R}^k$ we have:*

(i) $\quad \int \Pi^n(x_1, dx_2)(1 + |x_2|^p) \leq K_1(1 + |x_1|^p)$

(ii) $\quad |\Pi^n g(x_1) - \Pi^n g(x_2)| \leq K_2 \rho^n [g]_p |x_1 - x_2|(1 + |x_1|^p + |x_2|^p)$

Then for all $f(\theta, x)$ of class $\overline{Li}(\mathbb{R}^d, L_1, L_2, p_1, p_2)$, there exist functions $h(\theta)$, $\nu_\theta(x)$ and constants C_i depending only on the L_j, p_j such that $(I - \Pi)\nu_\theta = f_\theta - h(\theta)$, and satisfying for all $\theta, \theta' \in \mathbb{R}^d$

$$|h(\theta) - h(\theta')| \leq C_1|\theta - \theta'|$$
$$|\nu_\theta(x)| \leq C_2(1 + |\theta|)(1 + |x|^{p_1+1})$$
$$|\nu_\theta(x) - \nu_{\theta'}(x)| \leq C_3|\theta - \theta'|(1 + |x|^{p_2+1})$$
$$|\Pi\nu_\theta(x) - \Pi\nu_{\theta'}(x)| \leq C_3|\theta - \theta'|(1 + |x|^{p_2+1})$$

Proof. The upper bound on $\nu_\theta(x)$ comes from Lemma 1, that on $|\nu_\theta - \nu_{\theta'}|$ is obtained by considering $u_2(\theta, \theta', x)$ as in Subsection 2.2.5. $\square$

2.3 Linear Dynamical Processes

We consider processes (X_n) taking values in $\mathbb{R}^k$ whose dynamics are defined by

$$X_{n+1} = A(\theta)X_n + B(\theta)W_{n+1}$$

where:

- $A(\theta)$ is a $k \times k$ matrix,
- $B(\theta)$ is a $k \times k'$ matrix,
- $(W_n)_{n \geq 0}$ is a sequence of independent, identically distributed random variables taking values in $\mathbb{R}^k$.

We shall show that under suitable regularity assumptions, the family of transition probabilities (Π_θ) associated with such Markov chains satisfies the assumptions of Theorems 5 and 6. Thus the results of Chapter 1 (and as we shall see later) of Chapter 3 apply to these processes.

2.3.1 Assumptions and Notation

We suppose that for all $p \geq 0$, the (W_n) satisfy:
(W_n) is a sequence of independent identically distributed r.v. and

$$\|W_n\|_p \leq \mu_p < +\infty \tag{2.3.1}$$

Let Q denote a subset of $\mathbb{R}^k$, and suppose that there exist constants $\alpha_1, \alpha_2, \beta_1, \beta_2, M, \rho \in]0,1[$ such that for all $\theta, \theta' \in Q$, all $n \geq 0$

$$|A(\theta)| \leq \alpha_1, \qquad |A(\theta) - A(\theta')| \leq \alpha_2 |\theta - \theta'| \tag{2.3.2}$$
$$|A^n(\theta)| \leq M\rho^n \tag{2.3.3}$$
$$|B(\theta)| \leq \beta_1, \qquad |B(\theta) - B(\theta')| \leq \beta_2 |\theta - \theta'| \tag{2.3.4}$$

Set:

$$U_n(\theta) = \sum_{k=1}^n A^{n-k}(\theta) B(\theta) W_k \tag{2.3.5}$$

$$V_n(\theta) = \sum_{k=1}^n A^{k-1}(\theta) B(\theta) W_k \tag{2.3.6}$$

$$\Pi_\theta g(x) = E\{g(A(\theta)x + B(\theta)W_1)\} \tag{2.3.7}$$

Note that because of the symmetry of the distribution of $(W_1, \ldots, W_n)$, the r.v. $U_n(\theta)$ and $V_n(\theta)$ are identically distributed. It is now easy to show that:

$$\Pi_\theta^n g(x) = E\{g(A^n(\theta)x + U_n(\theta))\} = E\{g(A^n(\theta)x + V_n(\theta))\} \tag{2.3.8}$$

2.3.2 Preliminary Results

Lemma 8. *For all $p \geq 0$, there exist constants K_1, K_2 such that, for all $\theta, \theta' \in Q$, $n \geq 0$,*

(i) $\qquad \|U_n(\theta)\|_p \leq K_1$
(ii) $\qquad \|U_n(\theta) - U_n(\theta')\|_p \leq K_2 |\theta - \theta'|$

Moreover, for each θ, $V_n(\theta)$ converges a.s. and in L^p to $V_\infty(\theta)$.

2.3 Linear Dynamical Processes

Proof. We have

$$\begin{aligned}
\|U_n(\theta)\|_p &\leq \sum_{k=1}^n \|A^{n-k}(\theta)B(\theta)W_k\|_p \\
&\leq \sum_{k=1}^n |A^{n-k}(\theta)||B(\theta)|\|W_k\|_p \\
&\leq M\beta_1\mu_p \sum_{k=1}^n \rho^{n-k} \leq M\beta_1\mu_p(1-\rho)^{-1}
\end{aligned}$$

Next we note that if A_1 and A_2 are square matrices (not necessarily commuting) then

$$A_1^n - A_2^n = \sum_{k=1}^n A_1^{n-k}(A_1 - A_2)A_2^{k-1} \qquad (2.3.9)$$

Thus

$$\begin{aligned}
&U_n(\theta) - U_n(\theta') \\
&= \sum_{k=1}^n (A^{n-k}(\theta) - A^{n-k}(\theta'))B(\theta)W_k + \sum_{k=1}^n A^{n-k}(\theta)(B(\theta) - B(\theta'))W_k \\
&= Z'_n(\theta,\theta') + Z''_n(\theta,\theta')
\end{aligned}$$

Since

$$\begin{aligned}
|A^{n-k}(\theta) - A^{n-k}(\theta')| &= |\sum_{i=1}^{n-k} A^{n-k-i}(\theta)(A(\theta) - A(\theta'))A^{i-1}(\theta')| \\
&\leq |A(\theta) - A(\theta')| \sum_{i=1}^{n-k} |A^{n-k-i}(\theta)||A^{i-1}(\theta')| \\
&\leq \alpha_2|\theta - \theta'|M^2(n-k)\rho^{n-k-1}
\end{aligned}$$

it follows that

$$\begin{aligned}
\|Z'_n(\theta',\theta)\|_p &\leq \beta_1\mu_1\alpha_2|\theta - \theta'|M^2 \sum_{k=1}^n (n-k)\rho^{n-k-1} \\
&\leq K_3(p) \cdot |\theta - \theta'|
\end{aligned}$$

and that

$$\|Z''_n(\theta,\theta')\|_p \leq \beta_2|\theta - \theta'|\mu_p M \sum_{k=1}^n \rho^{n-k} \leq K_4(p)|\theta - \theta'|$$

Whence $\|U_n(\theta) - U_n(\theta')\|_p \leq K_2(p)|\theta - \theta'|$.

Now $V_n(\theta)$ is a sum of independent r.v. satisfying $\sup_n \|V_n(\theta)\|_p < +\infty$, thus $V_n(\theta)$ converges a.s. and in L^p to $V_\infty(\theta)$. □

As a result of this lemma, denoting the distribution of the r.v. $V_\infty(\theta)$ by m_θ, it follows from (2.3.8) and since $A^n(\theta)x \to 0$, that for any continuous bounded function $g(x)$, $\Pi_\theta^n g(x)$ tends to $\int g(x) dm_\theta(x) = \Gamma_\theta g$. It is easy to see that m_θ is the unique invariant probability of the Markov chain with transition probability Π_θ. This may also be deduced from Proposition 2, together with the properties of Π_θ^n to be established in the following subsection.

2.3.3 Properties of the Π_θ^n

Lemma 9. For all $p \geq 0$, there exist constants $C_1, C_2, C_3, C_4, \rho_1 \in]0,1[$, such that for all $\theta, \theta' \in Q$, all $x, x_1, x_2 \in \mathbb{R}^k$, all $n \geq 0$, all g:

(a) $\quad \int \Pi_\theta^n(x, dy)(1 + |y|^p) \leq C_1(1 + |x|^p)$

(b) $\quad |\Pi_\theta^n g(x_1) - \Pi_\theta^n g(x_2)| \leq C_2 \rho^n [g]_p |x_1 - x_2|(1 + |x_1|^p + |x_2|^p)$

(c) $\quad |\Pi_\theta^n g(x) - \Pi_{\theta'}^n g(x)| \leq C_3 [g]_p |\theta - \theta'|(1 + |x|^{p+1})$

(d) $\quad |\Pi_\theta^n g(x_1) - \Pi_\theta^n g(x_2) - \Pi_{\theta'}^n g(x_1) + \Pi_{\theta'}^n g(x_2)|$
$\quad \leq C_4 N_p(g') \rho_1^n |\theta - \theta'| |x_1 - x_2|(1 + |x_1|^{p+1} + |x_2|^{p+1})$

Proof. Inequality (a)

$$\int \Pi_\theta^n(x, dy)(1 + |y|^p) = E[1 + |A^m(\theta)x + U_n(\theta)|^p]$$
$$\leq C(p)[1 + M|x|^p + E|U_n(\theta)|^p] \leq C_1(p)(1 + |x|^p)$$

Inequality (b)

$$|\Pi_\theta^n g(x_1) - \Pi_\theta^n g(x_2)|$$
$$= |E[g(A^n(\theta)x_1 + U_n(\theta)) - g(A^n(\theta)x_2 + U_n(\theta))]|$$
$$\leq [g]_p |A^n(\theta)||x_1 - x_2| E[1 + |A^n(\theta)x_1 + U_n(\theta)|^p + |A^n(\theta)x_2 + U_n(\theta)|^p]$$
$$\leq M[g]_p \rho^n |x_1 - x_2| C(p)(1 + |x_1|^p + |x_2|^p)$$

Inequality (c)

$$|\Pi_\theta^n g(x) - \Pi_{\theta'}^n g(x)|$$
$$= |E[g(A^n(\theta)x + U_n(\theta)) - g(A^n(\theta')x + U_n(\theta'))]|$$
$$\leq [g]_p E\{[\|A^n(\theta) - A^n(\theta')\||x| + |U_n(\theta) - U_n(\theta')|]$$
$$\quad [1 + |A^n(\theta)x + U_n(\theta)|^p + |A^n(\theta')x + U_n(\theta')|^p]\}$$
$$\leq [g]_p [\|A^n(\theta) - A^n(\theta')\||x| + \|U_n(\theta) - U_n(\theta')\|_2 C(p)(1 + |x|^p)$$
$$\leq C(p)[g]_p [M^2 \alpha_2 |\theta - \theta'| n \rho^{n-1}|x| + \|U_n(\theta) - U_n(\theta')\|_2](1 + |x|^p)$$
$$\leq C_3(p)|\theta - \theta'|(1 + |x|^{p+1})$$

(Here we have used (2.3.9) to evaluate $|A^n(\theta) - A^n(\theta')|$.)

2.3 Linear Dynamical Processes

Inequality (d). We must evaluate

$$\begin{aligned}
a &= |E\{g(A^n(\theta)x_1 + U_n(\theta)) - g(A^n(\theta)x_2 + U_n(\theta)) \\
&\quad -g(A^n(\theta')x_1 + U_n(\theta')) - g(A^n(\theta')x_2 + U_n(\theta'))\}|
\end{aligned}$$

$$\begin{aligned}
&= |E\{A^n(\theta)(x_2 - x_1) \\
&\quad \int_0^1 g'[A^n(\theta)x_1 + U_n(\theta) + tA^n(\theta)(x_2 - x_1)]dt \\
&\quad -A^n(\theta')(x_2 - x_2) \\
&\quad \int_0^1 g'[A^n(\theta')x_1 + U_n(\theta') + tA^n(\theta')(x_2 - x_1)]dt\}|
\end{aligned}$$

$$\begin{aligned}
&\leq E\{|A^n(\theta) - A^n(\theta')||x_2 - x_1| \\
&\quad \int_0^1 |g'[A^n(\theta)x_1 + U_n(\theta) + tA^n(\theta)(x_2 - x_1)]|dt\} \\
&\quad +E\{|A^n(\theta')||x_2 - x_1| \\
&\quad \int_0^1 |g'[A^n(\theta)x_1 + U_n(\theta) + tA^n(\theta)(x_2 - x_1)] \\
&\quad -g'[A^n(\theta')x_1 + U_n(\theta') + tA^n(\theta')(x_2 - x_1)]|dt\} \\
&= a_1 + a_2
\end{aligned}$$

$$\begin{aligned}
a_1 &\leq M^2\alpha_2|\theta - \theta'|n\rho^{n-1}|x_2 - x_1|N_p(g') \\
&\quad \cdot E\{\int_0^1 [1 + |A^n(\theta)x_1 + U_n(\theta) + tA^n(\theta)(x_2 - x_1)|^{p+1}]dt\} \\
&\leq C(p)\rho_1^n|\theta - \theta'||x_2 - x_1|N_p(g')(1 + |x_1|^{p+1} + |x_2|^{p+1})
\end{aligned}$$

$$\begin{aligned}
a_2 &\leq M\rho^n|x_2 - x_1|N_p(g') \\
&\quad \int_0^1 E\{[|A^n(\theta) - A^n(\theta')||x_1| + |U_n(\theta) - U_n(\theta')| + t|A^n(\theta) - A^n(\theta')||x_2 - x_1|] \\
&\quad [1 + |A^n(\theta)x_1 + U_n(\theta) + tA^n(\theta)(x_2 - x_1)|^p \\
&\quad +|A^n(\theta')x_1 + U_n(\theta') + tA^n(\theta')(x_2 - x_1)|^p]\}dt \\
&\leq C(p)\rho^n|x_2 - x_1|N_p(g')|\theta - \theta'|(1 + |x_1|^{p+1} + |x_2|^{p+1})
\end{aligned}$$

This completes the proof of the lemma. □

2.3.4 Verification of Assumption (A.4)

Let us consider the algorithm

$$\begin{aligned}
\theta_{n+1} &= \theta_n + \gamma_{n+1}H(\theta_n, X_{n+1}) \\
X_{n+1} &= A(\theta_n)X_n + B(\theta_n)W_{n+1}
\end{aligned} \quad (2.3.10)$$

The properties of (Π_θ) established in Lemma 9 enable us to apply Theorems 5 and 6. We obtain,

Proposition 10. *Let D be an open subset of $\mathbb{R}^d$. Suppose that the sequence (W_n) satisfies (2.3.1) and that for any compact subset Q of D, the matrices $A(\theta)$ and $B(\theta)$ satisfy (2.3.2), (2.3.3) and (2.3.4); then if $H(\theta, x)$ or $\Pi_\theta H_\theta(x)$ is of class $Li(Q)$ —see Subsection 2.2.1— (or of class $Li^1(Q)$, respectively— see Subsection 2.2.4), for any compact subset Q of D, algorithm (2.3.10) satisfies Assumption (A.4) (see Chapter 1, Subsection 1.1.3) for all $\lambda < 1$ (for $\lambda = 1$, respectively).*

If now we consider the algorithm with dynamics independent of θ

$$\begin{aligned} \theta_{n+1} &= \theta_n + \gamma_{n+1} H(\theta_n, X_{n+1}) \\ X_{n+1} &= A \cdot X_n + B W_{n+1} \end{aligned} \quad (2.3.11)$$

we obtain the following algorithm by applying Theorem 7:

Proposition 11. *Suppose that the sequence (W_n) satisfies (2.3.1) and that $|A^n| \leq M\rho^n$ for $0 < \rho < 1$. Then if $H(\theta, x)$ or $\Pi_\theta H_\theta$ is of class $\overline{Li}(\mathbb{R}^d)$ (see Subsection 2.2.6), algorithm (2.3.11) satisfies assumptions (1.9.3), (1.9.4) and (1.9.5) of Chapter 1 for $\lambda = 1$.*

2.4 Examples

2.4.1 Markov Representation of a Received Signal

In several examples in Part I, we used

$$Y_n^T = (y_{n+N}, \ldots, y_{n-N}) \quad (2.4.1)$$

to denote the vector representing a signal received at times $n+N, \ldots, n-N$, where $N \geq 0$ is a fixed integer.

Suppose that the received signal y_n has a Markov representation (see Part I, Appendices 1 and 2)

(i) $\qquad U_n = AU_{n-1} + Ba_n$
(ii) $\qquad y_{n+N} = CU_{n-1} + \nu_n \qquad (2.4.2)$

In the above, A, B and C are matrices of suitable dimensions, and the eigenvalues of A have modulus strictly less than 1, and (a_n, ν_n) is a sequence of independent, identically distributed r.v., with a_n and ν_n mutually independent, taking values in $E \times \mathbb{R}$, for some finite subset E of $\mathbb{R}$. We suppose that $\text{var}(\nu_1) \neq 0$ and that for all $p \geq 0$, $E|\nu_1|^p < +\infty$. We set

(i) $\qquad \eta_n^T = (a_n, \ldots a_{n-N}) \qquad (2.4.3)$

2.4 Examples

and introduce the "state vector"

(ii) $$X_n = (U_n, Y_n, \eta_n)$$

Warning. There is a difference in notation from the Part I, so that the innovation indexed by n is a function of a_n and ν_n. X_n is a vector in $\mathbb{R}^M$ and we denote a generic point x of $\mathbb{R}^M$ by

(iii) $$x = (u, Y, \bar{a}_N, \ldots, \bar{a}_0)$$

It is easy to see that we may write

(i) $\quad X_{n+1} = D_1 X_n + D_2 W_{n+1}$ where
(ii) $\quad |D_1|^n \le M\rho^n, \quad 0 < \rho < 1$
(iii) $\quad W_n^T = (a_n, \nu_n)$ (2.4.4)

Then (2.4.4-ii) follows since D_1 has the same eigenvalues as A. The results of Section 2.3 imply that the Markov chain (X_n) has a unique invariant probability m and that

(i) $\quad \int g(x)dm(x) = E\{g(X_\infty)\}$

(ii) $\quad X_\infty = \sum_{k \ge 1} D^{k-1} D_2 W_k$

(iii) $\quad X_\infty = (U_\infty, Y_\infty, a_N^\infty m \ldots, a_0^\infty)$ (2.4.5)

Note finally that if we set

$$R_* = E\{Y_\infty \cdot Y_\infty^T\} \tag{2.4.6}$$

then the matrix R_* is positive definite. In fact, if we write W_n^T in the form $W_n^T = (a_n, 0) + (0, \nu_n)$, we see that $Y_\infty = Y_\infty^1 + Y_\infty^2$ where Y_∞^1 and Y_∞^2 are independent and $E\{Y_\infty^2 (Y_\infty^2)^T\} = \text{var}(\nu_1) \cdot I$.

2.4.2 Transversal Equaliser, Learning Phase

Here we consider the algorithm given in Part I, Chapter 1, Paragraph 1.2.1.2. Taking into account the change in the numbering of the time indices, this may be written as

$$\theta_{n+1} = \theta_n + \gamma_{n+1} Y_{n+1}(a_{n-N} - \theta_n^T \cdot Y_{n+1}) \tag{2.4.7}$$

where Y_n is given by (2.4.1) and satisfies the assumptions of Subsection 2.4.1. This is an algorithm with linear dynamics independent of θ, of type (2.3.11). With the notation of (2.4.3-iii) we have:

$$H(\theta, x) = Y \cdot (\bar{a}_0 - \theta^T \cdot Y)$$

Since $H(\theta, x)$ is of class $\overline{Li}(\mathbb{R}^d)$ (see Subsection 2.2.6), by Proposition 11, Theorem 7-b may be applied to it for $\lambda = 1$. Thus, taking (2.4.5) and (2.4.6) into account, we obtain

$$
\begin{align}
(i) & \quad h(\theta) = E\{Y_\infty \cdot a_o^\infty\} - R_* \theta \\
(ii) & \quad \theta_* = R_*^{-1} E\{Y_\infty \cdot a_0^\infty\} \\
(iii) & \quad U(\theta) = (\theta - \theta_*)^T \cdot R_* \cdot (\theta - \theta_*)
\end{align}
\tag{2.4.8}
$$

In conclusion θ_n converges to θ_* if $\sum \gamma_n^2 < +\infty$.

2.4.3 Least Squares Algorithm

Here we consider the algorithm given in Part I, Chapter 1, Paragraph 1.2.1.2. which may be rewritten using our new system for numbering the time as

$$
\begin{align}
(i) & \quad \theta_{n+1} = \theta_n + \gamma_{n+1} R_{n+1}^{-1} Y_{n+1} (a_{n-N} - Y_{n+1}^T \cdot \theta_n) \\
(ii) & \quad R_{n+1} = R_n + \gamma_{n+1} (Y_{n+1} Y_{n+1}^T - R_n)
\end{align}
\tag{2.4.9}
$$

where Y_n satisfies the assumptions of Subsection 2.4.1 and $\sum \gamma_n^2 < +\infty$.

If we choose R_0 to be symmetric, we may assume that R_n is also a symmetric matrix. Then algorithm (2.4.9) takes its values in $\mathbb{R}^{2N+1} \times \mathbb{R}^{(2N+1)(2N+2)/2}$, where the set M_s of symmetric matrices of order $(2N+1)$ is identified with $\mathbb{R}^{(2N+1)(2N+2)/2}$. We shall denote the subset of M_s of positive definite matrices by M_s^+.

Note firstly that R_n converges a.s. to R_* as defined in (2.4.6). In fact, Theorem 17-b of Chapter 1 may be applied with $h(R) = -(R - R_*)$ and $U(R) = \|R - R_*\|^2$.

Next note that since the moments of Y_n of all orders are bounded in n,

$$\gamma_n^{1/2} Y_n \cdot Y_n^T \to \quad \text{a.s.} \tag{2.4.10}$$

since $E\{\sum (\gamma_n^{1/2} |Y_n \cdot Y_n^T|)^4\} \leq K_1 \sum \gamma_n^2$.

Since $R_* \in M_s^+$, we can construct a regular function u from M_s to M_s^+ such that

$$
\begin{align}
(i) & \quad \forall R, \ |u(R)| \leq K_2 \leq +\infty \\
(ii) & \quad \{|R - R_*| \leq \beta\} \Rightarrow \{u(R) = R^{-1}\}
\end{align}
\tag{2.4.11}
$$

Similarly, we can construct a regular map ν from M_s to M_s satisfying:

$$
\begin{align}
(i) & \quad \forall R, \ |\nu(R)| \leq K_3 < +\infty \\
(ii) & \quad \{|R| \leq 1, \alpha \in [-\frac{1}{2}, \frac{1}{2}]\} \Rightarrow \{(I + \alpha R)^{-1} = I - \alpha R + \alpha^2 \nu(R)\}
\end{align}
\tag{2.4.12}
$$

2.4 Examples

We introduce the algorithm

(i) $\theta_{n+1} = \theta_n + \gamma_{n+1} u(R_n) S_{n+1}(a_{n-N} - Y_{n+1}^T \cdot \theta_n)$
(ii) $R_{n+1} = R_n + \gamma_{n+1}(Y_{n+1} Y_{n+1}^T - R_n)$
(iii) $S_{n+1} = I + \gamma_{n+1}(I - Y_{n+1} Y_{n+1}^T u(R_n)) + \gamma_{n+1} \nu [\gamma_{n+1}^{1/2}(Y_{n+1} Y_{n+1}^T u(R_n) - I)]$

$$(2.4.13)$$

Following (2.4.12-ii) for $\alpha = \gamma_{n+1}^{1/2}$ and $R = \gamma_{n+1}^{1/2}(Y_{n+1} Y_{n+1}^T u(R_n) - I)$, we have
$S_{n+1} = \{I + \gamma_{n+1}(Y_{n+1} Y_{n+1}^T u(R_n) - I)\}^{-1}$ when:

$$\gamma_{n+1}^{1/2} \leq \frac{1}{2} \text{ and } \gamma_{n+1}^T |Y_{n+1} Y_{n+1}^T| \leq \frac{1}{2K_2} \qquad (2.4.14)$$

Then, following (2.4.11-ii), if

$$|R_n - R_*| \leq \beta$$
$$u(R_n) S_{n+1} = R_n^{-1}(I + \gamma_{n+1}(Y_{n+1} Y_{n+1}^T R_n^{-1} - I)) = R_{n+1}^{-1} \qquad (2.4.15)$$

Then, in view of the convergence of R_n to R_*, and of (2.4.10), there exists a time n_0 (random) after which algorithms (2.4.9) and (2.4.13) coincide. Thus it is sufficient to show that (2.4.13) converges a.s. for any initial condition. But algorithm (2.4.13) takes the form (1.1.1) of Chapter 1 if we set

$$\Theta_n = (\theta_n, R_n) \qquad (2.4.16)$$

Using Theorem 7 (or a variant of Proposition 11 including a term in $\gamma_{n+1}^2 \rho_n(\theta_n, X_{n+1})$), we see immediately that assumptions (1.9.1) to (1.9.5) of Theorem 17 of Chapter 1 are satisfied by the function

(i) $h(\Theta) = (h_1(\Theta), h_2(\Theta))$
(ii) $h_1(\Theta) = -u(R)(R_* \theta - \zeta)$
(iii) $h_2(\Theta) = -(R - R_*)$
(iv) $\zeta = E\{Y_\infty \cdot a_0^\infty\}$ $\qquad (2.4.17)$

(see (2.4.5)). To enable us to apply Theorem 17-b of Chapter 1, it is sufficient to set

$$U(\Theta) = (\theta - \theta_*)^T R_*(\theta - \theta_*) + (R - R_*)^T(R - R_*) \qquad (2.4.18)$$

where $\theta_* = R_*^{-1} \cdot \zeta$. In fact, if U_1' denotes the gradient with respect to θ and U_2' denotes the gradient with respect to R, we have

$$\begin{aligned} U'(\Theta) \cdot h(\Theta) &= U_1'(\Theta) \cdot h_1(\Theta) + U_2'(\Theta) \cdot h_2(\Theta) \\ &= -2(\theta - \theta_*)^T R_*^T u(R) R_*(\theta - \theta_*) - 2(R - R_*)^T(R - R_*) \end{aligned}$$

Then, since $R_*^T \cdot u(R) \cdot R_* \in M_s^+$ for all R, $U'(\Theta) \cdot h(\Theta) < 0$ except if $\theta = \theta_*$ and $R = R_*$. We have proved that θ_n as defined by (2.4.9) converges a.s. to $\theta_* = R_*^{-1} \zeta$.

2.4.4 Decision-Feedback Loop

Let b'_n, b''_n be independent r.v. with values ± 1, and set

$$a_n = b'_n + ib''_n \tag{2.4.19}$$

Suppose that we have a complex received signal y_n with a complex Markov representation

(i) $\quad U_n = AU_{n-1} + Ba_n, \quad |A^n| \leq M_\rho^n, \quad 0 < \rho < 1$
(ii) $\quad y_n = CU_{n-1} + \nu_n$ $\hfill(2.4.20)$

where ν_n is a complex noise independent of b'_n and b''_n. This is the complex variant of Subsection 2.4.1 with $N = 0$. Suppose further that ν_n has a rapidly decreasing probability density function. If we set

$$X_n = (U_n, y_n) \tag{2.4.21}$$

then X_n is a Markov chain with values in $\mathbb{C}^k \times \mathbb{C}$. We shall denote the transition function of X_n by Π and $x = f(u, y)$ will denote a generic point of $\mathbb{C}^k \times \mathbb{C}$.

Consider a real algorithm of the form

$$\theta_{n+1} = \theta_n + \gamma_{n+1} H(\theta_n, X_{n+1}) \tag{2.4.22}$$

From Proposition 10, we see that the Assumption (A.4) of Chapter 1 is satisfied if ΠH_θ is of class $Li(Q)$ (see Subsection 2.2.1) for all compact subsets Q of $\mathbb{R}$. The decision-feedback loop is that of (Paragraph 1.2.2.2, Chapter 1, Part I) where

(i) $\quad H(\theta, x) = H(\theta, y) = \mathrm{Im}\{ye^{-i\theta}\bar{\hat{a}}(\theta, y)\}$
(ii) $\quad \hat{a}(\theta, y) = b_1(\theta, y) + ib_2(\theta, y)$
(iii) $\quad b_1(\theta, y) = \mathrm{sgn}[\mathrm{Re}\,(ye^{-i\theta})]$
(iv) $\quad b_2(\theta, y) = \mathrm{sgn}[\mathrm{Im}(ye^{-i\theta})]$ $\hfill(2.4.23)$

Then we have

$$\Pi H_\theta(x) = E\{H(\theta, Cu_0 + \nu_1)\} \tag{2.4.24}$$

The problem is to find an upper bound for $\Pi H_\theta(x) - \Pi H_{\theta'}(x')$. Setting

$$Z = (Cu + \nu_1)e^{-i\theta} \qquad Z' = (Cu' + \nu_1)e^{-i\theta}$$
$$\Pi H_\theta(x) = E\{Z \cdot [\mathrm{sgn}(\mathrm{Re}\,Z) + i\,\mathrm{sgn}(\mathrm{Im}\,Z)]\} \tag{2.4.25}$$

it is sufficient to find an upper bound for

$$\Delta = |E\{Z\,\mathrm{sgn}(\mathrm{Re}\,Z) - Z'\,\mathrm{sgn}(\mathrm{Re}\,Z')\}|$$

The term in $\mathrm{Im}\,Z$ may be treated similarly. We have

$$\Delta \leq E|Z - Z'| + E\{|Z||\mathrm{sgn}(\mathrm{Re}\,Z) - \mathrm{sgn}(\mathrm{Re}\,Z')|\} = \Delta_1 + \Delta_2$$

2.4 Examples

Using (2.4.25), we see immediately that
$$\Delta_1 \leq |C|\{|u||\theta - \theta'| + |u - u'|\} \tag{2.4.26}$$
Moreover, if we set $e_1 = (1,0) \in \mathbb{C}$, then
$$\begin{aligned}\Delta_2 &\leq E\{|Z| \cdot I(\operatorname{Re}|Z| \leq |\operatorname{Re}(Z - Z')|)\} \\ &\leq E\{|Z| \cdot I(|e_1 Z| \leq |e_1(Z - Z')|)\}\end{aligned}$$
and again from (2.4.25)
$$\begin{aligned}\Delta_2 \leq\ &|C|(1+|u|)E\{(1+|\nu_1|) \\ &I(|e^{-i\theta}e_1\nu_1 + e^{-i\theta}e_1 Cu| \leq |e^{-i\theta} - e^{-i\theta'}||\nu_1| + |C|(|ue^{-i\theta} - u'e^{-i\theta'}|))\}\end{aligned}$$
Noting that $|e^{-i\theta} \cdot e_1| = 1$, we may apply Lemma 12 of Subsection 2.4.5 (below). Thus, for some constant K_1, which depends only on the density of ν_1, we obtain
$$\begin{aligned}\Delta_2 &\leq K_1|C|(1+|u|)\{|C||ue^{-i\theta} - u'e^{-i\theta'}| + \\ &\quad |e^{-i\theta} - e^{-i\theta'}| + |ue^{-i\theta}| \cdot |e^{-i\theta} - e^{-i\theta'}|\} \\ &\leq K_2(1+|u|^2)(|\theta - \theta'| + |u - u'|)\end{aligned}$$
It is now easy to see that $\Pi H_\theta(x)$ is of class $Li(Q)$ for any compact subset Q of $\mathbb{R}$. Thus, the results of Chapter 1 apply to algorithm (2.4.22), (2.4.23).

2.4.5 A Technical Lemma

Lemma 12. *Let $p \geq 0$ and suppose that V is a r.v. with values in $\mathbb{R}^m$ with density function ϕ for which, for all $q \geq 0$, $\phi(t)(1 + |t|^q) \leq \mu_q < +\infty$. Then there exists a constant K, such that for all $\varepsilon > 0$, all $a > 0$, all $b > 0$ with $b < \varepsilon/2$, all $c \in \mathbb{R}$, all $\alpha \in \mathbb{R}^m$ with $|\alpha| \geq \varepsilon$ we have:*
$$\overline{E} = E\{(1+|V|^p)I(|\alpha^T \cdot V + c| \leq a + b|V|)\} \leq \frac{K}{\varepsilon}(a + b + \frac{b}{\varepsilon}|c|)$$

Proof. We can find an orthogonal matrix O such that if we set $U = O \cdot V$, then $U_1 = \frac{\alpha}{|\alpha|} \cdot V$. Then we set $U = (U_1, \bar{U})$. The density function ψ of U clearly has the same properties as ϕ, whence for all q,
$$\psi(u_1, \bar{u}) \leq K_1(q)(1+|u_1|^q)^{-1}(1+|\bar{u}|^q)^{-1} \tag{2.4.27}$$
$$\begin{aligned}\overline{E} &\leq E\{(1+|U|^p)I(|U_1 + \frac{c}{|\alpha|}| \leq \frac{a}{\varepsilon} + \frac{b}{\varepsilon}|U_1| + \frac{b}{\varepsilon}|\overline{U}|)\} \\ &\leq E\{(1+|U|^p)I(|U_1 + \frac{c}{|\alpha|}| \leq \frac{a}{\varepsilon} + \frac{b}{\varepsilon}|U_1 + \frac{c}{|\alpha|}| + \frac{b}{\varepsilon^2}|c| + \frac{b}{\varepsilon}|\overline{U}|) \\ &\leq E\{(1+|U|^p)I(|U_1 + \frac{c}{|\alpha|}| \leq \frac{2b}{\varepsilon}|U_1 + \frac{c}{|\alpha|}|)\} \\ &\quad + E\{(1+|U|^p)I(|U_1 + \frac{c}{|\alpha|}| \leq \frac{2a}{\varepsilon} + \frac{2b}{\varepsilon^2}|c| + \frac{2b}{\varepsilon}|\overline{U}|\} \\ &= \overline{E}_1 + \overline{E}_2\end{aligned}$$

In view of the assumption $b < \varepsilon/2$, $\overline{E}_1 = 0$, and using (2.4.27), for $q = p+m+1$ (which guarantees the convergence of the integrals below) we have:

$$\begin{aligned}
\overline{E}_2 &\leq K \int d\bar{u} \frac{1+|\bar{u}|^p}{1+|\bar{u}|^q} \int du_1 \frac{1+|u_1|^p}{1+|u_1|^q} I\{|u_1 + \frac{c}{|\alpha|}| \leq 2(\frac{a}{\varepsilon} + \frac{b}{\varepsilon^2}|c| + \frac{b}{\varepsilon}|\overline{U}|)\} \\
&\leq K \int d\bar{u} \frac{1+|\bar{u}|^p}{1+|\bar{u}|^q} 4(\frac{a}{\varepsilon} + \frac{b}{\varepsilon^2}|c| + \frac{b}{\varepsilon}|\overline{U}|) \\
&\leq K(\frac{a}{\varepsilon} + \frac{b}{\varepsilon^2}|c| + \frac{b}{\varepsilon})
\end{aligned}$$

□

2.5 Decision-Feedback Algorithms with Quantisation

2.5.1 The Algorithm. Assumptions

We consider the self-adaptive equaliser algorithm given in Part I, Chapter 1, Paragraph 1.2.1.2.

We reintroduce the notation. N and p are two fixed integers, and the vector θ is written in the form

$$\theta^T = (\alpha^T, \beta^T), \qquad \alpha \in \mathbb{R}^{2N+1}, \qquad \beta \in \mathbb{R}^p \qquad (2.5.1)$$

Then we have a process (U_n, Y_n) which has the same dynamics as in Section 2.4

(i) $\qquad U_n = AU_{n-1} + Ba_n$
(ii) $\qquad y_{n+N} = CU_{n-1} + \nu_n$
(iii) $\qquad Y_n^T = (y_{n+N}, \ldots, y_{n-N})$

where the matrix A satisfies:

(iv) $\qquad |A^n| \leq K \cdot \rho^n, \qquad 0 < \rho < 1 \qquad (2.5.2)$

The sequence $(a_n, \nu_n)_{\nu \geq 1}$ is a sequence of (setwise) independent, identically distributed r.v. with values in $A \times \mathbb{R}$ for some **finite** subset A of $\mathbb{R}$, over which the a_n are uniformly distributed. Further, for the "quantiser" Q (i.e. a mapping from $\mathbb{R}$ into A), we define a sequence of r.v. $\hat{a}_n, \eta_n, \theta_n$ by:

(i) $\qquad \eta_n^T = (\hat{a}_n, \hat{a}_{n-1}, \ldots, \hat{a}_{n-p+1})$
(ii) $\qquad \hat{a}_n = Q(\alpha_{n-1}^T Y_n + \beta_{n-1}^T \eta_{n-1})$
(iii) $\qquad \theta_n = \theta_{n-1} + \gamma_n H(\theta_{n-1}, Z_n), \quad \theta_n^T = (\alpha_n^T, \beta_n^T)$

where

(iv) $\qquad Z_n^T = (U_{n+N}^T, Y_n^T, \eta_n^T) \qquad (2.5.3)$

2.5 Decision-Feedback Algorithms with Quantisation

Q is assumed to satisfy

(i) $\quad \exists s_1, \ldots, s_q \in \mathbb{R}$ such that $\{|s - s_i| > |s - s'| \forall i\} \Rightarrow Q(s) = Q(s')$
(ii) $\quad \exists \bar{a} \in A, \bar{s} \in \mathbb{R}$ such that $s > \bar{s} \Rightarrow Q(s) = \bar{a}$ $\hfill (2.5.4)$

Remark 1. Thus we have

$$\{(s, s'); Q(s) \neq Q(s')\} \subset \cup_{i=1}^{q} \{|s - s_i| \leq |s - s'|\}$$

Remark 2. In the examples in Part I, we used $A = \{+1, -1\}$ or $A = \{\pm 1, \pm 3\}$, with $Q(s)$ taking the value in A closest to s. Then we have (2.5.4-i.ii) with respectively $s_1 = 0, \bar{a} = 1, \bar{s} = 0$ and $s_1 = -2, s_2 = 0, s_3 = 2$, $\bar{a} = 2, \bar{s} = 1$.

We denote $E = A^p$, $z = (u, Y, e)$, and $x = (u, Y)$ and suppose that for any compact subset K of $\mathbb{R}^{2N+1+p}$:

$$H(\theta, x, e) \text{ is of class } Li(K) \hfill (2.5.5)$$

(see Subsection 2.2.1). Furthermore, we suppose that the r.v. (ν_n) have density ϕ satisfying

$$\phi > 0 \text{ and } \forall m \geq 0: \sup_{t}(1 + |t|^m)\phi(t) = K(m) < +\infty \hfill (2.5.6)$$

2.5.2 Associated Markov Chain

For all θ, we may associate this algorithm with a Markov chain $Z_n(\theta) = (U_{n+N}, Y_n, \eta_n(\theta))$ defined by (2.5.1), (2.5.2) and

(i) $\quad \eta_n(\theta) = (\hat{a}_n(\theta), \ldots, \hat{a}_{n-p+1}(\theta))$
(ii) $\quad \hat{a}_n(\theta) = Q(\alpha^t \cdot Y_n + \beta^t \cdot \eta_{n-1}(\theta))$ $\hfill (2.5.7)$

We set $X_n = (U_{n+N}, Y_n)$ and let $\Pi_\theta(x, e; dx', de')$ denote the transition probability associated with this chain.

2.5.3 Results

We see immediately that algorithm (2.5.3–iii) satisfies Assumption (A.2) of Chapter 1 with the above transition probability Π_θ. Similarly, from (2.3.8) and (2.5.2-iv), we have

$$E_{x,e,a}(1 + |Z_n|^q) \leq \mu_q(1 + |x|^q)$$

where $P_{x,e,a}$ is the distribution of the algorithm with $Z_0 = (x, e)$ and $\theta_0 = a$. Thus Assumption (A.5) of Chapter 1, Subsection 1.2.1 is satisfied. In view of

(2.5.5), Assumption (A.3) is clearly satisfied. It remains to establish (A.4). We have

Theorem 13. *Under the assumptions (2.5.4), (2.5.5) and (2.5.6), the algorithm (2.5.2), (2.5.3) satisfies (A.4-i,ii) (Chapter 1, Subsection 1.1.3) relative to the open subset $D = \{\theta = (\alpha, \beta); |\alpha| > 0\}$ of $\mathbb{R}^{2N+1+p}$. If, further we suppose that for any integer r*

$$\liminf_n \frac{\gamma_{n+r}}{\gamma_n} > 0$$

then it satisfies (A.4-iii)' (Chapter 1, Subsection 1.7.1) for all $\mu < 1$, again relative to D. Thus the conclusions of Theorem 14 of Chapter 1 may be applied to it.

The remainder of this paragraph is given over to a proof of this theorem. First we shall establish the existence of a solution to the Poisson equation, using Proposition 2 and Theorem 5. Then we shall prove that this solution satisfies (A.4-i,ii) and (A.4-iii)'.

Remark. Actually, conclusions similar to those of Theorem 14, Chapter 1 would still be true for this algorithm without the assumption that

$$\liminf_n \gamma_{n+r}\gamma_n > 0$$

See the remark at the end of Section 2.5.

2.5.4 Notation and Preliminary Calculations

Formulae (2.5.2) and (2.5.7) which define the Markov chain with transition probability Π_θ, allow us to obtain Z_n as a function of θ and z_0 and of the a_p, ν_p for the initial condition $z_0 = (u_0, Y_0, \eta_0)$. More precisely we have

$$Z_n(\theta, z_0) = [U_n(u_0), Y_n(u_0, Y_0), \eta_n(\theta, z_0)] \tag{2.5.8}$$

If we set

$$U_n(0) = \sum_{k=1}^{n} A^{n-k} B a_k \tag{2.5.9}$$

we have

$$U_n(u_0) = A^n u_0 + U_n(0) \tag{2.5.10}$$

Then it is easy to see that if $n \geq 2N+1$, there exist matrices D_1, D_2 depending only on N such that

(i) $Y_n(u_0, Y_0) = Y_n(u_0) = D_1 A^{n-2N-1} u_0 + D_2 \overline{U}_n + V_n$
(ii) $\overline{U}_n = (U_{n-1}(0), \ldots, U_{n-2N-1}(0))$
(iii) $V_n = (\nu_n, \ldots, \nu_{n-2N})$ (2.5.11)

2.5 Decision-Feedback Algorithms with Quantisation

and if $n \leq 2N$, then for suitable matrices $D_1(n), D_2(n), D_3(n)$, we have

(i) $\qquad Y_n(u_0, Y_0) = D_1(n)u_0 + D_2(n)\overline{U}_n + V_n + D_3(n)Y_0$
(ii) $\qquad \overline{U}_n = (U_{n-1}(0), \ldots, U_1(0))$
(iii) $\qquad V_n = (\nu_n, \ldots, \nu_1)$ $\hfill (2.5.12)$

Then we have

(i) $\qquad \hat{a}_n(\theta, z_0) = Q(c_n(\theta, z_0))$
(ii) $\qquad c_n(\theta, z_0) = \alpha^T Y_n(u_0, Y_0) + \beta^T \eta_{n-1}(\theta, z_0)$ $\hfill (2.5.13)$

We shall denote the distribution of $(Z_n(\theta, z), Z_n(\theta', z'))$ on $(\mathbb{R}^{m_1} \times E)^2$ by $P_{z,z'}^{\theta;\theta'}$. By abuse of notation, we shall denote the canonical process on $\{(\mathbb{R}^{m_1} \times E)^2\}^N$ by $(Z_n, Z'_n) = (U_n, Y_n, \eta_n, U'_n, Y'_n, \eta'_n)$. We also recall the previous notation $(u, Y) = x$ and $X_n = (U_n, Y_n)$.

Then for any function g on $(\mathbb{R}^{m_1} \times E)^2$ we have:

$$E_{z,z'}^{\theta;\theta'}\{g(Z_n, Z'_n)\} = E\{g(Z_n(\theta, z), Z_n(\theta', z'))\}$$

and for any function g on $\mathbb{R}^{m_1} \times E$ we have

$$\Pi_\theta^n g(x, e) = E_{x,e,x',e'}^{\theta;\theta'}(g(Z_n)) = E\{g(Z_n(\theta, z))\} \qquad (2.5.14)$$

In particular

$$\Pi_\theta^n g(x, e) - \Pi_\theta^n g(x', e') = E_{x,e,x',e'}^{\theta;\theta}\{g(Z_n) - g(Z'_n)\} \qquad (2.5.15)$$

Finally, for given $\varepsilon > 0, \delta > 0, M > 0$, we set

$$K(\varepsilon, M) = \{\theta = (\alpha, \beta) \in \mathbb{R}^{2N+1+p} \ 0 < \varepsilon \leq |\alpha| \leq M, |\beta| \leq M\} \qquad (2.5.16)$$

$$\overline{K}(\varepsilon, M, \delta) = \{(\theta, \theta'); \theta, \theta' \in K(\varepsilon, M), |\theta - \theta'| \leq \delta\} \qquad (2.5.17)$$

2.5.5 A Fundamental Lemma

Lemma 14. *Given ε, M and $\bar{u} < 0$, there exist $\delta_1 > 0$ and $\rho_1 > 0$, such that for any pair $z = (u_0, Y_0, \eta_0), z' = (u'_0, Y'_0, \eta'_0)$ satisfying $|u_0| \leq \bar{u}, |u'_0| \leq \bar{u}$, for all $n \geq 2N + 1 + p$ and for all $(\theta, \theta') \in \overline{K}(\varepsilon, M, \delta_1)$*

$$P_{z,z'}^{\theta;\theta'}\{\eta_n = \eta'_n\} \geq \rho_1 > 0$$

Proof. Following (2.5-11-i), (2.5.13-ii) and (2.5.4-ii) and the fact that $\overline{U}_n$ is bounded, there exists $L > 0$ such that for all $\theta \in K(\varepsilon, M)$, and all z such that $|u_0| \leq \bar{u}$, $\alpha_T \cdot V_n > L$ implies $\hat{a}_n(\theta, z) = \bar{a}$. Thus we have

$$P_{z,z'}^{\theta;\theta'}\{\eta = \eta'\}$$
$$= P\{\eta_n(\theta, z) = \eta_n(\theta', z')\}$$
$$\geq P\{\alpha^T \cdot V_n > L, \alpha'^T \cdot V_n > L, \ldots, \alpha^T \cdot V_{n-p+1} > L, \alpha'^T \cdot V_{n-p+1} > L\}$$
$$= P\{W_n \in G(\theta, \theta', L)\}$$

where $W_n = (\nu_n, \ldots, \nu_{n-2N-p+1})$ and $G(\theta, \theta', L)$ is the open subset of $\mathbb{R}^{2N+p+1}$ given by $\{w; \alpha^{(1)} \cdot w > L, \alpha'^{(1)} \cdot w > L, \ldots, \alpha^{(p)} \cdot w > L, \alpha'^{(p)} \cdot w > L\}$ with

$$\alpha^{(1)} = (\alpha_1, \ldots, \alpha_{2N+1}, 0, \ldots 0)$$
$$\alpha^{(2)} = (0, \alpha_1, \ldots, \alpha_{2N+1}, 0, \ldots 0)$$
$$\alpha^{(p)} = (0, \ldots, 0, \alpha_1, \ldots, \alpha_{2N+1})$$

Since the vectors $\alpha^{(i)}, i = 1, \ldots, p$ are linearly independent, the system $\alpha^{(i)} \cdot w = 1$, $i = 1, \ldots, p$ has at least one solution $\bar{w} = \bar{w}(\alpha)$. We may choose $\bar{w}(\alpha)$ so that $\alpha \to \bar{w}(\alpha)$ is continuous, and then

$$\sup_{\theta \in K(\varepsilon, M)} |\bar{w}(\alpha)| \leq M_0$$

Let $\delta_1 = \frac{1}{2M_0}$. Then if θ' satisfies $|\theta - \theta'| \leq \delta_1$, we have

$$\alpha'^{(i)} \cdot \bar{w} = \alpha^{(i)} \bar{w} + (\alpha'^{(i)} - \alpha^{(i)})\bar{w} \geq 1 - \frac{1}{2} = \frac{1}{2}$$

Whence we deduce (by multiplying $\bar{w}$ by a scalar > 0) that, for $(\theta, \theta') \in \overline{K}(\varepsilon, M, \delta)$, $G(\theta, \theta', L)$ is non-empty. Since the random vector W_n has a strictly positive density function independent of n, $P(W_n \in G(\theta, \theta', L))$ is strictly positive and $(\theta, \theta') \to P(W_n \in G(\theta, \theta', L))$ is continuous and so bounded below by $\rho_1 > 0$ on the compact set $\overline{K}(\varepsilon, M, \bar{u})$.

2.5.6 Verification of Assumptions (X.3) and (X.4) of Proposition 4

Lemma 15. Let $A(n) = \{\eta_k \neq \eta'_k; k = 1, \ldots, n\}$. Given ε, M, there exist $\delta_2, C_2, \eta_2 \in]0, 1[$ such that, for all z, z', all $(\theta, \theta') \in K(\varepsilon, M, \delta_2)$ and all n

$$P_{z,z'}^{\theta;\theta'}(A(n)) \leq C_2 \rho_2^n (1 + |u_0| + |u'_0|)$$

Proof. Set $r = 2N + 1 + p$ and let $B(n) = A(nr)$. Since the r.v. $U_n(0)$ defined by (2.5.9) are bounded by R, fix $\bar{u} > 2R$. then

$$p_n = P_{z,z'}^{\theta;\theta'}(B(n))$$
$$= E_{z,z'}^{\theta;\theta'}\{I(B(n-1))I[|U_{(n-1)r}| \leq \bar{u}, |U'_{(n-1)r}| \leq \bar{u}]$$
$$\cdot P_{Z_{(n-1)r}, Z'_{(n-1)r}}^{\theta;\theta'}(B(1))\}$$
$$+ P_{z,z'}^{\theta;\theta'}\{(|U_{(n-1)r}| > \bar{u}) \cup (|U'_{(n-1)r}| > \bar{u})\}$$

2.5 Decision-Feedback Algorithms with Quantisation

Following Lemma 14, if $|\theta - \theta'| \leq \delta_2 = \delta_1(\bar{u})$, then for some $\rho_1 = \rho_1(\bar{u}) > 0$ we have

$$p_n \leq (1-\rho_1) \cdot p_{n-1} + P\{|U_{(n-1)r}(u_0)| > \bar{u}\} + P\{|U_{(n-1)r}(u_0')| > \bar{u}\}$$

Following (2.5.10) and Assumption (2.5.2-iv),

$$\{|U_n(u_0)| > \bar{u}\} \subset \{C\rho^n|u_0| + R > \bar{u}\}$$

Thus, since $\bar{u} > 2R$, $P(|U_n(u_0)| > \bar{u}) > 0$ implies that $C\rho^n|u_0| > \frac{\bar{u}}{2}$ and thus that

$$1 \leq \frac{2C\rho^n|u_0|}{\bar{u}} = \overline{C}\rho^n|u_0|$$

Whence

$$p_n \leq (1-\rho_1)p_{n-1} + \overline{C}\rho^n(|u_0| + |u_0'|)$$

It is easy to see that with $\rho_3 < 1$

$$p_n = P(B(n)) \leq C_3\rho_3^n(1 + |u_0| + |u_0'|)$$

Then

$$P(A(n)) \leq C_2\rho_3^{n/r}(1 + |u_0| + |u_0'|)$$

and the lemma follows. □

Lemma 16. *Given ε, M, there exists a constant C_3 such that, for all $\theta \in K(\varepsilon, M)$, all z, z', all $n \geq 2N + 1$,*

$$P_{z,z'}^{\theta;\theta'}\{\eta_{n-1} = \eta_{n-1}', \eta_n \neq \eta_n'\} \leq C_3\rho^n|u_0 - u_0'|$$

where $\rho < 1$ is given by (2.5.2-iv).

Proof. Following (2.5.13-i,ii) and assumption (2.5.4), we have

$$\begin{aligned}
p &= P_{z,z'}^{\theta;\theta'}\{\eta_{n-1} = \eta_{n-1}', \eta_n \neq \eta_n'\} \\
&= P\{\eta_{n-1}(\theta, z) = \eta_{n-1}(\theta, z'), \hat{a}_n(\theta, z) \neq \hat{a}_n(\theta, z')\} \\
&\leq \sum_{i=1}^{q}\sum_{e \in E} P\{\eta_{n-1}(\theta, z) = \eta_{n-1}(\theta, z') = e, \\
&\quad |c_n(\theta, z) - s_i| \leq |c_n(\theta, z) - c_n(\theta, z')|\}
\end{aligned}$$

For $n \geq 2N + 1$, it is easy to see using (2.5.11-i) that on $\{\eta_{n-1}(\theta, z) = \eta_{n-1}(\theta, z') = e\}$ we have, for all $\theta \in K(\varepsilon, M)$

$$|c_n(\theta, z) - c_n(\theta, z')| \leq C_4\rho^n|u_0 - u_0'|$$

Whence

$$p \le \sum_{i=1}^{q} \sum_{e \in E} P\{|\alpha^T \cdot D_1 A^{n-2N-1} u_0 + D_2 \overline{U}_n + \alpha^T \cdot V_n + \beta^T \cdot e - s_i| \le C_4 \rho^n |u_0 - u_0'|\}$$

Since the r.v. $\overline{U}_n$ and V_n are independent, in order to prove the lemma, it is sufficient to note that, following Lemma 12, there exists a constant C_5 such that for all scalars c, all α such that $|\alpha| \ge \varepsilon$,

$$P\{|\alpha^T \cdot V_n + c| \le C_4 \rho^n |u_0 - u_0'|\} \le \frac{C_5}{\varepsilon} \rho^n |u_0 - u_0'|$$

$\square$

2.5.7 Verification of Assumption (A.4-ii)

Fix $\theta \in K(\varepsilon, M)$. The constants below will be uniform in $\theta \in K(\varepsilon, M)$. We recall the previous notation $z = (u, Y, e), x = (u, Y), X_n = (U_n, Y_n)$. First note that following (2.5.11), (2.5.12) and (2.5.2-iv)

$$\sup_{x',e,e'} E_{z,z'}^{\theta;\theta'}(1 + |X_n|^p) \le B_1(1 + |x|^p) \qquad (2.5.18)$$

$$\sup_{x,e,e'} E_{z,z'}^{\theta;\theta'}(1 + |X_n'|^p) \le B_2(1 + |x'|^p) \qquad (2.5.19)$$

$$\sup_{e,e'} E_{z,z'}^{\theta;\theta'}(|X_n - X_n'|^2) \le B_3 \rho^n (1 + |x|^2 + |x'|^2) \qquad (2.5.20)$$

Furthermore, Lemmas 15 and 16 imply that the Markov chain $\{(Z_n, Z_n'), P_{z,z'}^{\theta;\theta}\}$ satisfies Assumptions (X.3) and (X.4) of Proposition 4, of Section 2.4, which may now be applied to it. We deduce that for all $g \in Li(p)$, there exist constants $C_4, q_1, \rho_4 < 1$ such that:

$$|\Pi_\theta^n g(x, e) - \Pi_\theta^n g(x', e')| \le C_4 \rho_4^n N_p(g)(1 + |x|^q + |x'|^q) \qquad (2.5.21)$$

The inequalities (2.5.18) and (2.5.21) allow us to apply Proposition 2. This implies that $h(\theta) = \lim_n \Pi_\theta^n H_\theta(x, e)$ exists (in view of (2.5.5)), and that

$$|\Pi_\theta^n H_\theta(x, e) - h(\theta)| \le C_5 \rho_4^n (1 + |x|^q) \qquad (2.5.22)$$

Moreover

$$\nu_\theta(x, e) = \sum_{n \ge 0} \{\Pi_\theta^n H_\theta(x, e) - h(\theta)\} \qquad (2.5.23)$$

exists and satisfies $(I - \Pi_\theta)\nu_\theta = H_\theta - h(\theta)$.

Finally, in view of the uniformity in $\theta \in K(\varepsilon, M)$ of the constants, from (2.5.22), we have

$$\sup_{\theta \in K(\varepsilon, M)} |\nu_\theta(x, e)| \le C_6 (1 + |x|^q) \qquad (2.5.24)$$

2.5.8 Verification of Assumption (A.4-i)

We recall that $\|g\|_{\infty,p}$ was defined in (2.1.6). Set

$$\psi^{\theta;\theta'}(z) = \Pi_\theta^n g(z) - \Pi_{\theta'}^n g(z) = E_{z,z}^{\theta;\theta'}\{g(X_n, \eta_n) - g(X'_n, \eta'_n)\} \quad (2.5.25)$$

Note that $X'_n = X_n$ $P_{z,z}^{\theta;\theta'}$ a.s. and that, following (2.5.18), if we set $G(X_n, \eta_n, \eta'_n) = g(X_n, \eta_n) - g(X_n, \eta'_n)$, then

$$E_{z,z}^{\theta;\theta'}|G(X_n, \eta_n, \eta'_n)|^m \leq \overline{C}_m \|g\|_{\infty,p}^m (1 + |x|^{pm}) \quad (2.5.26)$$

Then for $r \geq 1$

$$|\psi^{\theta;\theta'}(z)|^r$$
$$\leq \sum_{k=0}^{n-1} E_{z,z}^{\theta;\theta'}\{I(\eta_k = \eta'_k, \eta_j \neq \eta'_j, j = k+1, \ldots, n)$$
$$\cdot |G(X_n, \eta_n, \eta'_n)|^r\}$$
$$= \sum_{k=0}^{n-1} E_{z,z}^{\theta;\theta'}\{I(\eta_k = \eta'_k, \eta_{k+1} \neq \eta'_{k+1})$$
$$E_{Z_{k+1}, Z'_{k+1}}^{\theta;\theta'}(I(\eta_j \neq \eta'_j, j = 1, \ldots, n-k-1))$$
$$|G(X_{n-k-1}, \eta_{n-k-1}, \eta'_{n-k-1})|^r\}$$

The last expectation may be bounded above using Schwarz's inequality, Lemma 15 and (2.5.26). Thus, for some $\rho_4 < 1$ (in fact $\rho_4 = \rho_2^{1/2}$), and assuming $|\theta - \theta'| \leq \delta_2$ we have

$$|\psi^{\theta;\theta'}(z)|^r \leq B_r \cdot \|g\|_{\infty,p}^r \cdot \sum_{k=0}^{n-1} \rho_4^{n-k} \psi_{k,r}^{\theta;\theta'}(z) \quad (2.5.27)$$

$$\psi_{k,r}^{\theta;\theta'}(z) = E_{z,z}^{\theta;\theta'}\{I(\eta_k = \eta'_k, \eta_{k+1} \neq \eta'_{k+1})(1 + |X_{k+1}|^{1+pr})\} \quad (2.5.28)$$

Using (2.5.13) and (2.5.4-i) we have:

$$\psi_{k,r}^{\theta;\theta'}(z) \leq \sum_{i=1}^{q} \sum_{e \in E} E\{I[\eta_k(\theta, z) = \eta'_k(\theta, z) = e;$$
$$|c_{k+1}(\theta, z) - s_i| \leq |c_{k+1}(\theta, z) - c_{k+1}(\theta', z)|](1 + |X_{k+1}(x)|^{1+pr})\}$$
$$\leq \sum_{i=1}^{q} \sum_{e \in E} E\{(1 + |X_{k+1}(x)|^{1+pr})$$
$$I[|\alpha^T \cdot Y_{k+1}(x) + \beta^T e - s_i| \leq C_1|\theta - \theta'|(1 + |Y_{k+1}(x)|)]\}$$

Thus returning to the canonical space

$$\psi_{k,r}^{\theta;\theta'}(z) \leq \sum_{i=1}^{q} \sum_{e \in E} \psi_{k,r}^{\theta;\theta'}(z; i, e) \quad (2.5.29)$$

where

$$\psi_{k,r}^{\theta;\theta'}(z;i,e)$$
$$= E_{z,z}^{\theta;\theta'}\{(1+|X_{k+1}|^{1+pr})I[|\alpha^T Y_{k+1} + \beta^T e - s_i| \leq C_1|\theta - \theta'|(1+|Y_{k+1}|)]\} \quad (2.5.30)$$

Lemma 17 *Given ε, M and $p \geq 0$, there exists a constant C_4 such that for all $\theta, \theta' \in K(\varepsilon, M)$, all g, all z_0, all n, all $m \geq 2N + 1$*

$$E_{z_0,z_0}^{\theta;\theta'}\{|\Pi_\theta^n g(Z_m) - \Pi_{\theta'}^n g(Z_m)|\} \leq C_4\|g\|_{\infty,p}|\theta - \theta'|(1+|u_0|^{p+2}) \quad (2.5.31)$$

Proof. Suppose that ν denotes the distribution of Z_m for the initial condition $z_0 = (u_0, Y_0, \eta_0)$. Using the Markov property, then expressing $Y_{m+k+1}(x_0)$ explicitly for $m \geq 2N + 1$, taking into account that $\overline{U}_{m+k+1}$ is bounded, we have

$$\int \psi_{k,1}^{\theta;\theta'}(z;i,e)d\nu(z)$$
$$\leq B_2 E\{(1+|u_0|^{q+1}+|V_{m+k+1}|^{q+1})$$
$$I[|\alpha^T V_{m+k+1} + \alpha^T \cdot D_1 A^{m+k-2N} u_0 + \alpha^T \cdot D_2 \overline{U}_{m+k+1} + \beta^T \cdot e - s_i|$$
$$\leq C_2|\theta - \theta'|(1+|u_0|+|V_{m+k+1}|)]\}$$

Since the r.v. $\overline{U}_{m+k+1}$ and V_{m+k+1} are independent, and in view of (2.5.6), we may apply Lemma 12 for $|\theta - \theta'| \leq \frac{\varepsilon}{3C_2} = \delta_3$. Then, for $|\alpha| \geq \varepsilon$, we have

$$\int \psi_{k,1}^{\theta;\theta'}(z;i,e)d\nu(z) \leq C_3 \cdot |\theta - \theta'|(1+|u_0|^{p+2})$$

Using (2.5.29), (2.5.27) and (2.5.25), we deduce that the lemma is true for $|\theta - \theta'| \leq \delta_2 \wedge \delta_3$. But it is clearly true for $|\theta - \theta'| > \delta_2 \wedge \delta_3$, thanks to (2.5.18). □

We show that Lemma 17 implies that Assumption (A.4-i) is satisfied. Let $g \in Li(p)$. We saw in Subsection 2.5.7 that Proposition 2 is applicable, Thus $\Gamma_\theta g = \lim_n \Pi_\theta^n g(z)$ exists. Fixing an initial value z_0 and $m \geq 2N + 1$, we let n tend to $+\infty$ in (2.5.31). Since $|\Pi_\theta^n g(Z_n)| \leq C \cdot (1+|X_m|^{p+1})$, which is integrable, we obtain for all $\theta, \theta' \in K(\varepsilon, M)$:

$$|\Gamma_\theta g - \Gamma_{\theta'} g| \leq C_5 N_p(g)|\theta - \theta'| \quad (2.5.32)$$

It is easy to see now that if $H(\theta, z)$ satisfies (2.5.5), then $h(\theta)$ is Lipschitz on $K(\varepsilon, M)$.

2.5 Decision-Feedback Algorithms with Quantisation

2.5.9 Verification of Assumption (A.4-iii)′, and Completion of Proofs

We recall that $E_{z,a}$ denotes the distribution of the Markov chain defined by equations (2.5.2), (2.5.3) with the initial conditions $Z_0 = z, \theta_0 = a$. Let $K(\varepsilon, M)$ be the compact set as given in (2.5.16) and set

$$\tau = \tau(\varepsilon, M) = \inf(n; \theta_n \notin K(\varepsilon, M)) \qquad (2.5.33)$$

Lemma 18. *For all $p \geq 0$, $\lambda \in]0,1[$, there exist constants C_5, q_5 such that for all $m \geq 2N + 1 = r$, all n, all $a \in K(\varepsilon, M)$, all z, all g,*

$$E_{z,a}\{I(m+1 \leq \tau)|\Pi^n_{\theta_{m+1}} g(Z_{m+1}) - \Pi^n_{\theta_m} g(Z_{m+1})|^2\}$$
$$\leq C_5 \gamma^\lambda_{m+1-r} \|g\|_{\infty,p}(1+|x|^{q_5})$$

Proof. In view of (2.5.27) to (2.5.30), in order to prove the lemma, it is sufficient to show that

$$E_1 = E_{z,a}\{I(m+1 \leq \tau)\psi^{\theta_m;\theta_{m+1}}_{k,2}(Z_{m+1}, i, e)\} \leq A_1 \gamma^\lambda_{m+1-r}(1+|x|^{q_5})\} \qquad (2.5.34)$$

The function $\psi^{\theta;\theta'}_{k,2}(z; i, e)$ may be expressed using the distribution $P_{z,a}$ by the formula

$$\psi^{\theta;\theta'}_{k,2}(z; i, e) = E_{z,\theta'}\{\Phi(\theta, \theta', X_{k+1})\}$$

Applying the Markov property, we have

$$\begin{aligned}
E_1 &= E^\omega_{z,a}\{I[m+1 \leq \tau(\omega)] E^{\omega'}_{Z_{m+1}(\omega), \theta_{m+1}(\omega)}[\Phi(\theta_m(\omega), \theta_{m+1}(\omega), X_{k+1}(\omega'))]\} \\
&= E_{z,a}\{I[m+1 \leq \tau]\Phi(\theta_m, \theta_{m+1}, X_{m+k+2})\} \\
&= E_{z,a}\{I[m+1 \leq \tau](1+|X_t|^{2p+1}) \\
&\quad I[|\alpha^T_m \cdot Y_t + \beta^T_m e - s_i| \leq C_1|\theta_m - \theta_{m+1}|(1+|Y_t|)]\}
\end{aligned}$$

where we set $t = m + k + 2$.

Moreover, since $H(\theta, z)$ satisfies (2.5.5), we have the property: for all $q, q' \geq 0$, for all integers s, there exists a constant $C(q, q', r')$ such that, for all n, n', all z, all $a \in K(\varepsilon, M)$,

$$E_{z,a}\{I(n+s \leq \tau)|\theta_{n+s} - \theta_n|^q(1+|X_{n'}|^{q'})\} \leq C(q, q', s)\gamma^q_{n+1}(1+|x|)^{q+q'} \qquad (2.5.35)$$

If we choose q such that $q(1-\lambda) = \lambda$ then, following (2.5.35),

$$\begin{aligned}
&E_{z,a}\{I(m+1 \leq \tau)(1+|X_t|^{2p+1})I[C_1|\theta_m - \theta_{m+1}|(1+|Y_t|) > \gamma^\lambda_{m+1}]\} \\
&\leq \frac{C^q_1}{\gamma^{\lambda q}_{m+1}} E_{z,a}\{(1+|X_t|^{2p+q+1}) \cdot |\theta_m - \theta_{m+1}|^q)I(m+1 \leq \tau)\} \\
&\leq A_2 \gamma^{q(1-\lambda)}_{m+1}(1+|x|^{q_1}) \\
&= A_2 \gamma^\lambda_{m+1}(1+|x|^{q_1})
\end{aligned}$$

Thus it is sufficient to find an upper bound for

$$E_2 = E_{z,a}\{I(m+1 \leq \tau)(1+|X_t|^{2p+1})I(|\alpha_m^T Y_t + \beta_m^T e - s_i| \leq \gamma_{m+1}^\lambda)\}$$

Again following (2.5.35), for $r = 2N + 1$ we have

$$E_{z,a}\{I(m+1 \leq \tau)(1+|X_t|^{2p+1})$$
$$I[|(\alpha_m - \alpha_{m-r})^T \cdot Y_t + (\beta_m - \beta_{m-r})^T e - s_i| > \gamma_{m+1}^\lambda]\}$$
$$\leq \frac{A_3}{\gamma_{m+1}^{\lambda q}} E_{z,a}\{I(m+1 \leq \tau)(1+|X_t|^{2p+q+1})|\theta_m - \theta_{m-r}|^q\}$$
$$\leq A_4 \gamma_{m+1-r}^\lambda (1+|x|^{q_4})$$

Thus we must find an upper bound for

$$E_3 = E_{z,a}\{I(m+1 \leq \tau)(1+|X_t|^{2p+1})I(|\alpha_{m-r}^T \cdot Y_t + \beta_{m-r}^T \cdot e - s_i| \leq 2\gamma_{m+1}^\lambda)\}$$

On the initial probability space, from (2.5.11), since $t = m + k + 2 \geq 2N + 1$, and since $\{m+1 \leq \tau\} \subset \{m - r + 1 \leq \tau\}$, we have

$$E_3 \leq A_5 E\{I(m-r+1 \leq \tau)(1+|x|^{2p+1}+|V_{m+k+2}|^{2p+1})$$
$$I[|\alpha_{m-r}^T \cdot D_1 A^{m_1} u_0 + \alpha_{m-r}^T \cdot D_2 \overline{U}_{m+k+2} + \alpha_{m-r}^T \cdot V_{m+k+2}$$
$$+ \beta_{m-r}^T e - s_i| \leq 2\gamma_{m+1}^\lambda]\}$$

Since the r.v. V_{m+k+2} is independent of $\overline{U}_{m+k+2}$ and of the σ-field F_{m-r}, and since on $\{m - r + 1 \leq \tau\}$, $|\alpha_{m-r}| \geq \varepsilon$, Lemma 12 may be applied to the distribution of V_{m+k+2}. We obtain

$$E_3 \leq A_6 \cdot \gamma_{m+1}^\lambda \cdot (1+|x|^{2p+1})$$

This proves the lemma $\square$

Let us now complete the proof of Theorem 13. We must find an upper bound for

$$E(m) = E_{z,a}\{I(m+1 \leq r)|\Pi_{\theta_m} \nu_{\theta_m}(Z_{m+1}) - \Pi_{\theta_{m+1}} \nu_{\theta_{m+1}}(Z_{m+1})|^2\} \quad (2.5.36)$$

where, as we have already seen $\Pi_\theta \nu_\theta(z) = \sum_{k \geq 1}\{\Pi_\theta^k H_\theta(z) - h(\theta)\}$.

For all $\theta, \theta' \in K(\varepsilon, M)$, all z, all n

$$|\Pi_\theta \nu_\theta(z) - \Pi_{\theta'} \nu_{\theta'}(z)|$$
$$\leq |\sum_{k=1}^{n-1}(\Pi_\theta^k H_\theta(z) - \Pi_{\theta'}^k H_\theta(z)| + (n-1)|h(\theta) - h(\theta')|$$
$$+ |\sum_{k=1}^{n-1}(\Pi_{\theta'}^k H_\theta(z) - \Pi_{\theta'}^k H_{\theta'}(z))| + |\sum_{k=n}^{+\infty}(\Pi_\theta^k H_\theta(z) - h(\theta))|$$
$$+ |\sum_{k=n}^{+\infty}(\Pi_{\theta'}^k H_{\theta'}(z) - h(\theta'))|$$

2.5 Decision-Feedback Algorithms with Quantisation

Note that following (2.5.5) and Subsection 2.5.8, for all $\theta, \theta' \in K(\varepsilon, M)$

(i)   for all k, $|\Pi_{\theta'}^k(H_\theta - H_{\theta'})(z)| \leq B_1|\theta - \theta'|(1 + |x|^{q_1})$

(ii) $|h(\theta) - h(\theta')| \leq L_1|\theta - \theta'|$

(iii) $|\sum_{k=n}^{+\infty}(\Pi_\theta^k H_\theta(z) - h(\theta))| \leq B_2 \rho^n (1 + |x|^{q_2})$ (2.5.37)

Thus for $\theta, \theta' \in K(\varepsilon, M)$

$$|\Pi_\theta \nu_\theta(z) - \Pi_{\theta'}\nu_{\theta'}(z)|^2$$
$$\leq B_3\{n\sum_{k=1}^{n-1}|\Pi_\theta^k H_\theta(z) - \Pi_{\theta'}^k H_\theta(z)|^2 + n^2|\theta - \theta'|^2(1 + |x|^{2q_1})$$
$$+ n^2|\theta - \theta'|^2 + \rho^{2n}(1 + |x|^{2q_2})\}$$

Using (2.5.35) and Lemma 18, for $m \geq 2N + 1$ we obtain,

$$E(m) \leq B_4\{n^2 \gamma_{m+1-r}^\lambda + \rho^{2n}\}(1 + |x|^{2q_3})$$

If we choose $n = \log \gamma_{m+1-r}^\lambda \cdot (\log \rho^2)^{-1} + u$, $0 \leq u < 1$, then for all $\mu < \lambda$ we have,

$$n^2 \gamma_{m+1-r}^\lambda + \rho^{2n} \leq B_5\{|\log \gamma_{m+1-r}^\lambda|^2 + 1\}\gamma_{m+1-r}^\lambda + \gamma_{m+1-r}^\lambda$$
$$\leq B_6(\mu)\gamma_{m+1-r}^\mu$$

Whence, finally, using (2.5.37) if $m \leq 2N$, for $n = 1$ we have:

$$E(m) \leq B_6 \gamma_{m+1-r}^\mu (1 + |x|^{q_3}), \; m \geq 2N + 1$$
$$E(m) \leq B_7(1 + |x|^{q_2}), \; m \leq 2N$$

Under the assumption $\liminf_n \gamma_{n+r}/\gamma_n > 0$, it is easy to see that there exists a constant B such that for all m

$$E(m) \leq B\gamma_{m+1}^\mu(1 + |x|^{q_3 \vee q_2})$$

This implies (A.4-iii)', and so completes the proof of Theorem 13.   □

Remark As indicated in the remark following Theorem 13, we may dispense with the assumption $\liminf_n \gamma_{n+r}/\gamma_n > 0$. In order to prove this, it is sufficient to go back to the proof of Theorem 14 of Chapter 1, to see that Assumption (A.4-iii)' may be weakened to

$$E_{z,a}\{I(m+1 < \tau(Q))|\Pi_{\theta_m}\nu_{\theta_m}(Z_{m+1}) - \Pi_{\theta_{m+1}}\nu_{\theta_{m+1}}(Z_{m+1})|^2\}$$
$$\leq C_5(1 + |x|^{q_5})\gamma_{k+1-r}^\mu \wedge \gamma_1^\mu$$

for some constants $C_5, q_5 > 0, \mu \in]0,2[$ and $r \in \mathbb{N}$. We obtain conclusions similar to those of Theorem 14, where it is now sufficient to replace the sums $\sum_{k \geq n} \gamma_k^2$ by the sums $\sum_{k \geq n-r} \gamma_{n-k}^2$

2.6 Comments on the Literature

The case of the equaliser learning phase has been the object of many studies, and there are numerous references to this in (Eweda and Macchi 1983). The latter article includes the first complete proof of the almost sure convergence of the equaliser as an adaptive algorithm. The authors of the paper consider the case of a stationary perturbation and use a proof which entails the linearity in θ of $H(\theta, x)$. A direct investigation of the linear filter in liaison with the solution of a linear equation without Markov assumptions is also found in (Ljung 1984), (Walk 1985) and (Walk and Zsido 1989).

The book (Ljung and Söderstrom 1983) contains numerous examples on system identification, in which the mathematical analysis is essentially based on the result of (Ljung 1977b), see also (Goodwin and Sin 1984). Examples of infinite-dimensional parameter estimation (regression functions, probability densities) are also given in (Revesz 1973) and (Deheuvels 1973, 1974).

Further details of the applications of stochastic algorithms were given in the comments on the literature in Part I. However, we shall mention (Benveniste, Goursat and Ruget 1980a,b) and (Eweda and Macchi 1984a,b, 1986) which consider the blind equaliser. In fact, apart from (Benveniste, Goursat and Ruget 1980b) and (Kushner 1981, 1984), very few papers contain mathematical investigations of algorithms with discontinuities.

Chapter 3
Analysis of the Algorithm in the General Case

Introduction

In the stochastic algorithms studied in Chapter 1 there is a fairly strong condition on the moments of the process $(X_n)_{n\geq 0}$ (Assumption (A.5)): for any compact set Q, there exists a constant $\mu_q(\bar{Q}) < \infty$ such that for any initial condition (x, a)

$$E_{x,a}\{I(\theta_k \in Q, k \leq n)(1 + |X_{n+1}|^q)\} \leq \mu_q(Q)(1 + |x|^q)$$

As soon as the evolution of the process (X_n) depends on the sequence (θ_n) (unlike the equaliser discussed in Chapter 2), as in the algorithm with general linear dynamics, there is no reason for (A.5) to be satisfied.

This chapter picks up the analysis of Chapter 1, without Assumption (A.5) on the moments. The general procedure is as in Chapter 1. Firstly we shall give an upper bound for the sums $\sum_{i=1}^{k} \varepsilon_i(\phi)$ which in some way reflect the fluctuation of the algorithm around the solution of the mean differential equation (Section 3.2). Then we shall prove (Section 3.3) that the algorithm converges to this solution (finite horizon). In Section 3.4, we shall use the same upper bound, together with a Lyapunov function, to study the asymptotic behaviour of the algorithm for an initial condition near to an asymptotically stable equilibrium point of the ODE.

3.1 New Assumptions and Control of the Moments

3.1.1 The Linear Example

We consider an algorithm of the form (1.1.1), Chapter 1, with Assumptions (A.1), (A.2), (A.3) and (A.4). But, we wish to replace (A.5) by a weaker assumption which is particularly applicable to the linear case (Chapter 1, Subsection 1.1.4) in which $X_{n+1} = A(\theta_n)X_n + B(\theta_n)W_{n+1}$. As we have seen, the Assumptions (A.3), (A.4) and (A.5) are satisfied provided that $\sup\{|A(\theta)| : \theta \in Q\} \leq \rho < 1$. However, this condition is quite strong, and it is more natural to assume that for $\theta \in Q$, the eigenvalues of $A(\theta)$ have modulus less than $\rho < 1$; this implies that $|A^n| \leq K\rho^n$ and that the Markov chain with transition probability Π_θ has nice properties. On the other hand from equation (1.1.19) of Chapter 1 we have $X_{n+1} = A(\theta_n) \cdot A(\theta_{n-1}) \ldots A(\theta_0)x + \ldots$, and we have no control over $|X_{n+1}|$, at least at the start of the operation of the algorithm. In fact, when γ_n becomes small, θ_n varies slowly and $A(\theta_{n+p})\ldots A(\theta_n) \approx A^p(\theta_n)$ and we can expect to have control over $|X_{n+p}|$, given $|X_n|$. The assumptions given below take this situation into account.

3.1.2 The New Assumptions

Given a function $g(x)$ on $\mathbb{R}^k$, for $q \geq 0$, denote

$$[g]_q = \sup_{x_1 \neq x_2} \frac{|g(x_1) - g(x_2)|}{|x_1 - x_2| \cdot [1 + |x_1|^q + |x_2|^q]}$$

Note that if $[g]_0 < \infty$ then g is Lipschitz, and if, for $q \geq 1$ we set $\Psi_q(x) = |x|^q$, then $[\Psi_q]_{q-1} < +\infty$. Note also that if $[g]_q \infty$, we have

$$|g(x)| \leq (g(0) + 2[g]_q)(1 + |x|^{q+1})$$

Thus $\int \Pi_\theta(x,dy)|y|^{q+1} < \infty$ is a sufficient condition for any Borel function g such that $[g]_q < \infty$ to be integrable with respect to $\Pi_\theta(x,dy)$ for any θ and x.

Given an open subset D of $\mathbb{R}^d$, we assume

(A'.5) *For all $q \geq 1$ and for any compact subset Q of D, there exist $r \in \mathbb{N}_*$ and constants $\bar{\alpha} < 1, \beta, M_1, K_1, K_2$, such that*

(i) $$\sup_{\theta \in Q} \int \Pi_\theta^r(x,dy)|y|^q \leq \bar{\alpha}|x|^q + \beta$$

(i') $$\sup_{\theta \in Q} \int \Pi_\theta(x,dy)|y|^q \leq M_1|x|^q + \beta$$

For any Borel function g on $\mathbb{R}^k$ such that $[g]_q < \infty$:

(ii) $\sup_{\theta \in Q} |\Pi_\theta g(x_1) - \Pi_\theta g(x_2)| \leq K_1[g]_q|x_1 - x_2|(1 + |x_1|^q + |x_2|^q)$

For all $\theta, \theta' \in Q$ and for any Borel function g with $[g]_q < \infty$:

(iii) $$|\Pi_\theta g(x) - \Pi_{\theta'} g(x)| \leq K_2[g]_q|\theta - \theta'|(1 + |x|^{q+1})$$

Remark 1. It is not necessary to assume (A'.5) for all $q \geq 1$. It is sufficient in fact that it be valid for some q "sufficiently large" (the value of which depends upon the exponents $q_1, q_2, q_3, q_2\lambda$ in Assumptions (A.3) and (A.4)).

Example.

We return to the linear case (Chapter 1, Subsection 1.1.4) in which we have:

$$\Pi_\theta g(x) = E\{g(A(\theta)x + B(\theta)W_1)\}$$

$$\Pi_\theta^r g(x) = E\left\{g\left(A^r(\theta)x + \sum_{k=0}^{r-1} A^k(\theta)B(\theta)W_{r-k}\right)\right\}$$

3.1 New Assumptions and Control of the Moments

If $A(\theta)$ and $B(\theta)$ are locally Lipschitz then (ii) and (iii) are readily satisfied. Suppose that $\sup\{|A^n(\theta)| : \theta \in Q\} \leq C\rho^n$, $\rho < 1$, then for $\theta \in Q$,

$$\int \Pi_\theta^r(x, dy)|y|^q = E\{|A^r(\theta)x + \sum_{k=0}^{r-1} A^k(\theta)B(\theta)W_{r-k}|^q\}$$
$$\leq 2^q |A^r(\theta)|^q |x|^q + 2^q E|\sum_{k=0}^{r-1} A^k(\theta)B(\theta)W_{r-k}|^q$$
$$\leq \bar{\alpha}|x|^q + \beta$$

with $\bar{\alpha} < 1$ if r is sufficiently large that $2C\rho^r < 1$.

3.1.3 Proposition Regarding the Moments of X_n

For the compact subset Q of D, and $\varepsilon > 0$, we define

$$\tau(Q) = \inf(n; \theta_n \notin Q) \tag{3.1.1}$$
$$\sigma(\varepsilon) = \inf(n \geq 1, |\theta_n - \theta_{n-1}| > \varepsilon) \tag{3.1.2}$$
$$\nu(\varepsilon, Q) = \inf(\tau(Q), \sigma(\varepsilon)) \tag{3.1.3}$$

Proposition 1. *We assume (A.1) and (A'.5). For any compact subset Q of D and any $q \geq 1$, there exist constants ε_0, M, such that for all $\varepsilon \leq \varepsilon_0$, all $a \in Q$, all x:*

$$\sup_n E_{x,a}\{|X_n|^q I(n \leq \nu(\varepsilon, Q))\} \leq M \cdot (1 + |x|^q)$$

Proof of Proposition 1. We shall frequently use C to denote any constant which is independent of q and Q. We recall that $\{k+1 \leq \nu\}$ is F_k-measurable. The proof will be given in various stages. First we have the following lemmas.

Lemma 2. *There exist constants $C(l)$, such that for all $n, l \geq 1, x, a \in Q$,*

$$E_{x,a}\{|X_{n+l}|^q I(n+l \leq \nu(\varepsilon, Q))|F_n\} \leq C(l)[1 + |X_n|^q] \cdot I(n+1 \leq \nu(\varepsilon, Q))$$

Proof of Lemma 2. It suffices to note that

$$E\{|X_{k+1}|^q I(k+1 \leq \nu)|F_k\} = I(k+1 \leq \nu) \int \Pi_{\theta_k}(X_k, dy)|y|^q$$
$$\leq I(k+1 \leq \nu)[K_3 |X_k|^q + K_4]$$
$$\leq (K_4 + K_3 |X_k|^q) I(k \leq \nu)$$

This proves that the lemma is true for $l = 1$ and moreover allows us to carry out the recurrence over l. □

Lemma 3. *Suppose $\phi(\theta, x)$ is function on $\mathbb{R}^d \times \mathbb{R}^k$ such that for all $\theta \in Q$, $[\phi_\theta]_{q-1} \leq A < +\infty$, then for all $n, l \geq 1, a \in Q, x$:*

$$E_{x,a}\{\phi(\theta_n, X_{n+l})I(n+l \leq \nu(\varepsilon, Q))|F_n\}$$
$$\leq E_{x,a}\{I(n+l-1 \leq \nu(\varepsilon, Q))\int \phi(\theta_n, x)\Pi_{\theta_n}(X_{n+l-1}, dx)|F_n\}$$
$$+\varepsilon C(A, l)(1+|X_n|^q)I(n+1 \leq \nu(\varepsilon, Q))$$

Proof of Lemma 3.

$$E\{\phi(\theta_n, X_{n+l})I(n+l \leq \nu)|F_{n+l-1}\}$$
$$= I(n+l \leq \nu)\int \phi(\theta_n, x)\Pi_{\theta_{n+l-1}}(X_{n+l-1}, dx)$$
$$= I(n+l \leq \nu)\int \phi(\theta_n, x)\Pi_{\theta_n}(X_{n+l-1}, dx) + R_l$$

where

$$R_l = I(n+l \leq \nu)\int \phi(\theta_n, x)[\Pi_{\theta_{n+l-1}}(X_{n+l-1}, dx) - \Pi_{\theta_n}(X_{n+l-1}, dx)]$$

But on $\{n+l \leq \nu\}$, $|\theta_{n+l-1} - \theta_n| \leq l\varepsilon$, and, using (A'.5-iii), we have $|R_l| \leq I(n+l \leq \nu)K_2 A l\varepsilon(1+|X_{n+l-1}|^q)$. Following Lemma 2,

$$E\{|R_l||F_n\} \leq \varepsilon C(A, l)(1+|X_n|^q)I(n+1 \leq \nu)$$

The lemma is thus true. □

Lemma 4. *There exist constants $C'(l)$ such that for all $n, l \geq 1, a \in Q, x$:*

$$E_{x,a}\{|X_{n+l}|^q I(n+l \leq \nu(\varepsilon, Q))|F_n\}$$
$$\leq I(n+1 \leq \nu(\varepsilon, Q))\{\int \Pi_{\theta_n}^l(X_n, dx)|x|^q + \varepsilon C'(l)(1+|X_n|^q)\}$$

Proof of Lemma 4. Note that if $g(y) = |y|^q$, $[g]_{q-1} < +\infty$, then, after (A'.5-ii), $[\Pi_\theta g]_{q-1} \leq K_1 [g]_{q-1}, \ldots, [\Pi_\theta^{l-1} g]_{q-1} \leq K_1^{l-1}[g]_{q-1}$ for all $\theta \in Q$. Applying Lemma 3 successively to $g, \Pi_\theta g, \ldots, \Pi_\theta^{l-1} g$, we have:

$$E\{|X_{n+l}|^q I(n+l \leq \nu)|F_n\}$$
$$\leq E\{\Pi_{\theta_n} g(X_{n+l-1})I(n+l-1 \leq \nu)|F_n\} + \varepsilon \overline{C}_1(l)(1+|X_n|^q)I(n+1 \leq \nu)$$

$$E\{\Pi_{\theta_n} g(X_{n+l-1})I(n+l-1 \leq \nu)|F_n\}$$
$$\leq E\{\Pi_{\theta_n}^2 g(X_{n+l-2})I(n+l-2 \leq \nu)|F_n\} + \varepsilon \overline{C}_2(l)(1+|X_n|^q)I(n+1 \leq \nu)$$

...

$$E\{\Pi_{\theta_n}^{l-1} g(X_{n+1})I(n+1 \leq \nu)|F_n\} = I(n+1 \leq \nu)\Pi_{\theta_n}^l g(X_n)$$

It is sufficient to add these inequalities term by term. □

3.1 New Assumptions and Control of the Moments

Proof of Proposition 1 (Continued). We apply Lemma 4 for the r of (A'.5-i) to obtain

$$\begin{aligned}E\{|X_{n+r}|^q I(n+r \leq \nu)|F_n\} &\leq \{\bar{\alpha}|X_n|^q + \beta + \varepsilon C'(r)(1+|X_n|^q)\}I(n+1 \leq \nu)\\ &= \{(\bar{\alpha} + \varepsilon C'(r))|X_n|^q + C''(r)\}I(n+1 \leq \nu)\end{aligned}$$

r being fixed, choose ε_0 such that $\bar{\alpha} + \varepsilon_0 C'(r) = \rho < 1$, then for all $\varepsilon \leq \varepsilon_0$ we have:

$$E\{|X_{n+r}|^q I(n+r \leq \nu)|F_n\} \leq \rho |X_n|^q I(n \leq \nu) + C''$$

Iterating this we obtain

$$E\{|X_{lr}|^q I(lr \leq \nu)\} \leq \rho^l |x|^q + \frac{C''}{1-\rho} \leq M(1 + |x|^q)$$

for all l.

But for $j < r$, following Lemma 2, we have

$$\begin{aligned}E\{|X_{lr+j}|^q I(lr+j \leq \nu)|F_{lr}\} &\leq C(j)[1 + |X_{lr}|^q] I(lr \leq \nu)\\ &\leq C' r [1 + |X_{lr}|^q] I(lr \leq \nu)\end{aligned}$$

Proposition 1 now follows. □

3.2 L^q Estimates

This section is the analogue of Chapter 1, Section 1.4. However, since we wish to carry out our asymptotic analysis for the case in which $\sum \gamma_n^\alpha < \infty$ with $\alpha > 1$, but with α possibly different from 2, we shall require bounds in L^q and not just in L^2.

Note also that, clearly, the formulae use the stopping time $\nu(\varepsilon, Q)$ in place of $\tau(Q)$.

3.2.1 Main Inequality

We recall that given a function ϕ on $\mathbb{R}^d$ with bounded second derivatives, we have

$$\varepsilon_i(\phi) = \phi(\theta_{i+1}) - \phi(\theta_i) - \gamma_{i+1} \phi'(\theta_i) \cdot h(\theta_i) \qquad (3.2.1)$$

In this paragraph, we shall obtain upper bounds for

$$E\{\sup_{n < k \leq m(n,T)} I(k \leq \nu(\varepsilon, Q)) |\sum_{i=n}^{k-1} \varepsilon_i(\phi)|^q\}$$

where $m(n, T)$ was defined in Chapter 1, (1.1.6) and $\nu(\varepsilon, Q)$ in (2.1.4). We shall do this for $\varepsilon \leq \varepsilon_0(q, Q)$ as given in Proposition 1. For $q = 2$ and under Assumption (A.5), we obtained this upper bound in Chapter 1, Proposition 7.

Proposition 5. *We assume conditions (A.1), (A.2), (A.3), (A.4) and (A'.5). For any regular function ϕ with bounded second derivatives, any compact subset $Q \subset D$ and for all $q \geq 2$, there exist constants $B, s, \varepsilon_0 > 0$ (ε_0 independent of ϕ), such that for all $\varepsilon \leq \varepsilon_0, T > 0, x, a$:*

$$E_{x,a}\{\sup_{n<k\leq m(n,T)} I(k \leq \nu(\varepsilon,Q))|\sum_{i=n}^{k-1}\varepsilon_i(\phi)|^q\}$$
$$\leq B(1+T)^{q-1}(1+|x|^s)\sum_{i=n+1}^{m(n,T)}\gamma_i^{1+q/2}$$

The above inequality will be proved in Subsection 3.2.3 (below).

3.2.2 Statement of Lemmas

Before proving Proposition 5, we shall state, in the form of lemmas, two results which will be useful to us. The first is a classical result from the theory of martingales (Burkholder inequalties, see (Hall and Heyde 1980) Theorem 2.10 page 23). The second is a simple application of Hölder's inequality.

Lemma 6. *Let $(U_n; n \geq 0)$ be a sequence of real-valued random variables defined on a filtered probability space (Ω, F_n, F, P), satisfying for some $q \geq 2$, $E|U_n|^q < +\infty$ and $E[U_{n+1}|F_n] = 0$. Then*

$$E(\sup_{n\leq k\leq m}|\sum_{i=n}^{k}U_i|^q) \leq C(q) \cdot E(\sum_{i=n}^{m}U_i^2)^{q/2}$$

Lemma 7. *Let $a_i \geq 0, b_i \in \mathbb{R}, u > 1, 0 < \delta < 1$, then*

$$|\sum_{i=n}^{m}a_i b_i|^u \leq (\sum_{i=n}^{m}a_i^{\delta u/u-1})^{u-1} \cdot \sum_{i=n}^{m}a_i^{(1-\delta)u} \cdot |b_i|^u$$

Proof of Lemma 7. We write

$$|\sum_{i=n}^{m}a_i b_i| = |\sum_{i=n}^{m}a_i^{\delta} \cdot a_i^{1-\delta}b_i|$$

and we apply Hölder's inequality for the conjugate exponents $\frac{u}{u-1}$ and u; whence

$$|\sum_{i=n}^{m}a_i b_i| \leq (\sum_{i=n}^{m}a_i^{\delta u/u-1})^{\frac{u-1}{u}}(\sum_{i=n}^{m}a_i^{(1-\delta)u}|b_i|^u)^{\frac{1}{u}} \qquad \square$$

3.2 L^q Estimates

Note finally that for any sequence Z_i,

$$\sup_{n<k\leq m} I(k\leq \nu)\cdot |\sum_{i=n}^{k-1} Z_i| = \sup_{n<k\leq \nu\wedge m} |\sum_{i=n}^{k-1} Z_i| = \sup_{n<k\leq m} |\sum_{i=n}^{k\wedge\nu-1} Z_i|$$

$$= \sup_{n<k\leq m} |\sum_{i=n}^{k-1} Z_i I(i+1\leq \nu)| \leq \sum_{i=n}^{m-1} |Z_i| I(i+1\leq \nu)$$

$$= \sum_{i=n}^{m\wedge\nu} |Z_i| \tag{3.2.2}$$

3.2.3 Proof of Proposition 5

In order to establish Proposition 5, we shall prove it for each term of the decomposition given in Lemma 1, Chapter 1, Subsection 1.3.2. We shall write m for $m(n,T)$, ν for $\nu(\varepsilon,Q)$, K will denote any constant which depends only on q and Q and on the derivatives of ϕ, and s will denote an exponent which depends only on q and on the exponents in (A.3) and (A.4). The $\varepsilon_k^{(i)}$ are as defined in Chapter 1, Lemma 1.

1. Following (3.2.2)

$$S_1 = E\{\sup_{n<k\leq m} I(k\leq \nu)|\sum_{i=n}^{k-1} \varepsilon_i^{(1)}|^q\} = E\{\sup_{n<k\leq m} |\sum_{i=n}^{k-1} U_i|^q\}$$

where

$$U_i = \gamma_{i+1}\phi'(\theta_i)[\nu_{\theta_i}(X_{i+1}) - \Pi_{\theta_i}\nu_{\theta_i}(X_i)]I(i+1\leq \nu)$$
$$= \gamma_{i+1} V_i \tag{3.2.3}$$

Following (A4-ii) and Proposition 1,

$$E|\nu_{\theta_i}(X_{i+1})I(i+1\leq \nu)|^q \leq K(1+|x|^s)$$

whence, in view of the contractivity of the q-norms of the conditional expectation

$$E|V_i|^q \leq K(1+|x|^s)$$

Following (A.2), $E\{U_{i+1}|F_n\} = 0$. Then Lemma 6 implies that

$$S_1 \leq KE\left\{\sum_{i=n}^{m} \gamma_{i+1}^2 V_i^2\right\}^{q/2}$$

If $q=2$, this is the desired result. If $q>2$, we apply Lemma 7 with $a_i = \gamma_{i+1}^2, b_i = V_i^2, u = \frac{q}{2}, \delta = \frac{q-2}{2q}$ to obtain:

$$S_1 \leq K\left(\sum_{i=n}^{m-1} \gamma_{i+1}\right)^{\frac{q}{2}-1} \cdot \sum_{i=n}^{m-1} \gamma_{i+1}^{1+\frac{q}{2}} E(|V_i|^q)$$
$$\leq KT^{\frac{q}{2}-1} \cdot \sum_{i=n}^{m-1} \gamma_{i+1}^{1+\frac{q}{2}} \cdot (1+|x|^s)$$

2. Furthermore we have

$$S_2 = E\{\sup_{n<k\leq m} I(k \leq \nu)| \sum_{i=n+1}^{k-1} \varepsilon_i^{(2)}|^q\}$$
$$\leq E\{\sum_{i=n+1}^{m-1} |\varepsilon_i^{(2)}|I(i+1 \leq \nu)\}^q$$

Using formulae (1.3.13), (1.3.14) and (1.3.15) of Chapter 1, we have

$$|\varepsilon_i^{(2)}|I(i+1 \leq \nu)$$
$$\leq K\gamma_{i+1}(1+|X_i|^{q_4 \vee q_3})\{|\theta_i - \theta_{i-1}|^\lambda + |\theta_i - \theta_{i-1}|\}I(i+1 \leq \nu)$$
$$\leq K\gamma_i^{1+\lambda}(1+|X_i|^s)I(i+1 \leq \nu)$$

Thus

$$S_2 \leq KE\{\sum_{i=n+1}^{m-1} \gamma_i^{1+\lambda}(1+|X_i|^s)I(i+1 \leq \nu)\}^q$$

whence, applying Lemma 7 to

$$a_i = \gamma_{i+1}^{1+\lambda}, \qquad b_i = (1+|X_i|^s)I(i+1 \leq \nu)$$
$$u = q, \qquad \delta = \frac{q-1}{q}\frac{1}{1+\lambda}$$

we have

$$S_2 \leq K\{\sum_{i=n+1}^{m-1} \gamma_i\}^{q-1} \cdot \sum_{i=n+1}^{m-1} \gamma_i^{1+\lambda q} E\{(1+|X_i|^s)^q I(i+1 \leq \nu)\}$$

then, following Proposition 1, for some suitable s

$$S_2 \leq K \cdot T^{q-1} \cdot \sum_{i=n+1}^{m-1} \gamma_i^{1+\lambda q} \cdot (1+|x|^s)$$

3. Analysis of S_3.

$$S_3 = E\{\sum_{i=n+1}^{m-1} |\varepsilon_i^{(3)}|I(i+1 \leq \nu)\}^q$$

Using formula (1.3.14) of Chapter 1, we obtain

$$|\varepsilon_i^{(3)}|I(i+1 \leq \nu) \leq K(\gamma_i - \gamma_{i+1})(1+|X_i|^{q_3})I(i+1 \leq \nu)$$

Then, as above, we have

$$S_3 \leq K\{\sum_{i=n+1}^{m-1} (\gamma_i - \gamma_{i+1})\}^{q-1} \sum_{i=n+1}^{m-1} (\gamma_i - \gamma_{i+1})$$
$$\cdot E\{(1+|X_i|^{q_3})^q I(i+1 \leq \nu)\}$$
$$\leq K\gamma_{n+1}^q(1+|x|^s)$$

3.2 L^q Estimates

4. Analysis of S_4:

$$S_4 = E\{\sum_{i=n}^{m-1} |\varepsilon_i^{(4)}|I(i+1\leq\nu)\}^q$$

where (Chapter 1, Lemma 1)

$$|\varepsilon_i^{(4)}|I(i+1\leq\nu) \leq K\gamma_{i+1}^2(1+|X_{i+1}|^s)I(i+1\leq\nu)$$

whence, as in 2., but with $\lambda = 1$, we have

$$S_4 \leq K \cdot T^{q-1} \cdot \sum_{i=n}^{m-1} \gamma_{i+1}^{1+q}(1+|x|^s)$$

5. With the notation of Lemma 1, Chapter 1, we have

$$\begin{aligned}S_5 &= E\{\sup_{n<k\leq m} I(k\leq\nu)|\gamma_{n+1}\psi_{\theta_n}(X_n) - \gamma_k\psi_{\theta_{k-1}}(X_k)|^q\} \\ &\leq K_1\gamma_{n+1}^q E\{|\psi_{\theta_n}(X_n)|^q I(n+1\leq\nu)\} \\ &\quad + K_2 E\{\sum_{k=n+1}^{m} \gamma_k^q |\psi_{\theta_{k-1}}(X_k)|^q I(k\leq n)\} \\ &\leq K_1\gamma_{n+1}^q(1+|x|^s) + K_2\sum_{k=n+1}^{m}\gamma_k^q(1+|x|^s)\end{aligned}$$

Now Proposition 5 follows easily since $\lambda \geq \frac{1}{2}$ and $q \geq 1 + \frac{q}{2}$ as $q \geq 2$.

□

Remark. Adding together the various upper bounds, we obtain an upper bound which is slightly more exact than that given in Proposition 5, namely:

$$\{K_1 T^{q/2-1}\sum_{k=n}^{m-1}\gamma_{k+1}^{1+\frac{q}{2}} + K_2 T^{q-1}\sum_{k=n}^{m-1}\gamma_{k+1}^{1+\lambda q} + K_3\sum_{k=n}^{m-1}\gamma_{k+1}^q\}(1+|x|^s)$$

3.2.4 An Upper Bound Resulting from Proposition 5

Corollary 8. *With the assumptions of Proposition 5:*

(i)
$$E_{x,a}(\sup_n \sup_{n<k\leq m(n,T)} I(k\leq\nu(\varepsilon,Q))\cdot|\sum_{i=n}^{k-1}\varepsilon_i(\phi)|^q)$$
$$\leq 3^q B(1+T)^{q-1}(1+|x|^s)\sum_{i\geq 1}\gamma_i^{1+q/2}$$

(ii) If $\sum\gamma_i^{1+q/2} < +\infty$, then on $\{\nu(\varepsilon,Q) = +\infty\}$,

$$\sup_{n<k\leq m(n,T)}|\sum_{i=n}^{k-1}\varepsilon_i(\phi)| \text{ tends to } 0, P_{x,a} \text{ a.s. as } n \to \infty$$

Proof. We have seen (cf. (3.2.2)), that

$$\sup_{n<k\leq m(n,T)} I(k\leq \nu)\sum_{i=n}^{k-1}\varepsilon_i(\phi) \leq \sup_{n<k\leq m}|\sum_{i=n}^{k-1}Z_i|$$

where $Z_i = \varepsilon_i(\phi)I(i+1\leq \nu)$. Let $n_0 = 0, n_1 = m(0,T), \ldots, n_r = m(n_{r-1}, T)$; then for $n \in [n_r, n_{r+1})$ we have

$$\sum_{i=n}^{k-1} Z_i = \sum_{i=n_r}^{k-1} Z_i - \sum_{i=n_r}^{n-1} Z_i \text{ if } k \in [n_r, n_{r+1}[$$

$$\sum_{i=n}^{k-1} Z_i = \sum_{i=n_r}^{n_{r+1}-1} Z_i + \sum_{i=n_{r+1}}^{k-1} Z_i - \sum_{i=n_r}^{n-1} Z_i \text{ otherwise}$$

From which we deduce that

$$\sup_{n\geq n_p}\sup_{n<k\leq m(n,T)} I(k\leq n)|\sum_{i=n}^{k-1}\varepsilon_i(\phi)|^q \leq 3^q \sup_{r\geq p}\sup_{n_r<k\leq n_{r+1}} I(k\leq r)|\sum_{i=n_r}^{k-1}\varepsilon_i(\phi)|^q$$

From this, we deduce (i), since

$$E\{\sup_{r\geq 0}\sup_{n_r<k\leq n_{r+1}}(\ldots)\} \leq K \sum_{r\geq 0}\sum_{i=n_r}^{n_{r+1}-1}\gamma_{i+1}^{1+q/2} = K\sum_{i\geq 1}\gamma_i^{1+q/2}$$

and (ii), since

$$E\{\sum_{r\geq 0}\sup_{n_r<k\leq n_{r+1}} I(k\leq n)|\sum_{i=n_r}^{k-1}\varepsilon_i(\phi)|^q\} < +\infty$$

implies that

$$\sup_{n_r<k\leq n_{r+1}}|\sum_{i=n_r}^{k-1}\varepsilon_i(\phi)|$$

converges a.s. to 0 on $\{\nu = +\infty\}$

3.3 Convergence towards the Mean Trajectory

3.3.1 A Preliminary Result

For the stopping times $\tau(Q)$ and $\sigma(\varepsilon)$ as defined in (3.1.1) and (3.1.2), $\varepsilon_0(q,Q)$ as defined in Proposition 5 and $m(T)$ as defined in Chapter 1, (1.1.7), we have:

Lemma 9. *For any compact subset Q of D and $q \geq 2$, there exist constants A_1, s_1 such that for all $T > 0$, all $\varepsilon \leq \varepsilon_0(q,Q)$, all $a \in Q$, all x, we have:*

$$P_{x,a}(\sigma(\varepsilon) \leq \tau(Q), \sigma(\varepsilon) \leq m(T)) \leq \frac{A_1}{\varepsilon^q}(1+|x|^{s_1})\cdot \sum_{k=1}^{m(T)}\gamma_k^q$$

3.3 Convergence towards the Mean Trajectory

Proof. We have

$$P(\sigma \leq \tau, \sigma \leq m) = \sum_{k=1}^{m} P(\sigma = k \leq \tau)$$

$$= \sum_{k=1}^{m} P(\sigma = k \leq \tau, |\theta_k - \theta_{k-1}| > \varepsilon)$$

$$\leq \sum_{k=1}^{m} P(k \leq \sigma \wedge \tau, C_1 \gamma_k (1 + |X_k|^{q_1}) + C_2 \gamma_k^2 (1 + |X_k|^{q_2}) > \varepsilon)$$

(this follows from (A.3))

$$\leq \sum_{k=1}^{m} P(k \leq \sigma \wedge \tau, C' \gamma_k (1 + |X_k|^s) > \varepsilon)$$

$$\leq \sum_{k=1}^{m} \left(\frac{C'}{\varepsilon}\right)^q \gamma_k^q E\{(1 + |X_k|^s)^q I(k \leq \sigma \wedge \tau)\}$$

$$\leq \frac{A}{\varepsilon^q}(1 + |x|^{sq}) \sum_{k=1}^{m} \gamma_k^q \quad \text{(Proposition 1)}$$

□

Since this upper bound is valid for all T, we have:

Corollary 10. *With the assumptions of Lemma 9,*

$$P_{x,a}(\sigma(\varepsilon) \leq \tau(Q)) \leq \frac{A_1}{\varepsilon^q}(1 + |x|^{s_1}) \cdot \sum_{k=1}^{+\infty} \gamma_k^q$$

3.3.2

We return to the situation of Chapter 1, Subsection 1.5.1. Here $Q_1 \subset Q_2$ are two compact subsets of D satisfying, for any given $T > 0$, for all $a \in Q$, all $t \leq T$,

$$d(\bar{\theta}(t; 0, a), Q_2^c) \geq \delta_0 > 0 \tag{3.3.1}$$

Consider for $\delta \leq \delta_0$

$$\Omega(\delta) = \{\sup_{n \leq m(T)} |\theta_n - \bar{\theta}(t_n)| \geq \delta\}$$

If we introduce $\tau = \tau(Q_2), \sigma = \sigma(\varepsilon), \nu = \nu(\varepsilon, Q_2) = \tau(Q_2) \wedge \sigma(\varepsilon)$; then

$$\Omega(\delta) \subset \{\sup_{n \leq m(T)} |\theta_n - \bar{\theta}(t_n)| \geq \delta; m(T) \leq \nu\} \cup \{\nu < m(T)\}$$

But $\{\nu < m(T)\} \subset \{\sigma \leq \tau, \sigma < m(T)\} \cup \{\tau \leq \sigma, \tau < m(T)\}$ and following (3.3.1)

$$\{\tau \leq \sigma, \tau < m(T)\} \subset \{\sup_{n \leq \nu \wedge m(T)} |\theta_n - \bar{\theta}(t_n)| \geq \delta\}$$

Thus

$$\Omega(\delta) \subset \{\sup_{n \leq \nu \wedge m(T)} |\theta_n - \bar{\theta}(t_n)| \geq \delta\} \cup \{\sigma \leq \tau, \sigma < m(T)\}$$

The second of these sets is bounded above, as shown in Lemma 9 (above). To investigate the first set, we have:

Lemma 11. *Given compact subsets $Q_1 \subset Q_2$ of D and $q \geq 2$, there exist constants A_2, s_2, such that for all T satisfying (3.3.1), all $\varepsilon \leq \varepsilon_0(q, Q_2)$, all $a \in Q$, all x:*

$$E_{x,a}\{\sup_{n \leq \nu \wedge m(T)} |\theta_n - \bar{\theta}(t_n)|^q\}$$
$$\leq A_2(1 + |x|^{s_2})(1+T)^{q-1}\exp(qL_2T)\sum_{k=1}^{m(T)} \gamma_k^{1+q/2}$$

Proof. We do not need to change many of the details of Subsection 1.5.2 of Chapter 1. In the same way, on $\{n \leq \nu \wedge m(T)\}$, and for $r = 0, 1, \ldots, n$:

$$|\theta_r - \bar{\theta}(t_r)| \leq L_2 \cdot \sum_{k=0}^{r-1} \gamma_{k+1}|\theta_k - \bar{\theta}(t_k)| + \sup_{m \leq m(T)}\{I(m \leq \nu)|\sum_{k=0}^{m-1}\varepsilon_k|\} + L_2\sum_{k=1}^{m(T)}\gamma_k^2$$
$$= L_2 \sum_{k=0}^{r-1} \gamma_{k+1}|\theta_k - \bar{\theta}(t_k)| + U_1 + U_2$$

Whence, applying Lemma 8 of Chapter 1,

$$\sup_{n \leq \nu} |\theta_n - \bar{\theta}(t_n)|^q \leq 2^q \cdot \exp(qL_2T) \cdot (U_1^q + U_2^q)$$

Following Proposition 5, if $\varepsilon \leq \varepsilon_0(q, Q)$,

$$E\{U_1^q\} \leq B(1+T)^{q-1}(1+|x|^s)\sum_{k=1}^{m(T)} \gamma_k^{1+q/2}$$

and following Lemma 7

$$U_2^q \leq L_2^q\{\sum_{k=1}^{m(T)} \gamma_k^2\}^q \leq L_2^q T^{q-1} \cdot \sum_{k=1}^{m(T)} \gamma_k^{1+q}$$

This proves Lemma 11. □

3.3.3 Approximation by the Mean Trajectory

If we choose $\varepsilon = \varepsilon_0(q, Q_2)$, then Lemmas 9 and 11 give:

Theorem 12. *We assume the conditions (A.1), (A.2), (A.3), (A.4) and (A'.5). Suppose that $Q_1 \subset Q_2$ are compact subsets of D, and that $q \geq 2$; then there exist constants B_1, s_1, L_2 (L_2 is the Lipschitz constant for h on Q_2), such that for all $T > 0$ satisfying (3.3.1), all $\delta < \delta_0$, all $a \in Q_1$, all x:*

$$P_{x,a}\{\sup_{n \leq m(T)} |\theta_n - \bar{\theta}(t_n)| \geq \delta\}$$
$$\leq \frac{B_1}{\delta^q}(1 + |x|^{s_1})(1 + T)^{q-1} \cdot \exp(qL_2 T) \cdot \sum_{k=1}^{m(T)} \gamma_k^{1+q/2}$$

In particular if $\gamma_k = \gamma$ for all k:

$$P_{x,a}\{\sup_{n \leq T/\gamma} |\theta_n - \bar{\theta}(t_n)| \geq \delta\}$$
$$\leq \frac{B_1}{\delta^q}(1 + |x|^{s_1})T \cdot (1 + T)^{q-1} \cdot \exp(qL_2 T) \cdot \gamma^{q/2}$$

3.4 Asymptotic Analysis of the Algorithm

3.4.1

We shall now generalise the results of Chapter 1, Section 1.6. We shall replace Assumption (A.6) by:

(A'.6) *There exists $\alpha > 1$ such that $\sum \gamma_n^\alpha < +\infty$*

We retain assumption (A.7) and the notation $K(c), \tau(c)$ (Section 1.6, Chapter 1 (1.6.2) and (1.6.3)), and $\tau(Q), \sigma(\varepsilon), \nu(\varepsilon, Q)$ ((2.1.2), (2.1.3), (2.1.4)). We denote

$$q_0(\alpha) = \sup(2, 2(\alpha - 1)) \qquad (3.4.1)$$

Then if $q \geq q_0(\alpha)$ we have both $q \geq 2$ and $1 + \frac{q}{2} \geq \alpha$.

We shall suppose further that there exists a compact subset F of D satisfying

$$F = \{\theta; U(\theta) \leq c_0\} \supset \{\theta; U'(\theta) \cdot h(\theta) = 0\} \qquad (3.4.2)$$

for some $c_0 < C$ (where C is as defined in (A.7)).

Lemma 13. *Given c_1, c_2, q such that $c_0 < c_1 < c_2 < C$, $q \geq q_0(\alpha)$, there exist ε_0, B_2, s_2 such that for $\varepsilon \leq \varepsilon_0, a \in K(c_1), x$:*

$$P_{x,a}\{\tau(c_2) < +\infty, \sigma(\varepsilon) \geq \tau(c_2)\} \leq B_2(1 + |x|^{s_2})\sum_{k \geq 1} \gamma_k^{1+q/2}$$

Proof. Choose $\varepsilon_0 = \varepsilon_0(q, K(c_2))$ as in Proposition 5. From (A.7) and (3.4.2), there exists $\eta > 0$, such that, for all $\theta \in K(c_2) - K(c_1)$, $U'(\theta) \cdot h(\theta) \leq -\eta < 0$. We choose T such that $(T-1) \cdot \eta \geq c_2 - c_1$. Further, we let ϕ denote a function of class C^2 on $\mathbb{R}^d$ with bounded second order derivatives, which coincides with U on $K(c_2)$, and which is greater than or equal to c_2 outside $K(c_2)$. We define the integer valued r.v:

$$\rho = \sup\{n; n \leq \tau(c_2), \theta_n \in K(c_1)\}$$
$$\mu = \inf\{n; n > \rho, \gamma_{\rho+1} + \ldots + \gamma_{n+1} \geq T\}$$

ρ is the last time $K(c_1)$ is visited before the exit from $K(c_2)$.

Let $\Omega_1 = \{\tau(c_2) < +\infty, \sigma(\varepsilon) \geq \tau(c_2)\}$ and $\bar{\mu} = \mu \wedge \tau(c_2)$. On Ω_1, on the one hand $\bar{\mu} \leq m(\rho, T)$, $\bar{\mu} \leq \tau(c_2) \wedge \sigma(\varepsilon) = \nu(\varepsilon, K(c_2))$, and following (3.2.1)

$$\phi(\theta_{\bar{\mu}}) - \phi(\theta_\rho) - \sum_{i=\rho}^{\bar{\mu}-1} \gamma_{i+1} \phi'(\theta_i) \cdot h(\theta_i) = \sum_{i=\rho}^{\bar{\mu}-1} \varepsilon_i(\phi) \qquad (3.4.3)$$

The term on the left in (3.4.3) is bounded below on Ω_1 by $c_2 - c_1$ if $\bar{\mu} = \tau(c_2)$ and by $\eta(T-1) \geq c_2 - c_1$ if $\bar{\mu} < \tau(c_2)$ (since for $i \in [\rho, \bar{\mu}-1]$, $-\phi'(\theta_i) \cdot h(\theta_i) \geq \eta$). Thus, from Corollary 8, we have

$$(c_2 - c_1)^q \cdot P_{x,a}(\Omega_1) \leq E_{x,a}\{I(\Omega_1)|\sum_{i=\rho}^{\bar{\mu}-1} \varepsilon_i(\phi)|^q\}$$
$$\leq E_{x,a}\{\sup_n \sup_{n < k \leq m(n,T)} |\sum_{i=n}^{k-1} \varepsilon_i(\phi)|^q\}$$
$$\leq B_2(1 + |x|^{s_2}) \sum_{i \geq 1} \gamma_i^{1+q/2}$$

$\square$

From this lemma, it is easy to deduce

Proposition 14. *Given c_1, c_2, q satisfying $c_0 < c_1 < c_2 < C$, $q \geq q_0(\alpha)$, there exist ε_0, B_3, s_3 such that for all $a \in K(c_1)$, all x,*

$$P_{x,a}(\sigma(\varepsilon_0) = +\infty, \tau(c_2) = +\infty) \geq 1 - B_3(1 + |x|^{s_3}) \sum_{k \geq 1} \gamma_k^{1+q/2}$$

Proof. The set $\{\sigma(\varepsilon_0) = +\infty, \tau(c_2) = +\infty\}$ has complement

$$\{\tau(c_2) < +\infty, \sigma(\varepsilon_0) \geq \tau(c_2)\} \cup \{\sigma(\varepsilon_0) < +\infty, \tau(c_2) \geq \sigma(\varepsilon_0)\}$$

If we choose $\varepsilon_0 = \varepsilon_0(q, K(c_2))$ as in Proposition 5, the first of these sets is bounded above by Lemma 13, the second is bounded above by Corollary 10.

$\square$

Corollary 15. *With the assumptions of Proposition 14*

$$P_{x,a}(\tau(c_2) < +\infty) \leq B_3(1 + |x|^{s_3}) \sum_{k \geq 1} \gamma_k^{1+q/2}$$

3.4.2

Now we shall study the convergence of the algorithm towards the compact set F satisfying (3.4.2).

Proposition 16. *Suppose c and ε satisfy $c_0 < c < C$ and $\varepsilon \leq \varepsilon_0(q, K(c))$ for some $q \geq q_0(\alpha)$. Then for all x and all a in the interior of $K(c)$, θ_n converges $P_{x,a}$ a.s. towards F on $\{\tau(c) = +\infty, \sigma(\varepsilon) = +\infty\}$.*

Proof. Choose c_1 arbitrarily such that $c_0 < c_1 < c$ and

$$\Omega_2 = \{\tau(c) = +\infty, \sigma(\varepsilon) = +\infty\} \cap \{\limsup U(\theta_n) > c_1\}$$

Then we must show that $P_{x,a}(\Omega_2) = 0$.

Suppose c' is such that $c_0 < c' < c_1 < c$. Following (A.7) and (3.4.2), there exists $\eta > 0$ such that, for $\theta \in K(c) - K(c')$, $U'(\theta) \cdot h(\theta) \leq -\eta$. Choose T sufficiently large that $(T-1)\eta - c \geq c_1 - c'$ and $(T-2)\eta \geq c_1 - c'$.

We shall construct a sequence (V_r, W_r) of integer valued r.v. such that on Ω_2, $r < V_r < W_r \leq m(V_r, T)$ and :

$$U(\theta_{W_r}) - U(\theta_{V_r}) - \sum_{i=V_r}^{W_r - 1} \gamma_{i+1} U'(\theta_i) \cdot h(\theta_i) \geq c_1 - c' \quad (3.4.4)$$

We begin with $N \geq 1$. Let $\rho = \inf(n \geq N, \theta_n \in K(c'))$.

Case 1. $\rho = +\infty$ We set $V = N, W = m(N, T)$. Then (3.4.4) holds for the 2-tuple (V, W), since on $[V, W[$, $\theta_n \in K(c) - K(c')$ and $U(\theta_W) - U(\theta_V) \geq -c$, in view of the choice of T.

Case 2. $\rho < +\infty$. Here we define $\mu = \inf(n > \rho, \theta_n \notin K(c_1))$ (note that on Ω_2 we have $\mu < +\infty$), $\bar{\rho} = \sup(n \geq \rho, n \leq \mu, \theta_n \in K(c'))$ and $\bar{\mu} = \inf(n > \bar{\rho}, \gamma_{\bar{\rho}+1} + \ldots \gamma_{n+1} \geq T)$.

Let $V = \bar{\rho}, W = \mu \wedge \bar{\mu}$. If $\mu \leq \bar{\mu}$, $\theta_W \notin K(c_1), \theta_V \in K(c')$, and so $U(\theta_W) - U(\theta_V) \geq c_1 - c'$; if $\mu > \bar{\mu}$, then $\theta_W \notin K(c'), \theta_V \in K(c')$, whence $U(\theta_W) - U(\theta_V) \geq 0$ and $i \in]V, W[$, $\theta_i \in K(c) - K(c')$ whence

$$-\sum_{i=V}^{W-1} \gamma_{i+1} U'(\theta_i) h(\theta_i) \geq \eta \sum_{i=V+1}^{m(V,T)-1} \gamma_{i+1} \geq \eta \cdot (T-2)$$

and so (3.4.4) holds.

For $N = 1$, we apply this construction to obtain V_1, W_1, next for $N = W_1$, we obtain V_2, W_2, etc.

If ϕ is a regular extension of U outside $K(c)$, then on Ω_2 we have

$$c_1 - c' \leq \sum_{i=V_r}^{W_r - 1} \varepsilon_i(\phi) \leq \sup_{n \geq r} \sup_{n < k \leq m(n,T)} |\sum_{i=n}^{k-1} \varepsilon_i(\phi)|$$

However, in view of our assumptions, this last term tends to 0 a.s. on Ω_2 by Corollary 8-ii. Whence the proposition is true. □

3.4.3 Consequence of Propositions 14 and 16

Theorem 17. *We assume (A.1), (A.2), (A.3), (A.4), (A'.5), (A'.6) and (A.7), and suppose that F is a compact set satisfying (3.4.2). Then, for any compact subset Q of D, and $q \geq q_0(\alpha)$, there exist constants B_4, s_4, such that for all $a \in Q$ and all $x \in \mathbb{R}^k$:*

$$P_{x,a}(\theta_n \text{ converges to } F) \geq 1 - B_4(1 + |x|^{s_4}) \cdot \sum_{k \geq 1} \gamma_k^{1+q/2}$$

For the distributions $P_{n,x,a}$ (see Chapter 1, Subsection 1.1.1), we have

Corollary 18. *For all $n \geq 0$, all $a \in Q$, all $x \in \mathbb{R}^k$, we have*

$$P_{n,x,a}(\theta_n \text{ converges to } F) \geq 1 - B_4(1 + |x|^{s_4}) \cdot \sum_{k \geq n+1} \gamma_k^{1+q/2}$$

Now we have an alternative form of the convergence theorem.

Corollary 19. *Let Q be a compact subset of D and let Y be a finite positive r.v. Denote*

$$\tilde{\Omega}(Q,Y) = \{\omega : \text{ for infinitely many } n, \theta_n(\omega) \in Q \text{ and} |X_n(\omega)| \leq Y(\omega)\}$$

Then θ_n converges a.s. to F on $\tilde{\Omega}(Q,Y)$.

Proof. Denote

$$A = \{\theta_n \text{ converges to } F\}$$
$$\tilde{\Omega}_m = \{\omega; \text{ for infinitely many } n, \theta_n(\omega) \in Q \text{ and } |X_n(\omega)| \leq m\}$$

It is clear that $\tilde{\Omega}_m$ is increasing towards $\tilde{\Omega}(Q,Y)$. Define $\tau_0 = 0$ and $\tau_k = \inf(n > \tau_{k-1}, \theta_n \in Q \text{ and } |X_n| \leq m)$. By construction, the sequence τ_k is strictly increasing and finite on $\tilde{\Omega}_m$. Moreover, on $\{\tau_k < +\infty\}$, $A = \Theta_{\tau_k}(A)$, where, as usual, Θ_t denotes the translation onto the canonical space of the trajectories.

The Markov property then implies that for all k

$$\begin{aligned} P(A^c \cap \tilde{\Omega}_m) &\leq P(A^c \cap (\tau_k < +\infty)) \\ &\leq E\{I(\tau_k < +\infty)P_{\tau_k, X_{\tau_k}, \Theta_{\tau_k}}(A)\} \\ &\leq B_4(1 + |m|^{s_4}) \sum_{i \geq k} \gamma_{i+1}^{1+q/2} \end{aligned}$$

Thus $P(A^c \cap \tilde{\Omega}_m) = 0$, and so $P(A^c \cap \tilde{\Omega}) = 0$.

3.5 "Tube of Confidence" for an Infinite Horizon

We shall combine Theorems 17 and 12 to give a detailed description of the convergence of θ_n when D is the domain of attraction of a point θ_*. We replace (A.7) by (A'.7):

(A'.7) *There exists a positive function U of class C^2 on D, such that*
$$U(\theta) \to C \leq +\infty \text{ if } \theta \to \partial D \text{ or } |\theta| \to +\infty \text{ and } U(\theta) < C \text{ if } \theta \in D$$
which satisfies

(i) $\qquad\qquad U(\theta_*) = 0,\ U(\theta) > 0,\ \theta \in D,\ \theta \neq \theta_*$

(ii) $\qquad\qquad U'(\theta) \cdot h(\theta) < 0$ for all $\theta \in D, \theta \neq \theta_*$

Clearly $F = \{\theta_*\}$ satisfies (3.4.2) for $c_0 = 0$.

Theorem 20. *We assume (A.1), (A.2), (A.3), (A.4),(A'.5), (A'.6), (A'.7). Then for any compact subset Q of D, for all $q \geq q_0(\alpha)$, all $\delta > 0$, there exist constants B_5, s_5, such that for all $n \geq 0$, all $a \in Q$, all x:*

$$P_{n,x,a}(\sup_{k \geq n} |\theta_k - \bar{\theta}(t_k; t_n, a)| > \delta) \leq B_5 \cdot (1 + |x|^{s_5}) \cdot \sum_{k > n} \gamma_k^{1+q/2}$$

Proof. It is clearly sufficient to prove the theorem for $n = 0$ and Q of the form $K(c_1), c_1 < C$. If c_2 is such that $c_1 < c_2 < C$, we may suppose that $\delta < d(K(c_1), K(c_2)^c)$. We denote (see Proposition 14)

$$\Omega_1 = \{\tau(c_2) = +\infty, \sigma(\varepsilon_0(q, K(c_2))) = +\infty\}$$

Choose $\delta_2 > \delta_1$ and $\eta \in]0, \delta/4[$ such that

$$B(\theta_*, \delta/2) \supset K(\delta_2) \supset K(\delta_1) \supset B(\theta_*, \eta)$$

Choose $T \geq 2$ such that for all $a \in K(c_1)$ and $t \geq \frac{T}{2}$, $|\bar{\theta}(t; 0, a) - \theta_*| \leq \eta/2$. Then

$$P\{\sup_k |\theta_k - \bar{\theta}(t_k)| > \delta\} \leq P\{\Omega_1^c\} + P\{\sup_{k \leq m(T)} |\theta_k - \bar{\theta}(t_k)| \geq \eta/2\} + P\{\Omega_2\}$$

where

$$\Omega_2 = \Omega_1 \cap \{\sup_{k \leq m(T)} |\theta_k - \bar{\theta}(t_k)| \leq \frac{\eta}{2}\} \cap \{\sup_{k > m(T)} |\theta_k - \bar{\theta}(t_k)| > \delta\}$$

The first probability is bounded above by Proposition 14, the second by Theorem 12. In the latter case, note that on Ω_2, $\theta_{m(T)} \in B(\theta_*, \eta)$, since $\bar{\theta}_{m(T)} \in B(\theta_*, \eta/2)$ by virtue of the choice of T, whilst, on the other hand, there exists $k > m(T)$ such that $\theta_k \notin B(\theta_*, \delta/2)$, since for $k \geq m(T)$, we have $\bar{\theta}(t_k) \in B(\theta_*, \eta/2)$. Thus

$$\Omega_2 \subset \{\theta_{m(T)} \in K(\delta_1)\} \cap \{\nu(\varepsilon_0, K(c_2)) \geq m(T)\} \cap \cup_{k > m(T)} \{\theta_k \notin K(\delta_2)\}$$

whence

$$P(\Omega_2) \leq E\{I(\theta_{m(T)} \in K(\delta_1), \nu \geq m(T))P_{m(T),X_{m(T)},\theta_{m(T)}}(\tau(\delta_2) < +\infty)\}$$

Applying Corollary 15 we obtain

$$\begin{aligned}P(\Omega_2) &\leq B_3 E\{I(m(T) \geq \nu)(1 + |X_{m(T)}|^{s_3})\} \cdot \sum_{k > m(T)} \gamma_k^{1+q/2} \\ &\leq B_4(1 + |x|^{s_4}) \sum_{k \geq 1} \gamma_k^{1+q/2}\end{aligned}$$

and this completes the proof. □

3.6 Final Remark. Connections with the Results of Chapter 1

We end this chapter by observing that if we replace (A'.5) by (A.5) in Theorems 12, 17 and 20, all the results remain a fortiori true. We may replace $\nu(\varepsilon, Q)$ in all the proofs by $\tau(Q)$, and use (A.5) instead of Proposition 1 to control the moments. Thus the case in which we have only $\sum \gamma_n^\alpha < +\infty$ for some $\alpha > 1 + \lambda$ can be covered by the framework of Chapter 1.

3.7 Comments on the Literature

This chapter essentially follows (Métivier and Priouret 1987), with occasional extensions. The reader is referred to the comments on the literature in Chapter 1 for further details.

Chapter 4
Gaussian Approximations to the Algorithms

Introduction

In this chapter, we shall formally state and prove the "asymptotic normality" properties described in Chapter 3 of Part I.

Firstly we shall consider algorithms with constant step size, and in particular, the case in which the algorithm

$$\theta^\gamma_{n+1} = \theta^\gamma_n + \gamma H(\theta^\gamma_n, X_{n+1})$$

is such that the associated continuous-time step function $\theta^\gamma(t)$ converges to the solution $\bar\theta(t)$ of the mean differential equation, in the sense studied in the previous chapter:

$$\lim_{\gamma \to 0} P\{\sup_{0 \le t \le T} |\theta^\gamma(t) - \bar\theta(t)| > \eta\} = 0$$

for all $T > 0$ and $\eta > 0$.

In order to investigate the "quality" of the approximation of $\theta^\gamma(t)$ by $\bar\theta(t)$ (or the "rate of convergence" as $\gamma \to 0$), it was suggested in Part I, that we should study the process $\gamma^{-1/2}(\theta^\gamma(t) - \bar\theta(t))$. If we multiply the fluctuations of this process by $\sqrt\gamma$, we obtain the fluctuations of $\theta^\gamma(t)$ about $\bar\theta(t)$.

This process is in general difficult to study directly, whence, often a process with a "similar distribution" for small γ is substituted in its place. Chapter 3 of Part I expresses the fact that one such process is a "Gaussian diffusion".

Rigorously therefore, we are led to state a theorem describing the convergence of the process $\gamma^{-1/2}(\theta^\gamma(t) - \bar\theta(t))$ to the distribution of a Gaussian diffusion as γ tends to zero.

To state such a theorem, we require precise mathematical notions both about the convergence of process distributions and about diffusions.

Thus in Section 4.1, we recall some basic notions about the distributions of stochastic processes and their "weak convergence".

In Section 4.2, the basic concepts of diffusions are recalled, if only to characterise the distribution of a diffusion and to demonstrate how this characterisation may be used to calculate (in general by solving partial differential equations) moments of the process functions.

Section 4.3 studies some properties of the process $U^\gamma(t) = \frac{\theta^\gamma(t) - \bar\theta(t,a)}{\sqrt\gamma}$ associated with an algorithm with constant step size, including properties of the moments, and properties which are useful when passing to the limit as $\gamma \to 0$.

In Section 4.4, we state and prove a theorem giving the convergence of the distribution of U^γ to the distribution of a Gaussian diffusion when $\gamma \to 0$. In fact, this is the proof of Theorem 1 of Part I, Chapter 3.

In Section 4.5, we return to the problem of Gaussian approximation for algorithms with decreasing step size, as introduced in Chapter 3 of Part I. We essentially prove Theorem 3 of Chapter 3 of Part I.

In Section 4.6, we give a precise formulation of Theorem 2 of Chapter 3 of Part I, for the asymptotic Gaussian approximation of algorithms with constant step size.

4.1 Process Distributions and their Weak Convergence

In this section, we recall some of the notions and principal results relating to the (weak) convergence of the distributions of stochastic processes.

4.1.1 The Skorokhod Space $\bar{\Omega}_T = \mathbb{D}([0,T]; \mathbb{R}^d)$

$\mathbb{D}([0,t]; \mathbb{R}^d)$ will denote the set of functions from $[0,T]$ into $\mathbb{R}^d$ which are right-continuous and have a left limit for any point $t \in [0,T]$ (these are called rc–ll processes in current terminology). We shall denote $\tilde{\Omega}_T = \mathbb{D}([0,T]; \mathbb{R}^d)$, for simplicity.

Most stochastic process met in practice have trajectories which are elements of $\tilde{\Omega}_T$; the processes $(\theta(t))_{t \in [0,T]}$ considered in the previous chapter clearly have this property.

To obtain a sensible definition of "similar trajectories", we usually impose a metric structure on this set, thereby turning it into a complete metric space. This structure may be defined by the following distance function, which we mention only "for information", since we shall not use it explicitly in what follows. The distance $\bar{\delta}(\tilde{\omega}_1, \tilde{\omega}_2)$ between the trajectories $\tilde{\omega}_1$ and $\tilde{\omega}_2$ is given by (cf. (Billingsley 1968));

$$\bar{\delta}(\tilde{\omega}_1, \tilde{\omega}_2) = \inf_{\lambda \in \Lambda_T} \{ \sup_{0 \le t \le T} [|\lambda(t) - t| + |\tilde{\omega}_1(t) - \tilde{\omega}_2(t)|] + \sup_{s \ne t} |\log \frac{\lambda(t) - \lambda(s)}{t - s}| \}$$

(4.1.1)

where Λ_t denotes the set of all increasing homeomorphisms from $[0,T]$ to $[0,T]$. This metric induces the so-called **"Skorokhod topology"**.

Comments.

To interpret the notion of distance defined by (4.1.1), it is easy to see that the metric induced by restriction to the subset $\tilde{\Omega}_{T,c} = C([0,T]; \mathbb{R}^d)$ of continuous functions of $[0,T]$ into $\mathbb{R}^d$ is none other than the metric of uniform convergence on $[0,T]$.

4.1 Process Distributions and their Weak Convergence

Next we see, following (4.1.1), that two functions $\tilde{\omega}_1$ and $\tilde{\omega}_2$ are "similar" if for a "small" alteration in the time λ (i.e. a function λ uniformly similar to the identity with "derivative" uniformly similar to 1), $\tilde{\omega}_1(\lambda(t))$ and $\tilde{\omega}_2(t)$ are two functions which are uniformly similar on $[0,T]$. This is expressed intuitively by the statement that two functions are similar in $\tilde{\Omega}_T$ if, "in any interval in which they are both continuous, they are uniformly similar, and jumps in the functions occur at similar points and are similar in magnitude".

The Borel Field of $\tilde{\Omega}_T$.

It can be shown that the Borel field of the complete metric space $\tilde{\Omega}_T$ is the same as the σ-field generated by the functions $\tilde{\omega} \to \tilde{\omega}(t), t \in [0,T]$ (cf. (Billingsley 1968)). It is denoted by $\tilde{F}_T$.

It follows that if $(u(T))_{t \in [0,T]}$ is a stochastic process defined on a probability space (Ω, F, P), whose trajectories $t \to U(t, \omega)$ are rc-ll, then $\omega \to U(., \omega)$ defines a measurable function of $(\Omega.F)$ in $\tilde{\Omega}_T$.

Distribution of an rc-ll Process.

Consequently for such a process U, we can talk about the image measure $\tilde{P}$ of P under the mapping $\omega \to U(., \omega) \in \tilde{\Omega}_T$. This probability distribution $\tilde{P}$ on $\tilde{F}_T$ is called the **distribution of the random function** U.

Canonical Process and Canonical Filtration on $\tilde{\Omega}_T$.

If, for all $\tilde{\omega} \in \tilde{\Omega}_T$ we set

$$\xi_t(\tilde{\omega}) = \tilde{\omega}(t)$$

then $(\xi_t)_{t \in [0,T]}$ defines a process on $(\tilde{\Omega}_t, \tilde{F}_T)$.

We define the σ-field $\tilde{F}_t$ by

$$(\tilde{F}_t)_{t \in [0,T]} = \cap_{s>t} \{\sigma\text{-field generated by } \xi_u : u \le s\}$$

This increasing family of σ-fields $(\tilde{F}_t)_{t \in [0,T]}$ (which by construction has the following so-called right-continuity property: $\tilde{F}_t = \cap_{s>t} \tilde{F}_s$) is called the "canonical" filtration of $\tilde{\Omega}_T$. Note that if two distribution functions $\tilde{P}_1$ and $\tilde{P}_2$ on $(\tilde{\Omega}_T, \tilde{F}_T)$ are such that for any finite subset $\{t_0, t_1, \ldots, t_n\} \subset [0,T]$ the distributions of $(\xi_{t_1}, \ldots, \xi_{t_n})$ are identical for $\tilde{P}_1$ and $\tilde{P}_2$, then $\tilde{P}_1 = \tilde{P}_2$ (this follows since $\tilde{F}_T$ is generated by $\xi_t, t \in [0,T]$).

Note lastly that, in view of the definition of δ in (4.1.1), the function $\tilde{\omega} \to \xi_t(\tilde{\omega})$ is not continuous at any point $\tilde{\omega}_0$ such that $\tilde{\omega}_0$ is discontinuous in t. On the other hand, if $\tilde{\omega}_0 \in \tilde{\Omega}_{T,c}$, then the mapping $\tilde{\omega} \to \xi_t(\tilde{\omega})$ is continuous at the point $\tilde{\omega}_0$.

4.1.2 Weak Convergence of Probabilities on $\tilde{\Omega}_T$

If $(P_n)_{n \in \mathbb{N}}$ is a sequence of probability distributions on $(\tilde{\Omega}_T, \tilde{F}_T)$, we shall say that this sequence converges weakly to a distribution $\tilde{P}$ if, for any bounded function Ψ on $\tilde{\Omega}_T$

$$\lim_{n \to \infty} \int \Psi d\tilde{P}_n = \int \Psi d\tilde{P} \tag{4.1.2}$$

The sequence $(\tilde{P}_n)_{n \in \mathbb{N}}$ is said to be weakly compact if every subsequence of it contains a weakly convergent subsequence. It is then clear that the sequence $(\tilde{P}_n)_{n \in \mathbb{N}}$ converges weakly to $\tilde{P}$ if and only if:

(i) it is weakly compact

(ii) every convergent subsequence converges to $\tilde{P}$

The importance of weak convergence is clear from the defining formula (4.1.2): if a functional of a process X_n may be expressed as a continuous function Ψ of the trajectories of X_n, with a distribution similar to that of the process X, then the expectation of this functional of X_n is approximated by the expectation of the functional Ψ of the trajectories of X. The practical importance of this is clear if one or other of these expectations lends itself more readily to numerical evaluation.

We recall the following proposition (cf. (Billingsley 1968)) which shows that the approximation is valid for a much larger class of functionals other than continuous functions.

Proposition 1. *If $(\tilde{P}_n)$ converges weakly to $\tilde{P}$ and if Ψ is a bounded function on $\tilde{\Omega}_T$ such that $\tilde{P}\{\tilde{\omega} : \tilde{\omega} \in \tilde{\Omega}_T, \Psi \text{ is continuous at } \tilde{\omega}\} = 1$, then*

$$\lim_{n \to \infty} \int \Psi d\tilde{P}_n = \int \Psi d\tilde{P}$$

4.1.3 Criteria for Weak Compactness

For information, we give criteria for the weak compactness of a sequence of distributions $(\tilde{P}_n)$. In fact, in what follows, we shall only use the sufficient condition stated in Subsection 4.1.4 (below).

If $\tilde{\Omega}_T$ is a complete metric space, then there is a very general criterion (due to Prokhorov) for the weak compactness of a sequence $(\tilde{P}_n)$ of probability distributions on $\tilde{\Omega}_T$. This may be stated as follows: $(\tilde{P}_n)$ is weakly compact if it is **tight**, i.e. if for all $\varepsilon > 0$ there exists a compact subset K_ε of $\tilde{\Omega}_T$ such that

$$\sup_n \tilde{P}_n(\tilde{\Omega}_T - K_\varepsilon) \leq \varepsilon \tag{4.1.3}$$

Using a suitable characterisation of the compact sets of $\tilde{\Omega}_T$ (which in some ways extends Ascoli's Theorem for spaces of continuous functions), we obtain

4.1 Process Distributions and their Weak Convergence

the following criterion (Billingsley 1968):

Proposition 2. *The sequence* $(\tilde{P}_n)$ *of probability distributions on* $\tilde{\Omega}_T$ *is weakly compact (or tight) if and only if the following two conditions are satisfied:*

[T_1] *For all* $t \in [0,T]$, *the distributions of* ξ_t *for the probabilities* $\tilde{P}_n$ *form a tight sequence of probability distributions on* $\mathbb{R}^d$.

[T_2] *For all* $\eta > 0, \varepsilon > 0$, *there exist* $\delta > 0$ *and* $n_0 \in \mathbb{N}$, *such that*

$$\sup_{n \geq n_0} \tilde{P}_n \{\tilde{\omega} : W^T(\tilde{\omega}, \delta) > \eta\} \leq \varepsilon \qquad (4.1.4)$$

where $W^T(\tilde{\omega}, \delta)$ *is defined by*

$$W^T(\tilde{\omega}, \delta) = \inf_{\Pi_\delta} \max_{t_i \in \Pi_\delta} \sup_{t_i \leq s < t \leq t_{i+1}} |\tilde{\omega}(t) - \tilde{\omega}(s)|$$

(here $\{\Pi_\delta\}$ *denotes the set of all finite increasing sequences* $0 = t_0 < t_1 \ldots < t_n = T$ *such that* $\inf\{|t_{i+1} - t_i| : t_i \in \Pi_\delta\} \geq \delta$).

4.1.4 Sufficient Condition for Weak Convergence

Condition [T_2] is often difficult to verify directly. For this reason we use sufficient conditions which imply [T_2] and which are easier to handle in practice. One such condition which we shall use in the sequel is given by the following proposition.

Proposition 3 *If for all* $\varepsilon > 0, \eta > 0$ *there exist* $\delta > 0$ *with* $0 < \delta < 1$ *and* $\tilde{n}_\delta \in \mathbb{N}$ *such that for all* $t \in [0,T]$:

$$\sup_{n \geq \tilde{n}_\delta} \frac{1}{\delta} \tilde{P}_n \{ \sup_{t \leq s \leq t+\delta} |\xi_s - \xi_t| \geq \eta \} \leq \varepsilon \qquad (4.1.5)$$

then the sequence $(\tilde{P}_n)$ *satisfies condition* [T_2].

In fact, it is easy to see that if (4.1.5) is satisfied, consideration of the partition $\{0, \delta, 2\delta, \ldots, k\delta, \ldots, T\} \in \Pi_\delta$ shows that

$$\tilde{P}_n\{\tilde{\omega} : W^T(\tilde{\omega}, \delta) > 2\eta\} \leq \sum_{k \leq T/\delta} \tilde{P}_n\{ \sup_{k\delta \leq s \leq (k+1)\delta} |\xi_s - \xi_{k\delta}| \geq \eta \} \leq T\varepsilon$$

Remark. We can show (cf. (Billingsley 1968)) that if the sequence $(\tilde{P}_n)$ satisfies (4.1.5) then any weak limit $\tilde{P}$ of the sequence $(\tilde{P}_n)$ is "carried" by $\tilde{\Omega}_{T,c}$ the set of continuous trajectories, i.e. $\tilde{P}(\tilde{\Omega}_{T,c}) = 1$.

Consequently, following Proposition 1, if $(\tilde{P}_n)$ converges to $\tilde{P}$ we have

$$\lim_{n \to \infty} \int \Psi(\xi_{t_1}, \ldots, \xi_{t_k}) d\tilde{P}_n = \int \Psi(\xi_{t_1}, \ldots, \xi_{t_k}) d\tilde{P}$$

for any continuous bounded function Ψ on $(\mathbb{R}^d)^k$.

4.2 Diffusions. Gaussian Diffusions

4.2.1 Diffusions

For each $t \in \mathbb{R}_+$, let L_t denote a differential elliptic operator of the form

$$L_t \Psi(x) = \sum_{i=1}^{d} b^i(t,x) \partial_i \Psi(x) + \frac{1}{2} \sum_{i,j=1}^{d} R^{ij}(t,x) \partial_{ij}^2 \Psi(x) \qquad (4.2.1)$$

where Ψ is a twice-differentiable function. A diffusion $(X(t))_{t \in \mathbb{R}_+}$ in $\mathbb{R}^d$ associated with L_t, is by definition a Markov process defined on a probability space $(\Omega, F, P, (F_t)_{t \geq 0})$ of continuous trajectories such that for every function Ψ of class C^2 with compact support, the process

$$\left(\Psi(X_t) - \Psi(X_0) - \int_0^t L_s \Psi(X_s) ds\right)_{t \geq 0} \qquad (4.2.2)$$

is a martingale for the family $(F_t)_{t \geq 0}$ of σ-fields and the distribution P.

We observe that the martingale property of process (4.2.2) implies in particular that for all $s < t$ and for any function Ψ of class C^2

$$E(\Psi(X_t)) = E(\Psi(X_0)) + \int_0^t E L_s \Psi(X_s) ds \qquad (4.2.3)$$

In particular, if μ_t denotes the probability distribution of X_t and if L_t^* is the adjoint of L_t in the sense of the distributions, then formula (4.2.3) may be written as

$$\langle \Psi, \mu_t \rangle = \langle \Psi, \mu_0 \rangle + \int_0^t \langle \Psi, L_s^* \mu_s \rangle ds \qquad (4.2.4)$$

This equation shows that the family of probability distributions $(\mu_t)_{t \geq 0}$ is in a sense a "weak" solution of the equation of evolution

$$\begin{cases} \frac{d\mu_t}{dt} = L_t^* \mu_t \\ \mu_0 = \text{distribution of } X_0 \end{cases} \qquad (4.2.5)$$

If equation (4.2.5) has a unique solution $(\mu_t)_{t \geq 0}$, we see that for each fixed t, the distributions of the variables X_t are uniquely determined by equation (4.2.5). Moreover (cf. (Stroock and Varadhan 1969)), if the process X is such that (4.2.2) is a martingale, and if for all x the equation

$$\begin{cases} \frac{d\mu_t}{dt} = L_t^* \mu_t \\ \mu_0 = \varepsilon_x \end{cases} \qquad (4.2.6)$$

has a unique solution, then the distribution of the process $(X_t)_{t \geq 0}$ is uniquely determined by the martingale property (4.2.2) and by the given distribution μ_0 of X_0. This distribution $\tilde{P}_{\mu_0}$ is said to be the **unique solution** of the martingale problem $(\mu_0, (L_t)_{t \geq 0})$. It can also be shown that the trajectories of

4.2 Diffusions. Gaussian Diffusions

(X_t) are almost surely continuous; i.e. the distribution $\tilde{P}_{\mu_0}$ is carried by the subspace $C([0,T];\mathbb{R}^d)$ of $\mathbb{D}([0,T],\mathbb{R}^d)$.

In Chapter 3 of Part I (Theorem 1) some diffusions were expressed as solutions of stochastic differential equations. We recall (although we shall not use the result in what follows) that if $(X(t))_{t\geq 0}$ may be written in the form

$$X(t) = X(0) + \int_0^t b(s, X_s)ds + \int_0^t \sigma(s, X_s)dW_s$$

where

$$\sigma o \sigma^T = R$$

and W is a standard Wiener process in d dimensions, then X is the diffusion associated with the differential operator (4.2.1) and the initial condition X_0 (cf. for example (Priouret 1973)).

4.2.2 Gaussian Diffusions

If the coefficients R^{ij} and b_i for example are continuous in t and Lipschitz in x, then we have the previous case of uniqueness. A particular instance of this occurs when the b_i are linear in x and continuously dependent on t, and the a_{ij} are functions of t independent of x:

$$L_t \Psi(x) = \sum_{i=1}^d \partial_i \Psi(x)(\sum_i b_j^i x^j) + \frac{1}{2} \sum_{i,j}^d R^{ij}(t) \partial_{ij=1}^2 \Psi(x) \quad (4.2.7)$$

It can be shown in this case that for any initial condition x, the diffusion $(X_t)_{t\geq 0}$ with initial distribution ε_x, associated with L_t defined in (4.2.7) is such that for all $0 < t_1 \ldots < t_n$, the variables $(X_{t_1}, \ldots X_{t_n})$ have a Gaussian distribution. We say that (X_t) is a Gaussian diffusion.

4.2.3 Asymptotic Behaviour of some Homogeneous Gaussian Diffusions

Suppose that the matrix R is independent of t. Since R is positive definite, we denote by $R^{1/2}$ the positive definite matrix such that $R^{1/2} \circ R^{1/2} = R$. Suppose also that b is independent of t. This is the so-called homogeneous case.

The diffusion associated with (4.2.7) may be written as

$$X(t) = e^{-tB}X(0) + \int_0^t e^{-(t-s)B} \circ R^{1/2} dW_s \quad (4.2.8)$$

where B is the matrix b_j^i and W is a standard Wiener process (cf. for example (Métivier 1983) for a proof of this formula).

From formula (4.2.8), it is easy to calculate the second order moments of the Gaussian variables $X(t) - X(s)$. We have

$$E(X(t) - X(s)) = e^{-(t-s)B} X_s \qquad (4.2.9)$$

$$\mathrm{var}(X(t) - X(s)) = \int_s^t e^{(t-u)B} \circ R \circ e^{(t-u)B^T} du \qquad (4.2.10)$$

This has the following direct consequence: suppose that all the eigenvalues of B have a real part less than some number $\eta < 0$, then for all $t \to \infty$, the random variables $(X(t))_{t \geq 0}$ converge in distribution towards a Gaussian variable with mean zero and variance

$$C = \int_0^\infty e^{sB} \circ R \circ e^{sB^T} ds \qquad (4.2.11)$$

4.3 The Process $U^\gamma(t)$ for an Algorithm with Constant Step Size

4.3.1 Summary of the Notation and Assumptions of Chapter 3

We shall consider the algorithm

$$\begin{cases} \theta_{n+1}^\gamma = \theta_n^\gamma + \gamma H(\theta_n^\gamma, X_{n+1}) \\ \theta_0^\gamma = a, \ \theta_n^\gamma \in \mathbb{R}^d, \ X_n \in \mathbb{R}^k, \ \gamma > 0 \end{cases} \qquad (4.3.1)$$

We recall the notation (cf. Part I, Section 1.6)

$$t_n^\gamma = n\gamma \qquad (4.3.2)$$

$$m(n, T) = n + [T/\gamma] + 1, \quad m(T) = m(0, T) \qquad (4.3.3)$$

where $[\rho]$ denotes the integer part of the scalar ρ.

$$A_t^\gamma = \max\{t_r^\gamma : t_r^\gamma \leq t\} = \left[\frac{t}{\gamma}\right] \gamma \qquad (4.3.4)$$

The process $\theta^\gamma(t)$ is as defined in Chapter 1:

$$\theta^\gamma(t) = \theta_n^\gamma \text{ if } t_n^\gamma \leq t < t_{n+1}^\gamma$$

We assume that conditions (A.2), (A.3), (A.4) and (A'.5) of Chapter 3 are satisfied on some open set $D \subset \mathbb{R}^d$. We let $\bar{\theta}(t, a)$ denote the solution of the associated mean differential equation with initial condition a.

4.3.2 The Process U^γ

The process U^γ is defined by

$$U^\gamma(t) = \frac{\theta^\gamma(t) - \bar{\theta}(t, a)}{\sqrt{\gamma}} \qquad (4.3.5)$$

4.3 The Process $U^\gamma(t)$ for an Algorithm with Constant Step Size

Taking into account that

$$\theta^\gamma_{n+1} = \theta^\gamma_n + \gamma h(\theta^\gamma_n) + \gamma(H(\theta^\gamma_n, X_{n+1}) - h(\theta^\gamma_n))$$

using (A.4) and writing θ_n for θ^γ_n, for simplicity when θ_n appears as an index, we may write

$$\theta^\gamma_{n+1} = \theta^\gamma_n + \gamma h(\theta^\gamma_n) + \gamma[\nu_{\theta_n}(X_{n+1}) - \Pi_{\theta_n}\nu_{\theta_n}(X_{n+1})] \qquad (4.3.6)$$

If for all $t > 0$ we set

$$M^\gamma(t) = \sum_k^{t/\gamma} \sqrt{\gamma}(\nu_{\theta_k}(X_{k+1}) - \Pi_{\theta_k}\nu_{\theta_k})(X_k)) \qquad (4.3.7)$$

$$B^\gamma(t) = \sum_{k<t/\gamma} \sqrt{\gamma}(\Pi_{\theta_k}\nu_{\theta_k}(X_k) - \Pi_{\theta_k}\nu_{\theta_k}(X_{k+1})) \qquad (4.3.8)$$

$$\rho^\gamma(t) = \frac{1}{\sqrt{\gamma}} \int_{A^\gamma(t)}^t h(\theta^\gamma(u))du \qquad (4.3.9)$$

the definition of $\theta^\gamma(t)$ then gives

$$\frac{1}{\sqrt{\gamma}}\theta^\gamma(t) = a + \frac{1}{\sqrt{\gamma}} \int_0^t h(\theta^\gamma(u))du + M^\gamma(t) + B^\gamma(t) + \rho^\gamma(t)$$

Whence, taking into account (4.3.5) and the definition of $\bar{\theta}$ leads to

$$U^\gamma(t) = \frac{1}{\sqrt{\gamma}} \int_0^t [h(\theta^\gamma(u)) - h(\bar{\theta}(u,a))]du + M^\gamma(t) + B^\gamma(t) + \rho^\gamma(t) \qquad (4.3.10)$$

In what follows, **we shall consider a fixed compact set $Q \subset D$ and a fixed $T > 0$** such that

$$\{\bar{\theta}(t,a) : 0 \leq t \leq T\} \subset Q$$

We define

$$\tau^\gamma(Q) = \inf\{n : \theta^\gamma_n \notin Q\} \qquad (4.3.11)$$

$$\sigma^\gamma(\varepsilon) = \inf_n\{n : |\theta^\gamma_n - \theta^\gamma_{n-1}| > \varepsilon\} \qquad (4.3.12)$$

$$\nu^\gamma(\varepsilon, Q) = \inf(\tau^\gamma(Q), \sigma^\gamma(\varepsilon)) \qquad (4.3.13)$$

and

$$t^\gamma(Q) = t_{\tau^\gamma(Q)}, \quad t^\gamma(\varepsilon) = t_{\sigma^\gamma(\varepsilon)} \qquad (4.3.14)$$

$$\zeta^\gamma(\varepsilon, Q) = t_{\nu^\gamma(\varepsilon,Q)} \qquad (4.3.15)$$

We recall (Chapter 3, Proposition 1), that for all $q \geq 1$, there exist constants M and $\varepsilon_0 > 0$ such that for all $\varepsilon \leq \varepsilon_0, a \in Q$ all x, and all γ

$$\sup_n E_{x,a}\{|X_n|^q I(n \leq \nu^\gamma(\varepsilon, Q))\} M(1 + |x|^q) \qquad (4.3.16)$$

In what follows, we shall assume that h has continuous first and second derivatives (see Assumption (A.8) of the main theorem of this chapter). Note that if we then set

$$B_1^\gamma(t) = \frac{1}{\sqrt{\gamma}} \int_0^t [h(\theta^\gamma(s)) - h(\bar{\theta}(s,a)) - h'(\bar{\theta}(s,a)) \cdot (\theta(s) - \bar{\theta}(s,a))] ds \quad (4.3.17)$$

then (4.3.10) may be written as

$$U^\gamma(t) = \int_0^t h'(\bar{\theta}(s,a)) \cdot U^\gamma(s) ds + M^\gamma(t) + B^\gamma(t) + B_1^\gamma(t) + \rho^\gamma(t) \quad (4.3.18)$$

where the process M^γ is a martingale, following (A.2). If, for any function Ψ on $\mathbb{R}^d$ we denote

$$\|\Psi\|_Q = \sup_{\theta \in Q} |\Psi(\theta)| \quad (4.3.19)$$

then following (4.3.9) we have

$$I[t < \zeta^\gamma(\varepsilon, Q)] |\rho^\gamma(t)| \leq \|h\|_Q \gamma^{1/2} \quad (4.3.20)$$

and following (4.3.17)

$$I[t < \zeta^\gamma(\varepsilon, Q)] |B_1^\gamma(t)| \leq \gamma^{1/2} |U_t^\gamma|^2 \quad (4.3.21)$$

4.3.3 Upper Bounds on the Moments of U^γ

In this and the following paragraphs, we shall use an upper bound similar to that of Proposition 8, Chapter 1 (or Proposition 5, Chapter 3), in which we replace the sums

$$I[n \leq \tau(Q)] |\sum_{k=0}^{n-1} \varepsilon_k(\phi)|$$

where ϕ is a function of class C^2 on $\mathbb{R}^d$ and

$$\varepsilon_k(\phi) = \gamma_{k+1} \phi'(\theta_k) \cdot [\nu_{\theta_k}(X_{k+1}) - \Pi_{\theta_k} \nu_{\theta_k}(X_{k+1})] + R(\phi, \theta_k, \theta_{k+1})$$

(cf. Chapter 1, Section 1.3) by sums $\tilde{\varepsilon}_k$ which we shall now define.

Let $(\Psi_k)_{k \geq 0}$ be a sequence of random variables with values in $\mathbb{R}^{d'}$ which satisfy the following properties (4.3.22) to (4.3.24):

$$\Psi_k \text{ is } F_k\text{-measurable} \quad (4.3.22)$$

For all $q \geq 1$, and for any compact subset Q of D, there exist $K(Q,q), q' > 1$ and $\lambda \in [\frac{1}{2}, 1]$ such that for all $x \in \mathbb{R}^k, a \in Q$:

$$E_{x,a}\{|\Psi_k|^q I[k+1 \leq \nu^\gamma(\varepsilon, Q)]\} \leq K(Q,q)(1 + |x|^{q'}) \quad (4.3.23)$$

and

$$E_{x,a}\{|\Psi_k - \Psi_{k-1}|^q I[k+1 \leq \nu^\gamma(\varepsilon, Q)]\} \leq K(Q,q)(1 + |x|^q) \gamma^{\lambda q} \quad (4.3.24)$$

Let $u(\theta, x)$ be a function from $\mathbb{R}^d \times \mathbb{R}^k$ to $\mathbb{R}^{d'}$ which has the following property:

4.3 The Process $U^\gamma(t)$ for an Algorithm with Constant Step Size

(L) *For any compact subset Q of D, there exist constants C_3, C_4, q_3, q_4, $\lambda \in [\frac{1}{2}, 1]$ such that for all $x \in \mathbb{R}^k, \theta, \theta' \in Q$ we have*

(i) $\qquad |u_\theta(x)| \leq C_3(1 + |x|^{q_3})$

(ii) $\qquad |\Pi_\theta u_\theta(x) - \Pi_{\theta'} u_{\theta'}(x)| \leq C_4 |\theta - \theta'|^\lambda (1 + |x|^{q_4})$

For any decreasing sequence $(\gamma_k)_{k \geq 0}, \gamma_k > 0$, we define:

$$\tilde{\varepsilon}_k(\Psi, u) = \gamma_{k+1} \Psi_k \cdot [u_{\theta_k}(X_{k+1}) - \Pi_{\theta_k} u_{\theta_k}(X_{k+1})] \qquad (4.3.25)$$

$$\tilde{\varepsilon}_k^1(\Psi, u) = \gamma_{k+1} \Psi_k \cdot [u_{\theta_k}(X_{k+1}) - \Pi_{\theta_k} u_{\theta_k}(X_k)] \qquad (4.3.26)$$

$$\tilde{\varepsilon}_k^2(\Psi, u) = \gamma_{k+1}[\Psi_k \cdot \Pi_{\theta_k} u_{\theta_k}(X_k) - \Psi_{k-1} \cdot \Pi_{\theta_{k-1}} u_{\theta_{k-1}}(X_k)] \qquad (4.3.27)$$

$$\tilde{\varepsilon}_k^3(\Psi, u) = (\gamma_{k+1} - \gamma_k) \Psi_{k-1} \cdot \Pi_{\theta_{k-1}} u_{\theta_{k-1}}(X_k) \qquad (4.3.28)$$

$$\tilde{\eta}_{n,r}(\Psi, u) = \gamma_{r+1} \Psi_r \cdot \Pi_{\theta_r} u_{\theta_r}(X_r) - \gamma_n \Psi_{n-1} \cdot \Pi_{\theta_{n-1}} u_{\theta_{n-1}}(X_n) \qquad (4.3.29)$$

then clearly

$$\sum_{k=1}^{n-1} \tilde{\varepsilon}_k(\Psi, u) = \sum_{k=r}^{n-1} [\tilde{\varepsilon}_k^1(\Psi, u) + \tilde{\varepsilon}_k^2(\Psi, u) + \tilde{\varepsilon}_k^3(\Psi, u)] + \tilde{\eta}_{n,r}(\Psi, u)$$

Proposition 4. *For any compact subset Q of D, for all $q \geq 2$, there exist constants $B, s, \varepsilon_0 > 0$ such that for all $\varepsilon \leq \varepsilon_0$, $T > 0$, $x \in \mathbb{R}^k$, $a \in Q$, we have*

1. $\quad E_{x,a}\{|\sup_{n \leq k \leq m^\gamma(n,T)} I(k \leq \nu^\gamma(\varepsilon, Q)) \sum_{i=n}^{k-1} \tilde{\varepsilon}_i^1(\Psi, u)|^q\}$
$$\leq BT^{q/2-1}(1 + |x|^s) \sum_{i=n}^{m^\gamma(n,T)-1} \gamma_{i+1}^{1+q/2}$$

2. $\quad E_{x,a}\{|\sup_{n \leq k \leq m^\gamma(n,T)} I(k \leq \nu^\gamma(\varepsilon, Q)) \sum_{i=n}^{k-1} \tilde{\varepsilon}_i^2(\Psi, u)|^q\}$
$$\leq BT^{q-1}(1 + |x|^s) \sum_{i=n}^{m^\gamma(n,T)-1} \gamma_{i+1}^{1+\lambda q}$$

3. $\quad E_{x,a}\{|\sup_{n \leq k \leq m^\gamma(n,T)} I(k \leq \nu^\gamma(\varepsilon, Q))(\tilde{\eta}_{n,k}(\Psi, u) + \sum_{i=n}^{k-1} \tilde{\varepsilon}_i^3(\Psi, u))|^q\}$
$$\leq B(1 + |x|^s) \sum_{i=0}^{m^\gamma(n,T)-1} \gamma_{i+1}^q$$

4. $$E_{x,a}\{\sup_{n\leq k\leq m^{\gamma}(n,T)} I(k \leq \nu^{\gamma}(\varepsilon,Q))|\sum_{i=n}^{k-1}\tilde{\varepsilon}_i(\Psi,u)|^q\}$$
$$\leq B(1+|x|^s)\{T^{q/2-1}\sum_{k=n}^{m^{\gamma}(n,T)-1}\gamma_{k+1}^{1+q/2}$$
$$+T^{q-1}\sum_{k=n}^{m^{\gamma}(n,T)-1}\gamma_{k+1}^{1+\lambda q} + \sum_{k=n}^{m^{\gamma}(n,T)-1}\gamma_{k+1}^{q}\}$$
$$\leq B(1+|x|^s)\gamma_n^{q/2}(T^{q/2}+T^q)$$

Proof. Points 1 and 3 are proved by restatement of points 1, 3, and 5 of Subsection 3.2.3 of Chapter 3 (or equally of Lemmas 3, 5 and 7 of Chapter 1). To prove point 2, we write

$$\tilde{\varepsilon}_k^2(\Psi,u) = (\gamma_{k+1}(\Psi_k - \Psi_{k-1})\cdot\Pi_{\theta_k}u_{\theta_k}(X_k)$$
$$+\gamma_{k+1}\Psi_{k-1}\cdot(\Pi_{\theta_k}u_{\theta_k}(X_k)-\Pi_{\theta_{k-1}}u_{\theta_{k-1}}(X_k))$$
$$= V_k + W_k$$

Then we use (L) to give the upper bound:

$$E\{\sup_{n<k\leq m}(I(k\leq\nu^{\gamma}(\varepsilon,Q))\sum_{i=n}^{k-1}|V_k|^q\}$$
$$\leq C_3^q E\{|\sum_{i=n}^{m-1}\gamma_{i+1}|\Psi_i-\Psi_{i-1}|(1+|X_i|^{q_3})I(i+1\leq\nu^{\gamma}(\varepsilon,Q))|^q\}$$
$$\leq K(\sum_{i=n}^{m-1}\gamma_{i+1})^{q-1}\sum_{i=n}^{m-1}\gamma_{i+1}E\{|\Psi_i-\Psi_{i-1}|^q(1+|X_i|^{qq_3})I(i+1\leq\nu^{\gamma}(\varepsilon,Q))\}$$

Using (4.3.24), there exist s and B such that

$$E\{\sup_{n<k\leq m}|I(k\leq\nu^{\gamma}(\varepsilon,Q))(\sum_{i=n}^{k-1}V_k)|^q\} \leq B(1+|x|^s)(\sum_{i=n}^{m-1}\gamma_{i+1})^{q-1}\sum_{i=o}^{m-1}\gamma_{i+1}^{1+\lambda q}$$
(4.3.30)

Also following (L-ii) and (A.3) we have

$$E\{\sup_{n<k\leq m}[I(k\leq\nu^{\gamma}(\varepsilon,Q))\sum_{i=n}^{k-1}|W_k|]^q\}$$
$$\leq C_4^q E\{|\sum_{i=n}^{m-1}\gamma_{i+1}^{1+\lambda}|\Psi_{i-1}|(1+|X_i|^{q_4+q_1})|^q\}$$
$$\leq K(\sum_{i=n}^{m-1}\gamma_{i+1})^{q-1}(\sum_{i=n}^{m-1}\gamma_{i+1}^{1+\lambda q}E|\Psi_{i-1}|^q(1+|X_i|^{(q_4+q_1)q}))$$

4.3 The Process $U^\gamma(t)$ for an Algorithm with Constant Step Size

Using property (4.3.23), there exist s and B such that

$$E\{\sup_{n \leq k \leq m} |I(k \leq \nu^\gamma(\varepsilon, Q) \sum_{i=n}^{k-1} W_k|^q\}$$
$$\leq B(1 + |x|^s)(\sum_{i=n}^{m-1} \gamma_{i+1})^{q-1} \sum_{i=n}^{m-1} \gamma_{i+1}^{1+\lambda q} \qquad (4.3.31)$$

Point 2 of the proposition may now be deduced from (4.3.30) and (4.3.31). Point 4 is an immediate consequence of the decomposition of $\tilde{\varepsilon}_k(\Psi, u)$ and of points 1 and 3 of the proposition. $\square$

Proposition 5. *Let Q' be a compact subset containing Q, with*

$$\beta = \inf\{|x - x'| : x \in Q, x' \notin Q'\} > 0$$

Then for all $q \geq 1$, there exist s and a constant C depending only on Q, Q', q and ε, such that for all $\delta > 0, t \geq 0, a \in Q', x \in \mathbb{R}^k$

a.
$$E_{x,a}\{\sup_{t \leq u \leq t+\delta} |(M^\gamma + B^\gamma + \rho^\gamma)(u \wedge \zeta^\gamma(\varepsilon, Q'))$$
$$- (M^\gamma + B^\gamma + \rho^\gamma)(t \wedge \zeta^\gamma(\varepsilon, Q'))|^q\}$$
$$\leq C(1 + |x|^s)(\delta^{q/2} + \gamma^{q/2} + \delta^q)$$

b.
$$E_{x,a}\{\sup_{t \leq u \leq t+\delta} |U^\gamma(u \wedge \zeta^\gamma(\varepsilon, Q')) - U^\gamma(t \wedge \zeta^\gamma(\varepsilon, Q'))|^q\}$$
$$\leq C(1 + |x|^s)[(\delta \vee \gamma)^{q/2} + \delta^q]$$

Proof.

a. Applying Proposition 4-1 to the case $u(\theta, x) = \nu(\theta, x)$, where $\nu(\theta, x)$ is the function of (A.4) and $\Psi_k = 1$ gives the existence of C and s such that

$$E_{x,a}\{\sup_{t \leq u \leq t+\delta} |M^\gamma(u \wedge \sigma(\varepsilon, Q')) - M^\gamma(t \wedge \sigma(\varepsilon, Q'))|^q\}$$
$$\leq C(1 + |x|^s)\delta^{q/2} \qquad (4.3.32)$$

Proposition 4-4 gives

$$E_{x,a}\{\sup_{t \leq u \leq t+\delta} |(M^\gamma + B^\gamma)(u \wedge \zeta^\gamma(\varepsilon, Q')) - (M^\gamma + B^\gamma)(t \wedge \zeta^\gamma(\varepsilon, Q'))|^q\}$$
$$= \gamma^{-q/2} E_{x,a}\{\sup_{m^\gamma(t) \leq k \leq m^\gamma(t+\delta)} I[k+1 \leq \nu^\gamma(\varepsilon, Q')]|\sum_{i=m^\gamma(t)}^{m^\gamma(t+\delta)} \tilde{\varepsilon}_i(1, \nu)|^q\}$$
$$\leq B(1 + |x|^s)(\delta^{q/2} + \delta^q) \qquad (4.3.33)$$

where ν denotes the function $\nu(\theta, x) = \nu_\theta(x)$.

The first part of Proposition 5 follows from this inequality and from

$$\sup_{t \leq \zeta^\gamma(\varepsilon, Q')} |\rho^\gamma(t)| \leq \sqrt{\gamma} \sup_{\theta \in Q'} |h(\theta)| \qquad (4.3.34)$$

b. Using (4.3.10) and part a of the proposition, together with the fact that for $u < \zeta^\gamma(\varepsilon, Q')$:

$$|h(\theta^\gamma(u)) - h(\bar{\theta}(u, a))| \leq \gamma^{1/2} \sup_{\theta \in Q'} |h'(\theta)||U^\gamma(u)| \qquad (4.3.35)$$

for some suitable constant K we have:

$$E_{x,a}\{\sup_{t \leq u \leq t+\delta} |U^\gamma(u \wedge \zeta^\gamma) - U^\gamma(t \wedge \zeta^\gamma)|^q\}$$

$$\leq K\{\delta^{q-1} \int_t^{t+\delta} \sup_{t \leq u' \leq t+u} |U^\gamma(t \wedge \zeta^\gamma) - U^\gamma(u' \wedge \zeta^\gamma)|^q du$$

$$+ (1 + |x|^s)(\delta^{q/2} + \gamma^{q/2} + \delta^q)\}$$

Point b of the proposition follows using Gronwall's Lemma. □

Proposition 6. *Let Q' be a compact set containing Q with*

$$\beta = \inf\{|x - x'| : x \in Q, x' \notin Q'\} > 0$$

Then for all $q \geq 2, t > 0$ and $\varepsilon \leq \varepsilon_0(q, Q')$, there exist constants $K > 0$ and $s \geq 0$ such that for all $a \in Q$ and all x

$$P_{x,a}\{\zeta^\gamma(\varepsilon, Q') < T\} \leq K\gamma^{q/2}(1 + |x|^s)$$

Proof. Using Lemma 9 of Chapter 3, we have

$$P_{x,a}\{t^\gamma(\varepsilon) \leq t^\gamma(Q), t^\gamma(\varepsilon) < T\} \leq \frac{A_1}{\varepsilon^q}(1 + |x|^{s_1}) \cdot T\gamma^{q-1} \qquad (4.3.36)$$

then from Lemma 11 of the same chapter we have

$$E_{x,a}\{\sup_{t < T \wedge \zeta^\gamma(\varepsilon, Q')} |\theta^\gamma(t) - \bar{\theta}(t, a)|^q\}$$

$$\leq A_2(1 + |x|^{s_2})(1 + T)^{q-1} \exp(qL_2 T)T\gamma^{q/2}$$

Since $\bar{\theta}(t, a) \in Q$ for all $t \leq T$ and $a \in Q$ we have

$$P_{x,a}\{t^\gamma(Q') < t^\gamma(\varepsilon), t^\gamma(Q') < T\}$$

$$\leq \frac{A_2}{\beta^q}(1 + |x|^{s_2})(1 + T)^{q-1} \exp(qL_2 T)T\gamma^{q/2} \qquad (4.3.37)$$

The inequality in the proposition follows from (4.3.36) and (4.3.37). □

4.4 Gaussian Approximation of the Processes $U^\gamma(t)$

4.4.1 Assumptions

We shall add the following assumptions to those of Section 3.1.

(A.8) *The function h has continuous first and second order derivatives, for all $\theta \in D$, there exists a unique symmetric $d \times d$ matrix $(R^{ij}(\theta))$, and for all (θ, x), there exists a matrix $(w^{ij}(\theta, x))$ such that (where ν_θ is the function of (A.4))*
(i) R^{ij} is locally Lipschitz on D
(ii) $(I - \Pi_\theta)w_\theta^{ij}(x) = \Pi_\theta \nu_\theta^i \nu_\theta^j(x) - \Pi_\theta \nu_\theta^i(x)\Pi_\theta \nu_\theta^j(x) - R^{ij}(\theta)$
(iii) For any compact subset Q of D, there exist constants K_3, K_4, p_3, p_4, $\mu \in [1/2, 1]$ such that, for all $\theta, \theta' \in Q$ we have

$$|w_\theta^{ij}(x)| \leq K_3(1 + |x|^{p_3}) \qquad (4.4.1)$$
$$|\Pi_\theta w_\theta^{ij}(x) - \Pi_{\theta'} w_{\theta'}^{ij}(x)| \leq K_4|\theta - \theta'|^\mu (1 + |x|^{p_4}) \qquad (4.4.2)$$

4.4.2 Remark on Assumption (A.8) and Interpretation of $R(\theta)$

It is important to note that in the common situation in which the Markov chain with transition probability Π_θ has a distribution with an invariant probability Γ_θ, the matrix $R(\theta)$ of Assumption (A.8), is exactly the same matrix $R(\theta)$ as given in Part I, Chapter 3, Theorem 1.

Let us suppose now, as we did in Chapter 1, Section 1.3, comment c) that the Markov chain with transition probability Π_θ has an invariant probability Γ_θ for which we write as usual

$$h(\theta) = \int H(\theta, y)\Gamma_\theta(dy)$$

and that the solution ν_θ of the Poisson equation

$$(\nu_\theta - \Pi_\theta \nu_\theta) = H_\theta - h(\theta) \qquad (4.4.3)$$

is given by

$$\nu_\theta(y) = \sum_{k \geq 0} \Pi_\theta^k (H_\theta - h(\theta))(y) \qquad (4.4.4)$$

Since Γ_θ is invariant for Π_θ, equation (A.8-ii) shows that

$$R^{ij}(\theta) = \int [\nu_\theta^i \nu_\theta^j(x) - \Pi_\theta \nu_\theta^i(x)\Pi_\theta \nu_\theta^j(x)]\Gamma_\theta(dx) \qquad (4.4.5)$$

We may write

$$\begin{aligned}R^{ij}(\theta) &= \int \nu_\theta^i(x)(\nu_\theta^i(x) - \Pi_\theta \nu_\theta^i(x))\Gamma_\theta(dx) \\ &+ \int \nu_\theta^j(\nu_\theta^i(x) - \Pi_\theta \nu_\theta^i(x))\Gamma_\theta(dx) \\ &- \int (\nu_\theta^i(x) - \Pi_\theta \nu_\theta^i(x))(\nu_\theta^j(x) - \Pi_\theta \nu_\theta^j(x))\Gamma_\theta(dx)\end{aligned}$$

Then using (4.4.3) and (4.4.4) we have

$$\begin{aligned} R^{ij}(\theta) &= \sum_{k\geq 0} \int \Pi_\theta^k(H_\theta^i - h^i(\theta))(x)(H_\theta^j(x) - h^j(\theta))\Gamma_\theta(dx) \\ &+ \sum_{k\leq 0} \int \Pi_\theta^k(H_\theta^j - h^j(\theta))(x)(H_\theta^i(x) - h^i(\theta))\Gamma_\theta(dx) \\ &- \int (H_\theta^i(x) - h^i(\theta))(H_\theta^j(x) - h^j(\theta))\Gamma_\theta(dx) \end{aligned}$$

Suppose that $(X_n^\theta)_{n\in\mathbb{Z}}$ is a stationary Markov chain with transition probability Π_θ and invariant probability Γ_θ. Then

$$\int \Pi_\theta^k [H_\theta^i(x) - h^i(\theta)][H_\theta^j(x) - h^j(\theta)]\Gamma_\theta(dx) = \operatorname{cov}[H^i(\theta, X_k^\theta), H^j(\theta, X_0^\theta)]$$
$$= \operatorname{cov}[H^i(\theta, X_0^\theta), H^j(\theta, X_{-k}^\theta)]$$

Whence

$$R^{ij}(\theta) = \sum_{k=-\infty}^{\infty} \operatorname{cov}[H^i(\theta, X_k^\theta), H^j(\theta, X_0^\theta)] \qquad (4.4.6)$$

This is exactly the corresponding formula of Theorem 1 of Part I, Chapter 3.

4.4.3 Gaussian Approximation Theorem

Theorem 7. *Under Assumptions (A.2), (A.3), (A.4), (A'.5) and (A.8), the distributions $(\widetilde{P}_{x,a}^\gamma)_{\gamma>0}$ of the processes U^γ converge weakly as $\gamma \to \infty$ to the distribution of the Gaussian diffusion with initial condition 0 and generator L_t given by:*

$$L_t \Psi(x) = \sum_{i=1}^d \partial_i \Psi(x) [\sum_{j=1}^d \partial_j h^i(\bar\theta(t,a)) x^i] + \frac{1}{2} \sum_{i,j=1}^d \partial_{ij}^2 \Psi(x) R^{ij}(\bar\theta(t,a))$$

Proof of Theorem 7. The proof is divided into two stages.
Stage 1. The family of distributions $(\widetilde{P}_{x,a}^\gamma)_{\gamma>0}$ is tight.
Stage 2. If $\widetilde{P}$ is any limiting distribution for the family $(\widetilde{P}^\gamma)$, then $\widetilde{P}$ satisfies the martingale condition:

$$\left(\Psi(\xi_t) - \Psi(0) - \int_0^t L_s\Psi(\xi_s)ds\right)_{t\geq 0}$$

is a martingale for any function Ψ of class C^2 with compact support.

Following Section 4.2 (above), this characterises any limiting distribution of the sequence $(\widetilde{P}_{x,a}^\gamma)_{\gamma>0}$ as the unique distribution of the Gaussian diffusion with initial condition 0 and generator L_t.

4.4 Gaussian Approximation of the Processes $U^\gamma(t)$

Weak compactness of the distributions $\tilde{P}^\gamma_{x,a}$.

In order to prove the compactness, it is sufficient to prove that the following hold for some $q \geq 1$ and $\rho > 1$:

$$\lim_{\gamma \to 0} P_{x,a}\{\zeta^\gamma(\varepsilon, Q) < T\} = 0 \qquad (4.4.7)$$

$$\sup_\gamma E_{x,a}(\sup_{t \leq T \wedge \zeta^\gamma} |U^\gamma_t|^q) < \infty \qquad (4.4.8)$$

and for all $\delta > 0$ there exists n_δ such that

$$\sup_{t \in [0,T]} \sup_{n \geq n_\delta} E_{x,a}\{\sup_{t \leq s \leq t+\delta} |U^\gamma(s \wedge \zeta^\gamma) - U^\gamma(t \wedge \zeta^\gamma)|^q\} \leq \delta^\rho \qquad (4.4.9)$$

In fact, from conditions (4.4.7) and (4.4.8), it is easy to see that

$$\lim_{r \to \infty} \sup_\gamma P_{x,a}\{|U^\gamma_t| > r\} = 0$$

This is condition [T$_1$] of Subsection 4.1.3

Condition (4.4.9) implies that for all $n \leq n_\delta$ and $t \leq T$

$$\frac{1}{\delta} P_{x,a}\{\sup_{t \leq s \leq t+\delta} |U^\gamma(s) - U^\gamma(t)| \geq \eta\} \leq \frac{\delta^{\rho-1}}{\eta} + \frac{1}{\delta} P_{x,a}\{\zeta^\gamma < T\}$$

This inequality and (4.4.7) together imply condition (4.1.5) for the sequence $\tilde{P}^\gamma_{x,a}$. This proves that the distributions $\tilde{P}^\gamma_{x,a}$ are compact.

Identification of the Limit.

Following the remark in 4.1.4, we know that any limit $\tilde{P}$ of a convergent sequence $\tilde{P}^{\gamma_n}$ is carried by $C([0,T];\mathbb{R}^d)$. We shall show further that such a limit $\tilde{P}$ is such that, for any Ψ of class C^2 with compact support, the process

$$\left(\Psi(\xi_t) - \Psi(0) - \int_0^t L_u \Psi(\xi_u) du\right)_{t \geq 0}$$

is a martingale on $(\tilde{\Omega}_T, (\tilde{F}_t)_{t \leq T}, \tilde{P})$. For this, it is sufficient to show that for all $s \leq t \leq T$, for all functions f on $\tilde{\Omega}_T$ of the form

$$f(\tilde{\omega}) = h_1(\xi_{t_1}(\tilde{\omega})) \ldots h_r(\xi_{t_r}(\tilde{\omega})) \quad t_1 < \ldots < t_r \leq s$$

where the h_i are continuously bounded, we have

$$\tilde{E}(f \cdot [\Psi(\xi_t) - \Psi(\xi_s) - \int_s^t L_u \Psi(\xi_u) du]) = 0 \qquad (4.4.10)$$

Since by construction, the restriction of f to $C([0,T];\mathbb{R}^d)$ is continuous, we see that

$$\Phi(\tilde{\omega}) := f(\tilde{\omega})[\Psi(\xi_t(\tilde{\omega})) - \Psi(\xi_s(\tilde{\omega})) - \int_s^t L_u \Psi(\xi_u(\tilde{\omega})) du]$$

defines a continuous function in $C([0,T]; \mathbb{R}^d)$. Since $\tilde{P}[C([0,T]; \mathbb{R}^d)] = 1$, following the remark of Subsection 4.1.4, we have:

$$\tilde{E}(\Phi) = \lim_n \tilde{E}^{\gamma_n}(\Phi)$$
$$= \lim_n E_{x,a}\{([f(U^\gamma)][\Psi(U^\gamma_t) - \Psi(U^\gamma_s) - \int_s^t L_u \Psi(U^\gamma_u) du]\}$$

Since, by construction $f(U^\gamma)$ is F_s-measurable and bounded, we will have proved (4.4.10) if we can show that

$$\Psi(U^\gamma_t) = \Psi(U^\gamma_0) + \int_0^t L_u \Psi(U^\gamma_u) du + N^\gamma_t + R^\gamma_t \qquad (4.4.11)$$

where N^γ is a martingale and

$$\lim_{\gamma \to 0} E|R^\gamma_t \wedge \zeta^\gamma| = 0 \qquad (4.4.12)$$

since then we have

$$E_{x,a}\{([f(U^\gamma)][\Psi(U^\gamma_t) - \Psi(U^\gamma_s) - \int_s^t L_u \Psi(U^\gamma_u) du]\}$$
$$\leq \|f\|_\infty [E(|R^\gamma_{t\wedge\zeta^\gamma}| + |R^\gamma_{s\wedge\zeta^\gamma}|) + (2\|\Psi\|_\infty + |t-s|\|L_u\Psi\|_\infty)P_{x,a}\{\zeta^\gamma < T\}]$$

Thus in order to prove the theorem, it is sufficient to prove (4.4.11) and (4.4.12).

Note that

$$\Psi(U^\gamma(t)) = \Psi(U^\gamma(0)) + \sum_{k \leq t/\gamma} (\Psi(U^\gamma(t_k)) - \Psi(U^\gamma(t_{k-1}))) + r(t) \qquad (4.4.13)$$

where

$$|r(t)| = \frac{1}{\sqrt{\gamma}} |\int_{A^\gamma(t)}^t h(\bar{\theta}(s)) ds| \leq \sqrt{\gamma} \sup_{x \in Q} |h(x)| \qquad (4.4.14)$$

We write

$$U^\gamma(t_k) - U^\gamma(t_{k-1}) = \Delta^\gamma_1(k) + \Delta^\gamma_2(k) \qquad (4.4.15)$$

where

$$\Delta^\gamma_1(k) = \int_{t_{k-1}}^{t_k} \frac{h(\theta^\gamma(s)) - h(\bar{\theta}(s,a))}{\sqrt{\gamma}} ds \qquad (4.4.16)$$

and

$$\Delta^\gamma_2(k) = \sqrt{\gamma}[H(\theta_{k-1}, X_k) - h(\theta_{k-1})] \qquad (4.4.17)$$

Notation.

1. For simplicity, in what follows, we shall denote any process Z_t such that

$$E|Z_{t\wedge\zeta^\gamma}| < K\gamma^\alpha \text{ for all } t \leq T$$

for some constant K, by $O_t(\gamma^\alpha)$.

4.4 Gaussian Approximation of the Processes $U^\gamma(t)$

2. For any matrix α^{ij} and any vector ν we denote
$$\alpha \circ \nu^{\otimes 2} = \sum \alpha^{ij} \nu^i \nu^j$$

We then see that
$$\begin{aligned}
\Psi(U^\gamma(t)) &= \Psi(U^\gamma(0)) + \sum_{k \leq t/\gamma} \Psi'(U^\gamma(t_{k-1})) \cdot (\Delta_1^\gamma(k) + \Delta_2^\gamma(k)) \\
&\quad + \frac{1}{2} \sum_{k \leq t/\gamma} \Psi''(U^\gamma(t_{k-1})) \circ (\Delta_2^\gamma(k))^{\otimes 2} \\
&\quad + O_t(\sqrt{\gamma})
\end{aligned} \tag{4.4.18}$$

Lemma 8.
$$\sum_{k \leq t/\gamma} \Psi'(U^\gamma(t_{k+1})) \cdot \Delta_1^\gamma(k) = \int_0^t \Psi'(U^\gamma(s)) \cdot [h'(\bar{\theta}(s,a)) \circ U^\gamma(s)] ds + O_t(\sqrt{\gamma})$$

Proof of Lemma 8. This lemma follows from the definition of
$$\Delta_1^\gamma(k) = \int_{t_{k-1}}^{t_k} h'(\bar{\theta}(s,a)) \circ U^\gamma(s) ds + \rho_t^\gamma$$

with
$$|\rho_{t \wedge \zeta^\gamma}^\gamma| \leq \frac{1}{2} \sqrt{\gamma} \sup_{\theta \in Q'} |h''(\theta)| \sup_{t_{k-1} \leq s \leq t_k} |U^\gamma(s)|^2$$

Lemma 8 now follows from Proposition 5-b with $t = 0$ and $\delta = T$. $\square$

Lemma 9.
$$\sum_{k \leq t/\gamma} \Psi'[U^\gamma(t_{k-1})] \cdot \Delta_2^\gamma(h)$$
$$= M_1^\gamma(t) + \sum_{k \leq t/\gamma} \sum_{i,j=1}^d \partial_{ij}^2 \Psi(U_{t_{k-1}}^\gamma) \cdot \Delta_2^{\gamma,i}(k) \cdot \Pi_{\theta_{k-1}} \nu_{\theta_{k-1}}^j(X_k) + O_t(\sqrt{\gamma})$$

where M_1^γ is a martingale

Proof of Lemma 9. We write
$$\Delta_2^\gamma(k) = \sqrt{\gamma}(\nu_{\theta_{k-1}}(X_k) - \Pi_{\theta_{k-1}} \nu_{\theta_{k-1}}(X_k)) \tag{4.4.19}$$

and decompose
$$\sum_{k \leq t/\gamma} \Psi'(U^\gamma(t_{k-1})) \cdot \Delta_2^\gamma(k)$$
$$= \sum_{k \leq t/\gamma} \sqrt{\gamma} \Psi'(U^\gamma(t_{k-1})) \cdot [\nu_{\theta_{k-1}}(X_k) - \Pi_{\theta_{k-1}} \nu_{\theta_{k-1}}(X_{k-1})]$$
$$+ \sum_{k \leq t/\gamma} \sqrt{\gamma} [\Psi'(U^\gamma(t_{k-1})) \cdot \Pi_{\theta_{k-1}} \nu_{\theta_{k-1}}(X_{k-1}) - \Psi'(U^\gamma(t_k)) \cdot \Pi_{\theta_{k-1}} \nu_{\theta_{k-1}}(X_k)]$$
$$+ \sum_{k \leq t/\gamma} \sqrt{\gamma} [\Psi'(U^\gamma(t_k)) - \Psi'(U^\gamma(t_{k-1}))] \cdot \Pi_{\theta_{k-1}} \nu_{\theta_{k-1}}(X_k)$$

The first term on the right hand side of this equation defines the martingale M_1^γ.

The upper bounds of points 2 and 3 of Proposition 4 may be applied to the second term to show that this term is $O_t(\gamma^\lambda)$

Finally, the third term may be written as

$$\sum_{k \leq t/\gamma} \sqrt{\gamma} \sum_{i,j=1}^{d} \partial_{ij}^2 \Psi(U^\gamma(t_{k-1})) \Delta_2^{\gamma,i}(k) \Pi_{\theta_{k-1}} \nu_{\theta_{k-1}}^j(X_k) + O_t(\sqrt{\gamma})$$

□

Lemma 10.

$$\frac{1}{2} \sum_{k \leq t/\gamma} \sum_{i,j=1}^{d} \partial_{ij}^2 \Psi[U^\gamma(t_{k-1})] \cdot [\Delta_2^{\gamma,i}(k)\Delta_2^{\gamma,j}(k) + 2\sqrt{\gamma}\Delta_2^{\gamma,i}(k)\Pi_{\theta_{k-1}}\nu_{\theta_{k-1}}^j(X_k)]$$
$$= \int_0^t \text{Trace } \Psi''[U^\gamma(s)] \circ R[\theta(s)]ds + M_2^\gamma(t) + O_t(\sqrt{\gamma})$$

where M_2^γ is a martingale.

Proof of Lemma 10. If we write $\Delta_2^\gamma(k)$ in the form (4.4.19), then using Assumption (A.8), the expression in the lemma may be written as

$$\frac{1}{2} \sum_{k \leq t/\gamma} \gamma \sum_{i,j=1}^{d} \partial_{ij}^2 \Psi(U^\gamma(t_{k-1}))$$
$$[\nu_{\theta_{k-1}}^i(X_k)\nu_{\theta_{k-1}}^j(X_k) - \Pi_{\theta_{k-1}}\nu_{\theta_{k-1}}^i(X_k)\Pi_{\theta_{k-1}}\nu_{\theta_{k-1}}^j(X_k)]$$
$$= \frac{1}{2} \sum_{k \leq t/\gamma} \gamma \text{ Trace } \Psi''(U^\gamma(t_{k-1})) \circ R(\theta(t_{k-1}))$$
$$+ \sum_{k \leq t/\gamma} \gamma \Psi''(U_{t_{k-1}}^\gamma) \cdot [w_{\theta_{k-1}}(X_k) - \Pi_{\theta_{k-1}}w_{\theta_{k-1}}(X_k)]^{\otimes 2}$$

Lemma 10 then follows using Proposition 4-4. □

Lemma 11.

$$\int_0^t \text{Trace } \Psi''(U^\gamma(s)) \circ R(\theta(s))ds$$
$$= \int_0^t \text{Trace } \Psi''(U^\gamma(s)) \circ R(\bar{\theta}(s,a))ds + O_t(\sqrt{\gamma})$$

Proof of Lemma 11. Since h is Lipschitz

$$|\int_0^t \text{Trace } \Psi''(U^\gamma(s)) \circ (R(\theta(s)) - R(\bar{\theta}(s,a)))ds| \leq K\sqrt{\gamma}T \sup_{s \leq t} |U^\gamma(s)|$$

Lemma 11 follows from Proposition 5. □

4.5 Gaussian Approximation for Algorithms with Decreasing Step Size

4.5.1 The Problem—Assumptions

Proof of Theorem 7 (Continued). If we now rewrite (4.4.18,) using Lemmas 8 to 11, we see that

$$\Psi(U^\gamma(t)) = \Psi(U^\gamma(0)) + \int_0^t L_s \Psi(U^\gamma(s))ds + M_1^\gamma(t) + M_2^\gamma(t) + O_t(\sqrt{\gamma})$$

This is formula (4.4.11) and thus Theorem 7 is proved. □

We now consider an algorithm with decreasing step size (γ_n):

$$\begin{cases} \theta_{n+1} = \theta_n + \gamma_{n+1} H(\theta_n, X_{n+1}) \\ \theta_0 = a \end{cases} \quad (4.5.1)$$

In Chapter 1, Subsection 1.5.3, we introduced the algorithms $(\theta_{N+n})_{n\geq 0}$ where $P_{N,x,a}$ denotes the distribution of the sequence $(\theta_{N+n})_{n\geq 0}$ satisfying

$$\begin{cases} \theta_{N+n+1} = \theta_{N+n} + \gamma_{N+n+1} H(\theta_{N+n}, X_{N+n+1}) \\ \theta_N = a \quad X_N = x \end{cases} \quad (4.5.2)$$

We studied the behaviour of the algorithms (4.5.2) as a function of N.

More generally, we might also consider a sequence $(\theta_n^N)_{n\geq 0}$ $N=0,\ldots,k,\ldots$ of algorithms of the form

$$\begin{cases} \theta_{n+1}^N = \theta_n^N + \gamma_{n+1}^N H(\theta_n^N, X_{n+1}^N) \\ \theta_0^N = a \quad X_0^N = x \end{cases} \quad (4.5.3)$$

where the X^N are associated with the **same** transition probability $\Pi_\theta(x, dy)$, $\theta \in \mathbb{R}^d$. Then, if as in Subsection 1.1.1, we set

$$t_n^N = \sum_{k\geq 0}^n \gamma_i^N \quad (4.5.4)$$

$$\theta^N(t) = \sum_{k\geq 0} I(t_k^N \leq t < t_{k+1}^N) \theta_k^N \quad (4.5.5)$$

$$m^N(n,t) = \inf\{k : k \geq n, \gamma_{n+1}^N + \ldots + \gamma_{k+1}^N \geq T\} \quad (4.5.6)$$

$$m^N(T) = m^N(0,T) \quad (4.5.7)$$

then under the assumptions of Theorem 9 of Chapter 1, we have

$$\lim_{N\to\infty} P_{x,a}^N \{ \sup_{n\leq m^N(T)} |\theta_n^N - \bar\theta(t_n^N, a)| \geq \delta \} = 0$$

for all $\delta > 0$, where $\bar{\theta}(t_n^N; a)$ denotes the solution of the ODE

$$\begin{cases} \frac{d\bar{\theta}(t;a)}{dt} = h(\bar{\theta}(t;a)) \\ \bar{\theta}(0;a) = a \end{cases} \qquad (4.5.8)$$

associated with H and $(\Pi_\theta)_{\theta \in \mathbb{R}^d}$. Thus, analogously to Section 4.4, it is possible to study the weak convergence as $N \to \infty$ of the sequence of processes U^N defined by

$$U^N(t) = \frac{\theta^N(t) - \bar{\theta}(t, a)}{\sqrt{\gamma^N(t)}} \qquad (4.5.9)$$

where

$$\gamma^N(t) = \sum_{k \geq 0} I(t_K^N \leq t < t_{k+1}^N) \gamma_k^N \qquad (4.5.10)$$

This problem is discussed in Subsection 4.5.2

Following on from this, in Subsection 4.5.3, we shall obtain a Gaussian approximation result for the vectors $\gamma_n^{-1/2}(\theta(n) - \theta_*)$ in the case of an algorithm (4.5.1) converging to θ_*. As far as the sequences $(\gamma_n^N)_{n \geq 0}$ are concerned, we shall assume that they are decreasing, that $\lim_{N \to \infty} \gamma_n^N = 0$ and that there exists $\bar{\alpha} \geq 0$ such that

$$\lim_{N \to \infty} \sup_k \left| \frac{\sqrt{\gamma_k^N} - \sqrt{\gamma_{k+1}^N}}{(\gamma_{k+1}^N)^{3/2}} - \bar{\alpha} \right| = 0 \qquad (4.5.11)$$

Remark. If $\gamma_n^N = \frac{1}{(N+n)^\beta}$, it is easy to see that if $\beta = 1$ then $\bar{\alpha} = 1/2$ and that if $\beta < 1$ then $\bar{\alpha} = 0$.

4.5.2 Sequence of Algorithms with Decreasing Step Size. Gaussian Approximation Theorem

We shall prove the following result.

Theorem 12. *We assume that the function H, the sequences $(\gamma_n^N)_{n \geq 0} N \in \mathbb{N}$ and the transition probabilities $(\Pi_\theta)_{\theta \in \mathbb{R}^d}$ satisfy (A.1) to (A.4). ($\bar{A}'.5$) and (A.8).*

1. *The distributions $(\tilde{P}_{x,a}^N)_{N \geq 0}$ of the processes U^N converge weakly as $N \to \infty$ to the distribution of the Gaussian diffusion with initial condition 0 and generator L_t given by*

$$L_t \Psi(x) = \sum_{i=1}^d \partial_i \Psi(x) \left[\left[\sum_{j=1}^d \partial_j h^i(\bar{\theta}(t,a)) x^j \right] + \bar{\alpha} x^i \right]$$

$$+ \frac{1}{2} \sum_{i,j=1}^d \partial_{ij}^2 \Psi(x) R^{ij}(\bar{\theta}(t,a))$$

4.5 Gaussian Approximation for Algorithms with Decreasing Step Size

2. *We suppose that the random variables $\theta_0^N, N \geq 0$ are such that the distribution of the sequence $(U_0^N)_{N \geq 0} = [(\gamma_0^N)^{-1/2}.(\theta_0^N - \bar{\theta}_0)]_{N \geq 0}$ converges weakly towards a distribution ν. We also suppose that for any compact set Q, we have*

$$\sup_N E\{|U_0^N|^2 I(\theta_0^N \in Q)\} < \infty$$

Then the sequence of distributions of the processes $U^N, N \geq 0$ converges weakly as $N \to \infty$ to the distribution of the diffusion with initial condition ν and generator L_t.

Proof. Point 1 is clearly a special case of point 2. The proof of point 2 follows the same strategy as the proof of Theorem 7. We modify equations (4.3.5) to (4.3.10) as follows

$$\theta_{n+1}^N = \theta_n^N + \gamma_{n+1}^N h(\theta_n^N) + \gamma_{n+1}^N [\nu_{\theta_n^N}(X_{n+1}^N) - \Pi_{\theta_n^N}\nu_{\theta_n^N}(X_{n+1}^N)]$$

$$M^N(t) = \sum_{k \leq m^N(t)} \sqrt{\gamma_{n+1}^N}[\nu_{\theta_k}(X_{k+1}^N) - \Pi_{\theta_k}\nu_{\theta_k}(X_k^N)] \quad (4.5.12)$$

$$B^N(t) = \sum_{k \leq m^N(t)} \sqrt{\gamma_{n+1}^N}[\Pi_{\theta_k}\nu_{\theta_k}(X_k^N) - \Pi_{\theta_k}\nu_{\theta_k}(X_{k+1}^N)] \quad (4.5.13)$$

$$\rho^N(t) = \int_{A^N(t)}^t \left(\frac{1}{\sqrt{\gamma^N(u)}} h(\theta^N(u)) + \alpha^N(u) U^N(u)\right) du \quad (4.5.14)$$

where $A^N(t) = \max\{t_k^N; t_k^N \leq t\}$. If we set

$$Z_n^N = \theta_n^N - \bar{\theta}(t_n^N, a) \quad (4.5.15)$$

then we have

$$\begin{aligned}
Z_{n+1}^N &= Z_n^N + \gamma_{n+1}^N[h(\theta_n^N) - h(\bar{\theta}(t_n^N, a)] \\
&\quad + \sqrt{\gamma_{n+1}^N}[M^N(t_{n+1}^N) + B^N(t_{n+1}^N) - M^N(t_n^N) - B^N(t_n^N)] \\
&\quad - \int_{t_n^N}^{t_{n+1}^N} [h(\bar{\theta}(s,a)) - h(\bar{\theta}(t_n^N,a))]ds
\end{aligned}$$

whence

$$\begin{aligned}
U^N(t_{k+1}^N) &= U^N(t_k^N)\left(1 + \frac{\sqrt{\gamma_k^N} - \sqrt{\gamma_{k+1}^N}}{\sqrt{\gamma_{k+1}^N}}\right) \\
&\quad + \sqrt{\gamma_{k+1}^N}[h(\theta_k^N) - h(\bar{\theta}(t_k^N,a))] \\
&\quad + [M^N(t_{k+1}^N) + B^N(t_{k+1}^N) - M^N(t_k^N) - B^N(t_k^N)] \\
&\quad - \frac{1}{\sqrt{\gamma_{k+1}^N}} \int_{t_k^N}^{t_{k+1}^N} [h(\bar{\theta}(s,a)) - h(\bar{\theta}(t_k^N,a))]ds
\end{aligned}$$

If we set
$$\alpha^N(t) = \sum_k I(t_k^N \le t < t_{k+1}^N) \frac{\sqrt{\gamma_k^N} - \sqrt{\gamma_{k+1}^N}}{(\gamma_{k+1}^N)^{3/2}} \qquad (4.5.16)$$

then formula (4.3.10) for algorithms with constant step size is replaced here by

$$\begin{aligned} U^N(t) &= U^N(0) + \int_0^t \alpha^N(s) U^N(s) ds \\ &+ \int_0^t \frac{1}{\sqrt{\gamma^N(s)}} [h(\theta^n(s)) - h(\bar{\theta}(s,a))] ds \\ &+ M^N(t) + B^N(t) + \rho^N(t) \end{aligned}$$

This gives the formula analogous to (4.3.18);

$$\begin{aligned} U^N(t) &= U^N(0) + \int_0^t \alpha^N(s) U^N(s) ds + \int_0^t h'[\bar{\theta}(s,a)] \circ U^N(s) ds \\ &+ M^N(t) + B^N(t) + B_1^N(t) + \rho^N(t) \end{aligned} \qquad (4.5.17)$$

where

$$B_1^N(t) = \int_0^t \frac{1}{\sqrt{\gamma^N(s)}} [h(\theta^N(s)) - h(\bar{\theta}(s,a)) - h'(\bar{\theta}(s,a)).(\theta^N(s) - \bar{\theta}(s,a))] ds \qquad (4.5.18)$$

Next we give definitions analogous to (4.3.11) to (4.3.15)

$$\begin{aligned} \tau^N(Q) &= \inf\{n; \theta_n^N \notin Q\} & (4.5.19) \\ \sigma^N(\varepsilon) &= \inf\{n; |\theta_n^N - \theta_{n-1}^N| > \varepsilon\} & (4.5.20) \\ \nu^N(\varepsilon, Q) &= \inf(\tau^N(Q), \sigma^N(\varepsilon)) & (4.5.21) \\ \zeta^N(\varepsilon, Q) &= t_{\nu^N(\varepsilon,Q)} & (4.5.22) \end{aligned}$$

The definitions of ρ^N and B_1^N show that

$$\begin{aligned} I\{t < \zeta^N(\varepsilon, Q)\} \cdot |\rho^N(t)| \\ \le \|h\|_Q (\gamma(t))^{1/2} + \gamma(t) \sup_{s \le t} |I[s < \zeta^N(\varepsilon, Q)] \alpha_n(s) U^\gamma(s)| \end{aligned} \qquad (4.5.23)$$

and

$$I\{t < \zeta^N(\varepsilon, Q)\} \cdot |B_1^\gamma(t)| \le \|h''\|_Q \sup_{s \le t} I[s < \zeta^N(\varepsilon, Q) | U_s^\gamma|^2] \qquad (4.5.24)$$

We note here that the upper bounds in Proposition 4 were derived (cf. definitions (4.3.25) to (4.3.29)) for algorithms with decreasing gain. Thus the same inequalities as in Propositions 5 and 6 may be directly inferred

4.5 Gaussian Approximation for Algorithms with Decreasing Step Size 331

by replacing $M^\gamma, U^\gamma, B^\gamma$ and ρ^γ by M^N, U^N, B^N and ρ^N (respectively). In particular, if Q' is a compact set containing Q for which

$$\beta = \inf\{|x - x'| : x \in Q, x' \notin Q'\} > 0$$

there exist constants $C > 0, s > 0$ such that for all N, $t > 0$, $\delta > 0$ and $q \geq 1$:

$$E\{\sup_{t \leq u \leq t+\delta} |U^N(u \wedge \zeta^N(\varepsilon, Q')) - U^N(t \wedge \zeta^N(\varepsilon, Q'))|^q\}$$
$$\leq C(1 + |x|^s)(\delta \vee \gamma_1^N)^{q/2} \qquad (4.5.25)$$

As in the proof of Theorem 7, we now deduce that there exist $q \geq 1$ and $\rho > 1$ such that for all $\delta > 0$ there exists n_δ such that for all N

$$\sup_{t \in [0,T]} \sup_{n \geq n_\delta} E\{\sup_{t \leq s \leq t+\delta} |U^N(s \wedge \zeta^N) - U^N(t \wedge \zeta^N)|^q\} \leq \delta^\rho \qquad (4.5.26)$$

As in Proposition 6, we have

$$\lim_{N \to \infty} P\{\zeta^N(\varepsilon, Q) < T\} = 0 \qquad (4.5.27)$$

The assumption

$$\sup_N E\{|U_0^N|^2/(\theta_0^N \in Q)\} < \infty$$

together with (4.5.25) shows that

$$\sup_N E\{\sup_{0 \leq t \leq T} |U^N(t \wedge \zeta^N(\varepsilon, Q'))|^2\} < \infty$$

As in the proof of Theorem 7, the weak compactness of the sequence $\tilde{P}^N$ of the distributions of processes U^N follows immediately.

The limit is determined as in the proof of Theorem 7 by writing:

$$\Psi(U^N(t)) = \Psi(U^N(0)) + \sum_{k \leq m^N(t)} \Psi'[U^N(t_{k-1}^N)] \cdot [\Delta_1^N(k) + \Delta_2^N(k)]$$
$$+ \frac{1}{2} \sum_{k \leq m^N(t)} \Psi''[U^N(t_{k-1}^N)] \circ [\Delta_2^N(k)]^{\otimes 2} + O_t(\sqrt{\gamma_1^N})$$

where

$$\Delta_1^N(k) = \int_{t_{k-1}}^{t_k} \alpha^N(s) U^N(s) ds + \int_{t_{k-1}}^{t_k} \frac{h(\theta^N(s)) - h(\bar{\theta}(s,a)) ds}{\sqrt{\gamma^N(s)}}$$

and $\Delta_2^N(k) = \sqrt{\gamma^N}[H(\theta_{k-1}, X_k) - h(\theta_{k-1})]$.

We see that Lemmas 8 to 11 may be applied word for word to show that

$$\Psi(U_t^N) = \Psi(U_0^N) + \int_0^t L_u \Psi(U_u^N) du + N_t^N + R_t^N$$

where N^N is a martingale and

$$\lim_{N \to \infty} E|R_{t \wedge \zeta^N}^N| = 0$$

The proof is now completed as for Theorem 7. □

4.5.3 Gaussian Asymptotic Approximation of a Convergent Algorithm

In this paragraph, we shall consider an algorithm (θ_n) which converges almost surely, under the assumptions at the end of Chapter 1. More precisely, we shall suppose that the algorithm (4.5.1) satisfies Assumptions (A.1) to (A.5) of Chapter 1, and that if τ_R denotes the time at which θ_n leaves the compact set $\{|\theta| \leq R\}$, then

$$\inf_{R>0} P_{x,a}\{\tau_R < \infty\} = 0 \tag{4.5.28}$$

In particular, this condition is satisfied by any system whose assumptions include the a.s. boundedness of (θ_n) (c.f. Section 1.9 of Chapter 1) We shall also suppose that the following two conditions are satisfied

$$\lim_{n\to\infty} \frac{\sqrt{\gamma_n} - \sqrt{\gamma_{n+1}}}{\gamma_{n+1}^{3/2}} = \bar{\alpha} \tag{4.5.29}$$

$$(\theta - \theta_*).h(\theta) \leq -\delta|\theta - \theta_*|^2 \tag{4.5.30}$$

with

$$\liminf_{n\to\infty} 2\delta\frac{\gamma_n}{\gamma_{n+1}} + \frac{\gamma_{n+1} - \gamma_n}{\gamma_{n+1}^2} > 0 \tag{4.5.31}$$

Then we have the theorem:

Theorem 13 *We assume the conditions listed above hold, and that, in addition, the eigenvalues of the $d \times d$ matrix B with terms $B_{ij} = \bar{\alpha}\delta_{ij} + \partial_j h^i(\theta_*)$, $i,j = 1,\ldots,d$ all have strictly negative real parts. Then we have*

1. *The sequence $\widetilde{P}^N$ of the distributions of the processes defined by*

$$U^N(t) = \frac{\theta(t_N + t) - \theta_*}{\sqrt{\gamma(t_N + t)}}$$

   converges weakly towards the distribution of a stationary Gaussian diffusion with generator

$$L\Psi(x) = Bx \cdot \nabla\Psi(x) + \frac{1}{2}\sum_{i,j=1}^{d} \partial_{ij}^2 \Psi(x) R^{ij}(\theta_*) \tag{4.5.32}$$

2. *The sequence of random variables*

$$\frac{\theta_n - \theta_*}{\sqrt{\gamma_n}}, \quad n \in \mathbb{N}$$

   converges in distribution to a zero-mean Gaussian variable with covariance

$$C = \int_0^\infty e^{sB} \circ R \circ e^{sB^T} ds \tag{4.5.33}$$

4.5 Gaussian Approximation for Algorithms with Decreasing Step Size 333

Proof. Point 2 is clearly a special case of point 1 since $U^n(0) = \frac{\theta_n - \theta_*}{\sqrt{\gamma}}$ is the initial value of the process U^n which converges in distribution to a stationary Gaussian process with stationary distribution $N(0,C)$. We observe that, by virtue of Theorem 25 of Chapter 1, we have

$$\sup_N E_{x,a}\{|U^N(0)|^2 I(\theta_0^N \in Q)\} < \infty \qquad (4.5.34)$$

Thus, if the sequence U_0^N converges in distribution to a random variable with distribution ν, then the conclusions of Theorem 12 apply to the sequence $(U^N)_{N \geq 0}$. Conditions (4.5.34) and (4.5.28) together imply the weak convergence of the sequence $U^N(0)$. Thus any subsequence $(U^{N_k})_{k \in \mathbb{N}}$ of the sequence of processes $(U^N)_{N \in \mathbb{N}}$ contains a subsequence which converges to a diffusion with generator (4.5.32) and initial distribution ν_0 where ν_0 is the weak limit of the distributions of the $U^{N_k}(0)$.

Thus, in order to prove the theorem, it is sufficient to prove the following lemma.

Lemma 14 *Let $(U(t))_{t \geq 0}$ be a process with values in $\mathbb{R}^d$ such that any increasing sequence $(t_N)_{N \in \mathbb{N}}$ in $\mathbb{R}_+$ contains a subsequence $(t_{N_k})_{k \geq 0}$ for which the processes $(U^k(t))_{t \geq 0} = (U(t_{N_k} + t))_{t \geq 0}$ converge in distribution, as $k \to \infty$, to a diffusion with generator (4.5.32). Then, for any increasing sequence $(t_N)_{N \in \mathbb{N}}$, the distributions of the variables $U(t_N)$ converge weakly to $N(0,C)$, the stationary distribution of this diffusion.*

Proof of Lemma 14. It is clearly sufficient to show that, if any sequence (t_N) contains a subsequence (t'_k) such that the variables $U(t'_k)$ converge in distribution to a probability ν_∞, then, for any continuous function ϕ with compact support on $\mathbb{R}^d$ we have

$$\langle \phi, \nu_\infty \rangle = \int \phi(y) \nu_\infty(dy) = \langle \phi, g \rangle \qquad (4.5.35)$$

where g is the Gaussian distribution N(0,C). For any probability distribution ν on $\mathbb{R}^d$, we shall use $P_\nu(t)$ to denote the distribution of $V(t)$ where $(V(t))_{t \geq 0}$ is the diffusion with generator (4.5.32) and initial distribution ν. We note that for any weakly compact set K, of probability distributions on $\mathbb{R}^d$ we have the following property: for all ε, there exists $T > 0$ such that for all $t > T$

$$|\langle \phi, P_\nu(t) \rangle - \langle \phi, g \rangle| \leq \varepsilon \quad \forall \nu \in K$$

We fix such a T and extract a subsequence (t'_k) such that the sequences $(U(t'_k))$ and $(U(t'_k) - T)$ converge in distribution to limits ν_∞ and ν'_∞ (respectively). From the above, the convergence of the processes V^k to the Gaussian diffusion with initial distribution ν'_∞ implies that

$$\begin{aligned}|\langle \phi, \nu_\infty \rangle - \langle \phi, g \rangle| &\leq |\langle \phi, \nu_\infty \rangle - E\phi(U(t'_k))| \\ &+ |E\phi(U(t'_k)) - \langle \phi, P_{\nu'_\infty}(T) \rangle| + |\langle \phi, P_{\nu'_\infty}(T) \rangle - \langle \phi, g \rangle|\end{aligned}$$

It follows, letting t'_k tend to $+\infty$, that

$$|\langle \phi, \nu_\infty \rangle - \langle \phi, g \rangle| \leq \varepsilon$$

for all $\varepsilon > 0$. Condition (4.5.35) now follows, as do the lemma and Theorem 13.
□

Remark. It can be shown by elementary integration that

$$BC + CB^T = \int_0^\infty [B \circ e^{sB} \circ R \circ e^{sB^T} + e^{sB} \circ R \circ e^{sB^T} \circ B^T] ds = -R$$

The matrix C defined in (4.5.33) is thus a positive symmetric solution of the Lyapunov equation

$$BC + CB^T + R = 0$$

as indicated in the statement of Theorem 3 of Chapter 3, Part I.

4.6 Gaussian Approximation and Asymptotic Behaviour of Algorithms with Constant Steps

Theorem 13 says that for a class of algorithms with decreasing step size, the random variable $\gamma_n^{-1/2}(\theta_n - \theta_*)$ converges in distribution to a Gaussian variable. In the case of algorithms (θ_n^γ) with constant step size γ, such as those considered in Subsection 4.4.3, it is reasonable to wonder whether, when the ODE has an asymptotically stable equilibrium point θ_* (with additional assumptions in some cases), the random variable $\gamma^{-1/2}(\theta_n^\gamma - \theta_*)$ behaves asymptotically, for n large and γ small, like a Gaussian variable. We shall formalise the property of this type cited in Theorem 2 of Chapter 3, Part I in the statement of Theorem 15 (below).

4.6.1 Assumptions and Statement of the Asymptotic Theorem

We assume the conditions of Sections 4.3 and 4.4 hold, with the following reinforcements:

(A-i). For all $a \in \mathbb{R}^d$

$$\lim_{t \to \infty} \bar{\theta}(t, a) = \theta_*$$

(A-ii). The matrix $(\partial_j h^i(\theta) : i, j = 1, \ldots, d)$ is Lipschitz in θ and the eigenvalues of the matrix B defined by

$$B_{ij} = \partial_j h^i(\theta_*)$$

all have strictly negative real parts.

4.6 Gaussian Approximation for Algorithms with Constant Step Size

(B). *There exist* $q_1, q_2, q_3 \geq 0$ *and for all* $q > 0$ *and all compact sets* Q, *a constant* $\mu(q, Q)$ *such that for all* $\gamma \leq 1$, $x \in \mathbb{R}^d, a \in Q$:

(i) $\quad\quad\quad\quad \sup_n E_{x,a}(1 + |X_n^\gamma|^q) \leq \mu(1 + |x|^q)$

(ii) $\quad\quad\quad\quad \sup_n E_{x,a}|H(\theta_n^\gamma, X_{n+1}^\gamma)|^2 \leq \mu(1 + |x|^{q_1})$

(iii) $\quad\quad\quad\quad \sup_n E_{x,a}|\nu_{\theta_n^\gamma}(X_{n+1}^\gamma)|^2 \leq \mu(1 + |x|^{q_2})$

(iv) $\quad\quad\quad\quad \sup_n E_{x,a}|\theta_n^\gamma|^2 \leq \mu(1 + |x|^{q_3})$

Theorem 15. *We consider the stochastic algorithms* (θ_n^γ) *defined in (4.3.1) with the assumptions of Sections 4.3 and 4.4 reinforced by (A) and (B) (above). Then, for any sequence* $\tau_n \uparrow \infty$ *and any sequence* $\gamma_n \downarrow \infty$, *the sequence of random variables* $(U^{\gamma_n}(\tau_n))_{n \geq 0}$ *converges in distribution to a zero-mean Gaussian variable with covariance* C *where* C *is the matrix*

$$C = \int_0^\infty \exp^{sB} R e^{sB^T} ds$$

where R *is the matrix defined in Assumption (A.8).*

Proof. This will be carried out in several stages. After examining several simple consequences of Assumptions (A) and (B), in (4.6.3) we shall derive an upper bound for $E|U_t^\gamma|^2$ involving the weak compactness of the distributions of the $(U^{\gamma_n}(\tau_n))_{n \geq 0}$; we shall complete the proof in 4.6.4 using an argument similar to that of Lemma 14.

4.6.2 Initial Consequences of Assumptions (A) and (B)

Let $\{\Gamma(t, s) : s \leq t\}$ be the resolvent, for fixed a, of the linear system

$$\frac{dx(t)}{dt} = h'(\bar{\theta}(t,a))x(t)$$

(cf. (Reinhard 1982)). We recall that $\Gamma(t, s)$ is a solution of the differential equation

$$\frac{d\Gamma(t,s)}{dt} = h'(\bar{\theta}(t,a))\Gamma(t,s), \quad t \geq s$$

$$\Gamma(s,s) = \text{Id}$$

We also recall that the solution of the equation with second term

$$dx(t) = h'(\bar{\theta}(t,a))x(t)dt + dF(t), \quad t \geq s$$

$$x(s) = x_s$$

is given by

$$x(t) = \Gamma(t,s)x_s + \int_s^t \Gamma(t,u)dF(u) \quad\quad (4.6.1)$$

and that this formula is valid not only when F is a function of finite variation for which the integral of the second term of (4.6.1) exists, but also for a stochastic differential equation in which $F(t)$ is the sum of a martingale and a process whose trajectories are of finite variation. Thus it follows from (A-i) and (A-ii) that there exist numbers $\alpha > 0$ and t_0 such that for all $t > t_0$, the eigenvalues of $h'(\bar{\theta}(t, a))$ all have negative real parts $\leq -\alpha$ and that

$$\|\Gamma(t, s)\| \leq e^{-\alpha(t-s)} \tag{4.6.2}$$

We now return to equation (4.3.18), writing, in conformity with equation (4.6.1):

$$U^\gamma(t) = \int_0^t h'(\bar{\theta}(s, a))U^\gamma(s)ds + M^\gamma(t) + N^\gamma(t) \tag{4.6.3}$$

where

$$N^\gamma(t) = B^\gamma(t) + B_1^\gamma(t) + \rho^\gamma(t) \tag{4.6.4}$$

with the notation of Subsection 4.3.2. The process M^γ is a martingale (it is also a process with interval-wise constant trajectories, thus it is of finite variation). We also have

$$B_1^\gamma(t) = \int_0^t b_1^\gamma(s)ds$$

with, following (4.3.17), for some suitable constant K,

$$|b_1^\gamma(s)| \leq K|U^\gamma(s)| \tag{4.6.5}$$

From the definitions of M^γ, B^γ and ρ^γ, we deduce using Assumption (B) and denoting the jump in t of the process Z by $\Delta_t Z$:

$$\sup_{t \geq 0}(E|\Delta_t M^\gamma|^2 + E|\Delta_t B^\gamma|^2 + E|\Delta_t \rho^\gamma|^2) \leq K\gamma(1 + |x|^\beta) \tag{4.6.6}$$

for some suitable constant K, some β (in the remainder of this paragraph, we shall use K to denote any constant independent of γ).

Given a right-continuous function f on $\mathbb{R}_+$, we denote

$$w(f, \delta, s) = \sup_{s \leq s' \leq s+\delta} |f(s') - f(s)|$$

From the definition of ρ^γ, we have

$$\int_{]a,t]} f(s_-)d\rho^\gamma(s)$$

$$= \sum_{a \leq t_{n-1}^\gamma \leq t_n^\gamma \leq t} \int_{t_{n-1}^\gamma}^{t_n^\gamma} [f(s) - f(t_{n-1}^\gamma)] \frac{1}{\sqrt{\gamma}} h(\theta_s^\gamma)ds + \int_{A_t^\gamma} f(s) \frac{1}{\sqrt{\gamma}} h(\theta_s^\gamma)ds$$

whence

$$E|\int_{]a,t]} f(s_-)d\rho^\gamma(s)|$$

$$\leq \frac{1}{\sqrt{\gamma}} \int_a^t E|w(f, \delta, s)h(\theta_s^\gamma)|ds + \sqrt{\gamma} \sup_{A_t^\gamma \leq s < t} E|f(s)h(\theta_s^\gamma)| \tag{4.6.7}$$

4.6 Gaussian Approximation for Algorithms with Constant Step Size

We shall use $[M^\gamma]$ and $[N^\gamma]$ to denote the so-called processes of "square-variation" defined by

$$[M^\gamma]_t = \sum_{s \leq t} |\Delta_s M^\gamma|^2 \tag{4.6.8}$$

and

$$[N^\gamma]_t = \sum_{s \leq t} |\Delta_s N^\gamma|^2 \tag{4.6.9}$$

Lastly, we recall that if V is an adapted process whose trajectories have a left limit V_{s-} at all points and are integrable with respect to dM_s^γ and dN_s^γ then we have

$$E\left(\int_{]0,t]} V_{s-} dM_s^\gamma\right) = 0 \tag{4.6.10}$$

(martingale property) and

$$E|\int_{]0,t]} V_{s-} dN_s^\gamma|$$
$$\leq \int_0^t (E|V(s)|^2)^{1/2} (E|b_1^\gamma(s)|^2)^{1/2} ds + \sum_{s \leq t} [E|V(s_-)|^2]^{1/2} [E|\Delta_s N^\gamma|^2]^{1/2}$$
$$\tag{4.6.11}$$

4.6.3 Upper Bounds for the Process U^γ

The strengthening of the assumptions on the moments afforded by condition (B) leads directly to the following reinforcement of Proposition 4:

Proposition 16 *Let $\tilde{\varepsilon}_k(\nu), \tilde{\varepsilon}_k^i(\nu), i = 1, 2, 3, \tilde{\eta}_{n,k}(\nu)$ be as defined in 4.3.3, taking $\Phi = 1$ and u to be the solution v of the Poisson equation in Assumption (A.4). Under the assumptions of 4.6.1 there exist constants B, β (depending on a) such that for all $T > 0$*

(i) $\quad E_{x,a}(|\sup_{n \leq k \leq m^\gamma(n,T)} \sum_{i=n}^{k-1} \tilde{\varepsilon}_i^1(\nu)|^2) \leq BT(1+|x|^\beta)\gamma$

(ii) $\quad E_{x,a}(|\sup_{n \leq k \leq m^\gamma(n,T)} \sum_{i=n}^{k-1} \tilde{\varepsilon}_i^2(\nu)|^2) \leq BT^2(1+|x|^\beta)\gamma^2$

(iii) $\quad E_{x,a}(|\sup_{n \leq k \leq m^\gamma(n,T)} \sum_{i=n}^{k-1} \tilde{\varepsilon}_i^3(\nu)|^2 + |\tilde{\eta}_{n,k}(\nu)|^2) \leq B(1+|x|^\beta)\gamma^2$

(iv) $\quad E_{x,a}(|\sup_{n \leq k \leq m^\gamma(n,T)} \sum_{i=n}^{k-1} \tilde{\varepsilon}_i(\nu)|^2) \leq B(1+|x|^\beta)(T+\gamma+\gamma T^2)$

Proof. It is sufficient to follow the proof of Proposition 4 step by step, or even better Proposition 8 of Chapter 1 and the lemmas which precede it, together with all the intervening simplifications resulting from the assumptions in this case. Thus we have the following analogue of Proposition 5-a:

$$E_{x,a} \sup_{t \leq u \leq t+T} |M^\gamma(u) - M^\gamma(t)|^2 \leq BT(1 + |x|^\beta) \quad (4.6.12)$$

$$E_{x,a} \sup_{t \leq u \leq t+T} |A^\gamma(u) - A^\gamma(t)|^2 \leq B(1 + |x|^\beta)\gamma(1 + T^2) \quad (4.6.13)$$

and as in Proposition 5-b

$$E_{x,a} \sup_{t \leq u \leq t+T} |U^\gamma(u) - U^\gamma(t)|^2 \leq C(1 + |x|^\beta)(T \vee \gamma) \quad (4.6.14)$$

□

We shall now improve this estimate of $E|U^\gamma(t)|^2$, making it independent of T, using Assumption (A).

Proposition 17. *Under the assumptions of this section, there exists γ_0 such that*

$$\sup_{\gamma \leq \gamma_0} \sup_{t \geq 0} E_{x,a}|U_t^\gamma|^2 < \infty$$

Proof. A simple argument based on integration by parts (or on a classical formula for linear differential equations with a second term) gives from (4.6.3):

$$U^\gamma(t) = \Gamma(t,t_0)U^\gamma(t_0) + \int_{]t_0,t]} \Gamma(t,s)dM^\gamma(s) + \int_{]t_0,t]} \Gamma(t,s)dN^\gamma(s) \quad (4.6.15)$$

(Here we have $U^\gamma(0) = 0$). For $u \geq t$, we set

$$W^\gamma(u,t) = U^\gamma(t_0) + \int_{]t_0,t]} \Gamma(u,s)dM^\gamma(s) + \int_{]t_0,t]} \Gamma(u,s)dN^\gamma(s)$$

Applying Ito's formula (Métivier 1983) we have:

$$E|W^\gamma(u,t)|^2 = E|U^\gamma(t_0)|^2 + 2E\{\int_{]t_0,t]} \langle W^\gamma(u,s), \Gamma(u,s)dN^\gamma(s)\rangle\}$$

$$+ 2E\{\int_{]t_0,t]} \text{Trace}\,[\Gamma(u,s) \cdot \Gamma^T(u,s)](d[M^\gamma]_s + d[N^\gamma]_s)\}$$

Next we choose t_0 so that (4.6.2) is true (in fact $u = t$) and apply (4.6.4) and (4.6.11). Then we obtain

$$E|U^\gamma(t)|^2 \leq E|U^\gamma(t_0)|^2$$
$$+ 2E\{|\int_{]t_0,t]} [U^\gamma(s)]^T \circ \Gamma(t,s) \circ b_1^\gamma(s)ds|\}$$
$$+ 2E\{|\int_{]t_0,t]} [U^\gamma(s)]^T \circ \Gamma(t,s) \circ d\rho_s^\gamma|\}$$
$$+ 2E\{|\int_{]t_0,t]} [U^\gamma(s_-)]^T \circ \Gamma(t,s) \circ dB_s^\gamma|\}$$
$$+ 2E\{|\int_{]t_0,t]} e^{-\alpha(t-s)}(d[M^\gamma]_s + d[N^\gamma]_s)\} \quad (4.6.16)$$

4.6 Gaussian Approximation for Algorithms with Constant Step Size

Following (4.6.5)

$$E\{|\int_{]t_0,t]} [U^\gamma(s)]^T . \Gamma(t,s).b_1^\gamma(s)ds|\} \leq K \int_{]t_0,t]} e^{-\alpha(t-s)} E|U_s^\gamma|^2 ds \qquad (4.6.17)$$

We now note that, following (4.6.14), for $s \geq t_0$

$$E \sup_{s \leq s' < s+\delta < t} |[U^\gamma(s')]^T \cdot \Gamma(t,s') - [U^\gamma(s)]^T \cdot \Gamma(t,s)|^2$$
$$\leq K(1+|x|^\beta)(\gamma \vee \delta)e^{-2\alpha(t-s)} \qquad (4.6.18)$$

Following (4.6.14) and using the Lipschitz property of h, we have

$$\sup_{A_t^\gamma \leq s < t} E|[U^\gamma(s)]^T \cdot \Gamma(t,s) \cdot h(\theta_s^\gamma)| \leq K(1+|x|^\beta)(1+\sqrt{\gamma}E|U_t^\gamma|^2) \quad (4.6.19)$$

Thus, following (4.6.7), (4.6.18) and (4.6.19) we have

$$E\{|\int_{]t_0,t]} [U^\gamma(s)]^T \cdot \Gamma(t,s) \cdot d\rho_s^\gamma|\} \leq K(1+|x|^\beta) \int_{]t_0,t]} e^{-\alpha(t-s)} (E|h(\theta_s^\gamma)|^2)^{1/2} ds$$
$$+ K(1+|x|^\beta)(1+\sqrt{\gamma}E|U_t^\gamma|^2)$$

or

$$E\{|\int_{]t_0,t]} [U^\gamma(s)]^T \cdot \Gamma(t,s) \cdot d\rho_s^\gamma|\}$$
$$\leq K(1+|x|^\beta)\{\int_{]t_0,t]} e^{-\alpha(t-s)}(1+E|\theta_s^\gamma|^2)^{1/2} ds(1+\sqrt{\gamma}E|U_t^\gamma|^2)\} \qquad (4.6.20)$$

Following (4.6.6) we have

$$2E\{|\int_{]t_0,t]} [U^\gamma(s_-)]^T \cdot \Gamma(t,s)dB_s^\gamma|\} \leq \sum_{t_0 \leq s \leq t} e^{-\alpha|t-s|} E(|U^\gamma(s_-)||\Delta_s B^\gamma|)$$
$$\leq \sum_{t_0 \leq s \leq t} e^{-\alpha|t-s|}(E|\theta_s^\gamma|^2)^{1/2}(E|\Delta_s B^\gamma|^2)^{1/2}$$
$$\leq K(1+|x|^\beta)\sqrt{\gamma}\int_{t_0}^t e^{-\alpha(t-s)} ds \quad (4.6.21)$$

Lastly, also from (4.6.6), we obtain

$$E\{|\int_{]t_0,t]} e^{-\alpha(t-s)}(d[M^\gamma]_s + d[N^\gamma]_s)\} \leq K(1+|x|^\beta)\int_{t_0}^t e^{-\alpha(t-s)} ds \qquad (4.6.22)$$

Regrouping the inequalities (4.6.16), (4.6.17) and (4.6.20) to (4.6.22) we have

$$E|U^\gamma(t)|^2 \leq E|U^\gamma(t_0)|^2 K(1+|x|^\beta)(1+\int_{t_0}^t e^{t-s} ds)$$
$$+ K(1+|x|^\beta)\sqrt{\gamma}E|U^\gamma(t)|^2$$

Thus if we choose γ_0 such that
$$1 - K(1+|x|^\beta)\sqrt{\gamma_0} := \nu > 0$$
then for all $\gamma \leq \gamma_0$ and all t, we have:
$$\nu \cdot E|U^\gamma(t)|^2 \leq \sup_\gamma E|U^\gamma(t_0)|^2 + K(1+|x|^\beta)$$
and Proposition 17 is proved. □

4.6.4 End of the Proof of Theorem 15

We now consider sequences $t_n \uparrow \infty, \gamma_n \downarrow \infty$ and the processes
$$V^n(t) = U^{\gamma_n}(t_n + t) \tag{4.6.23}$$
Following (4.6.3) we have
$$V^n(t) = U^{\gamma_n}(t_n) + \int_0^t h'(\theta^*)V^n(s)ds + M^{\gamma_n}(t_n+t) - M^{\gamma_n}(t_n)$$
$$+ N^{\gamma_n}(t_n+t) - N^{\gamma_n}(t_n) + \int_0^t (h'[\bar{\theta}(t_n+s,a)] - h'[\theta_*])ds$$

The upper bounds on $M^{\gamma_n}(t_n+t) - M^{\gamma_n}(t_n)$ and $N^{\gamma_n}(t_n+t) - N^{\gamma_n}(t_n)$ derived in Section 4.3 enable us to carry over the arguments of Sections 4.4 and 4.5 almost word for word. Thus, if the sequence of random variables $U^{\gamma_n}(t_n)$ (which is weakly compact by Proposition 17) converges in distribution to a distribution ν, then the sequence of processes (V^n) converges in distribution to a stationary Gaussian diffusion with generator
$$L\phi(x) = Bx \cdot \nabla\phi(x) + \frac{1}{2}\sum_{i,j=1}^d \partial^2_{ij}\phi(x)R^{ij}(\theta_*) \tag{4.6.24}$$
and initial condition ν where the matrix B is as defined in 4.6.1, Assumption (A-ii) and $R(\theta_*)$ is as defined in (A.8).

As in the proof of Lemma 14, for all $\varepsilon > 0$ and all $\Phi \in C_K(\mathbb{R}^d)$, we can determine T so that for any ν in the weak closure of the distributions of the variables $\{U^\gamma(t); \gamma \leq \gamma_0, t > 0\}$ the distribution $P_\nu(t)$ at time t of the diffusion (4.6.24) with initial condition ν satisfies for all $t > T$
$$|\langle \Phi, P_\nu(t)\rangle - \langle \Phi, g\rangle| \leq \varepsilon \tag{4.6.25}$$
where g is the Gaussian distribution $N(0,C)$, C being the matrix defined in Theorem 15. We now consider an arbitrary convergent subsequence $(U^{\gamma_{n_k}}(t_{n_k}))$ with limiting distribution ν. We merely have to show that for all $\varepsilon > 0$
$$|\langle \Phi, \nu_\infty\rangle - \langle \Phi, g\rangle| \leq \varepsilon$$
Without loss of generality, without extracting a subsequence, we may suppose that $(U^{\gamma_{n_k}}(t_{n_k} - T))_{k\geq 0}$ is itself weakly convergent towards a distribution ν. This implies that $\nu_\infty = P_\nu(t)$. The result now follows from (4.6.25). □

4.7 Remark on Weak Convergence Techniques

Kushner and Shwartz (1984) used a method of weak convergence to establish the approximation by the ODE. If we consider an algorithm with constant gain, as in (4.3.1), together with the corresponding process $\theta^\gamma(t)$ in continuous time, we see that the latter may be written as:

$$\theta^\gamma(t) = \theta^\gamma(0) + \int_0^t h(\theta^\gamma(u))du + \sqrt{\gamma}[B^\gamma(t) + \rho^\gamma(t)] + \text{martingale} \quad (4.7.1)$$

where the processes B^γ, and ρ^γ are those defined in formulae (4.3.8) and (4.3.9).

Kushner and Shwartz (1984) showed that the distributions of the processes (θ^γ) are weakly compact as $\gamma \downarrow 0$ and that any limiting distribution is a solution of the martingale condition

$$\phi(\theta(t)) = \phi(\theta(0)) + \int_0^t h(\theta(u)) \cdot \phi'(\theta(u))du + \text{martingale} \quad (4.7.2)$$

Such a limiting process is associated with the differential operator

$$L\phi(x) = h(x) \cdot \phi'(x)$$

which contains no diffusion terms. This is thus a process with deterministic trajectories which are solutions of the equation

$$\theta(t) = \theta(0) + \int_0^t h(\theta(u))du$$

This brings us back to the ODE. If $\theta(0) = a$, and if the ODE has a unique solution, then the convergence in distribution of the random function θ^γ to a uniquely determined function θ implies convergence in probability. Thus

$$\limsup_{\gamma \downarrow 0} \{|\theta^\gamma(t) - \theta(t)| \geq \eta\} = 0 \text{ for all } T > 0 \text{ and } \eta > 0$$

This method is subtle since it does not rely on the calculation of detailed upper bounds on $\theta^\gamma(t) - \theta(t)$ and allows us to obtain convergence with slightly weaker assumptions than those used in Chapter 3.

Thus the approximation of the algorithm in a finite horizon (deterministic approximation with Gaussian fluctuations) may be carried out using only weak methods based on weak compactness criteria and on the martingale methods described at the start of this chapter. (Métivier 1988) takes a more systematic view of this.

4.8 Comments on the Literature

It is natural to think of extending the Gaussian approximation theorems which accompany the classical laws of large numbers to the convergence of stochastic algorithms. Results concerning what we have termed the "Robbins–Monro" case are given in (Gladyshev 1965), (Fabian 1968) and (Sacks 1958)

The convergence in distribution of the "renormalised algorithm", viewed as a process, to a limiting Gaussian distribution was apparently studied for the first time in (Khas'minskii 1966). More recent papers on the same topic include (Kushner and Huang 1979), (Kushner 1984) and (Kushner and Shwartz 1984). An invariance principle for iterative procedures is also found in (Berger 1986). The proof of such invariance principles uses not martingale methods, but earlier results on "triangular arrays" (cf. (Wald 1972), (Lai and Robbins 1978) and (Kersting 1977)). A variant is found in the theorems on mixingales in (McLeish 1976).

The results given in this chapter are an abridged version of (Bouton 1985); see also (Delyon 1986) for similar results with mixing vector fields and a discontinuous function $H(\theta, x)$.

As in the previous chapters, we have not considered algorithms with constraints. Results about Gaussian approximations for these algorithms are given for example in (Kushner and Clark 1978) and (Pflug 1986).

Chapter 5
Appendix to Part II

A Simple Theorem in the "Robbins–Monro" Case

5.1 The Algorithm, the Assumptions and the Theorem

This section was principally intended to provide course support. It was meant to be self-contained and all technical difficulties were to be eliminated through over-realistic assumptions. In fact, the particular instance which we shall describe has already been seen in substance in Example 4 of Subsection 1.1.2 (cf. also 1.1.4), and studied in particular in Subsection 1.10.1. Here once again is the algorithm with its assumptions.

$$\theta_{n+1} = \theta_n + \gamma_{n+1} H(\theta_n, X_{n+1})$$

The main, or so-called *Robbins–Monro*, assumption is the following: for any positive Borel function g,

$$E[g(\theta_n, X_{n+1})|F_n] = \int g(\theta_n, x) \mu_{\theta_n}(dx)$$

where μ_θ is a probability distribution on $\mathbb{R}^k$, and F_n is the σ-field generated by the variables $X_n, X_{n-1}, \ldots, \theta_n, \theta_{n-1}, \ldots$. In other words, the conditional distribution of X_{n+1}, knowing the past, depends only on θ_n. We suppose that

$$\sigma^2(\theta) = \int |H(\theta, x)|^2 \mu_\theta(dx) \leq C(1 + |\theta|^2) \quad (5.1.1)$$

for some constant C, which implies that

$$h(\theta) = \int H(\theta, x) \mu_\theta(dx)$$

exists. We also introduce the following *stability condition*:

$$\exists\, \theta_* : \sup_{\epsilon \leq |\theta - \theta_*| \leq \frac{1}{\epsilon}} (\theta - \theta_*)^T h(\theta) < 0 \quad (5.1.2)$$

for all $\epsilon > 0$: the mean vector field is "inward".

Theorem 1. *With the above assumptions, if the sequence of gains (γ_n) satisfies $\sum \sigma_n = \infty$, $\sum \sigma_n^2 = \infty$, then the sequence $(\theta_n)_{n \geq 0}$ converges almost surely to θ_*.*

This is a fundamental result, and we hope to convince the reader that its proof is extremely simple.

5.2 Proof of the Theorem

The proof depends on a lemma due to (Robbins and Siegmund 1971), see also (Neveu 1972).

5.2.1 A "Super-martingale" Type Lemma

We assume we have a probability space with an increasing family of σ-fields $\{\Omega, F, F_n, P\}$.

Lemma 2. *Suppose that Z_n, B_n, C_n and D_n are finite, non-negative random variables, adapted to the σ-field F_n, which satisfy*

$$E(Z_{n+1}|F_n) \leq (1 + B_n)Z_n + C_n - D_n \tag{5.2.1}$$

Then on the set $\{\sum B_n < \infty, \sum C_n < \infty\}$, we have

$$\sum D_n < \infty \text{ a.s.}; \qquad Z_n \to Z < \infty \text{ a.s.}$$

Proof of Lemma 2. We set

$$U_n = \frac{Z_n}{\prod_{k=1}^{n-1}(1 + B_k)} - \sum_{m=1}^{n-1} \frac{C_m - D_m}{\prod_{k=1}^{m}(1 + B_k)}$$

Following (5.2.1), (U_n) is a super-martingale. We introduce the stopping time

$$\nu_a = \inf\left\{ n : \sum_{m=1}^{n} \frac{C_m}{\prod_{k=1}^{m}(1 + B_k)} > a \right\}$$

where $a > 0$. Then $(a + U_{n \wedge \nu_a})$ is a finite, positive super-martingale, whence it converges a.s. to $U^a < \infty$. Thus, on the set $\{\nu_a = \infty\}$, (U_n) converges a.s. to a finite random variable, as also do the sequences

$$\frac{Z_n}{\prod_{k=1}^{n-1}(1 + B_k)} \quad \text{and} \quad \sum_{m=1}^{n-1} \frac{D_m}{\prod_{k=1}^{m}(1 + B_k)}$$

But

$$\{\sum C_n < \infty\} \subset \cup_{a_i} \{\nu_{a_i} = \infty\}$$

where (a_i) is a sequence which tends to $+\infty$. The proof of Lemma 2 follows directly. □

5.2 Proof of the Theorem

5.2.2 Proof of Theorem 1

Set
$$T_n = \theta_n - \theta_* \text{ and } Z_n = |T_n|^2$$

We have
$$Z_{n+1} = Z_n + 2\gamma_{n+1} T_n^T H(\theta_n, X_{n+1}) + \gamma_{n+1}^2 |H(\theta_n, X_{n+1})|^2 \qquad (5.2.2)$$

whence, using the Robbins–Monro property,
$$E[Z_{n+1}|F_n] = Z_n + 2\gamma_{n+1} T_n^T h(\theta_n) + \gamma_{n+1}^2 \sigma^2(\theta_n) \qquad (5.2.3)$$

Next we shall show that (Z_n) satisfies the assumptions of Lemma 2. Thanks to (5.2.3), we have
$$E[Z_{n+1}|F_n] \leq Z_n(1 + C\gamma_{n+1}^2) + \overline{C}\gamma_{n+1}^2 + 2\gamma_{n+1} T_n^T h(\theta_n)$$

with $T_n^T h(\theta_n) \leq 0$. Thus we may apply the lemma, to obtain
$$Z_n \to Z < \infty \text{ a.s. and } -\sum_n \gamma_{n+1} T_n^T h(\theta_n) < \infty \text{ a.s.} \qquad (5.2.4)$$

Now we need only show that the limit Z is actually equal to 0. Thus, suppose ω is such that $Z(\omega) \neq 0$. Then, there exist $\epsilon > 0$ and $N < \infty$, such that for all $n > N$, we have $\frac{1}{\epsilon} \geq |T_n(\omega)| \geq \epsilon$; whence it follows that $\liminf -T_n^T(\omega)h(\theta_n(\omega)) > 0$, and finally that $-\sum \gamma_{n+1} T_n^T(\omega) h_n(\theta_n(\omega)) = \infty$. Theorem 1 now follows, by virtue of (5.2.4). □

5.3 Variants

The variants involve a modification or slight weakening of the assumptions.

5.3.1 Time-varying Mean Field

The same proof also covers the case in which the random field is of the form $H_n(\theta_n, X_{n+1})$, which of course gives a time-varying mean field $h_n(\theta)$ in the statement of the Robbins–Monro condition. In order that we may copy the proof in its entirety, it is sufficient to introduce this modification into (5.1.1) and (5.1.2), maintaining the upper bounds *uniformly* in n.

5.3.2 A Lyapunov Function

Here we shall reintroduce the ODE more explicitly. We know (cf. Subsection 1.6.1) that the existence of a stable global equilibrium of the ODE $\dot{\theta} = h(\theta)$ is associated with the existence of a Lyapunov function $U(\theta)$. Thus we shall introduce such a Lyapunov function, without imposing any regularity conditions on the mean field $h(\theta)$ (in particular, we do not necessarily assume the existence or uniqueness of solutions of the ODE).

Thus we shall assume that there exists a Lyapunov function U, of class C^2, with a uniformly bounded second derivative, such that

$$\exists\, \theta_* : \sup_{\epsilon \leq |\theta - \theta_*| \leq \frac{1}{\epsilon}} U'(\theta)^T h(\theta) < 0$$

and

$$\sigma^2(\theta) = \int |H(\theta, x)|^2 \mu_\theta(dx) \leq C(1 + U(\theta))$$

In other words, $U(\theta)$ replaces $|\theta - \theta_*|^2$. Using a Taylor expansion of order 2, we obtain

$$U(\theta_{n+1}) = U(\theta_n) + 2\gamma_{n+1} U'^T(\theta_n) H(\theta_n, X_{n+1}) + r_{n+1}$$

where

$$|r_{n+1}| \leq \gamma_{n+1}^2 \|U''\|_\infty |H(\theta_n, X_{n+1})|^2$$

Thus we have

$$E(|r_{n+1}||F_n) \leq C\gamma_{n+1}^2 \|U''\|_\infty (1 + U(\theta_n))$$

But then, it is clear that the residue r_{n+1} may be treated in exactly the same way as the last term of right hand side of (5.2.2), with, in this case, $Z_n = U(\theta_n)$. The proof now follows as before, thus showing that θ_n converges a.s. to θ_*.

5.3.3 Small Additive Perturbation

The assumptions are as before, but now we suppose that the algorithm is of the form

$$\theta_{n+1} = \theta_n + \gamma_{n+1} H(\theta_n, X_{n+1}) + \gamma_{n+1}^2 \mathcal{E}(\theta_n, X_{n+1})$$

where the perturbation satisfies

$$\sigma_{\mathcal{E}}(\theta) = \int |\mathcal{E}(\theta, x)|^2 \mu_\theta(dx) \leq \tilde{C}(1 + U(\theta))$$

and

$$U'(\theta)^T \epsilon(\theta) \leq \tilde{C}(1 + U(\theta))$$

where

$$\epsilon(\theta) = \int \mathcal{E}(\theta, x) \mu_\theta(dx)$$

Once again the additive perturbation may be treated in the same way as the residue r_n if we expand $Z_n = U(\theta_n)$.

5.3 Variants

5.3.4 A Weakened Convergence Result

If we replace assumption (5.1.1) by

$$\forall\, r > 0,\ \sup_{|\theta|\le r} \int |H(\theta,x)|^2 \mu_\theta(dx) < \infty$$

and continue to assume (5.1.2) unchanged, then θ_n *converges a.s. to* θ_* *on the set* $\{\sup_n |\theta_n| < \infty\}$.

In fact, if we let ϕ be a function from $\mathbb{R}_+$ into $[0,1]$ which is equal to 1 on $[0,R]$ and to 0 on $[R+1, +\infty]$, and set

$$\widetilde{H}(\theta,x) = \phi(|\theta - \theta_*|) H(\theta,x) - (1 - \phi(|\theta - \theta_*|))(\theta - \theta_*)$$

then the algorithm

$$\tilde{\theta}_{n+1} = \tilde{\theta}_n + \gamma_{n+1} \widetilde{H}(\tilde{\theta}_n, X_{n+1})$$

satisfies the assumptions of Theorem 1. Thus $\tilde{\theta}_n$ converges a.s. to θ_*. But for all n, $\theta_n = \tilde{\theta}_n$ on $\{\tau = +\infty\}$, where $\tau = \inf\{n \ge 0 : |\theta_n - \theta_*| > R\}$, which proves the result stated above.

Bibliography

Anderson, B.D.O., Moore, J.B. (1979) Optimal Filtering. Prentice Hall, Englewood Cliffs, New Jersey

Anderson, B.D.O., Bitmead, R.R., Johnson, C.R. Jr., Kokotovic, P.V., Kosut, R.L., Mareels, L., Praly, L., Riedle, B. (1986) Stability of Adaptive Systems: Passivity and Averaging Analysis. MIT Press, Cambridge

Appel, U., Brandt, A.V. (1983) Adaptive sequential segmentation of piecewise stationary time series. Information Sciences 29 (1983)

Aström, K.J., Borisson, U., Ljung, L., Wittenmark, B. (1977) Theory and applications of self-tuning regulators. Automatica 13 (1977) 457–476

Azencott, R. (1980) Grandes Déviations et Applications. Lect. Notes in Math., vol. 774. Springer Verlag, Berlin Heidelberg New York (in French)

Basseville, M. (1986) Detection of Changes in Signals and Systems. 2nd IFAC Workshop on Adaptive Systems in Control and Signal Processing, Lund, Sweden, 7-12

Basseville, M. (1988) Detection of abrupt changes in signals and systems: a survey. Automatica 24 3 (1988) 309–326

Basseville, M., Benveniste, A. (eds.) (1986) Detection of Abrupt Changes in Signals and Dynamical Systems. LNCIS 77. Springer Verlag, Berlin Heidelberg New York

Basseville, M., Benveniste, A., Moustakides, G. (1986) Detection and diagnosis of abrupt changes in modal characteristics of non-stationary digital signals. IEEE Trans. on Information Theory IT–32 3 (1986) 412–417

Basseville, M., Benveniste, A., Moustakides, G., Rougée, A. (1987) Detection and diagnosis of changes in the eigenstructure of non-stationary multivariable systems. Automatica 23 (1987) 479–489

Benveniste, A. (1981) Introduction à la méthode de l'équation différentielle moyenne pour l'étude des algorithmes récursifs, exemples. In: Outils et Modèles Mathématiques pour l'Automatique, l'Analyse de Systèmes, et le Traitement du Signal, vol. 1, pp. 459–494. Editions du CNRS, Paris (in French)

Benveniste, A. (1982a) Estimation et factorisation spectrale: quelques points de vue féconds. In: Outils et Modèles Mathématiques pour l'Automatique, l'Analyse de Systèmes, et le Traitement du Signal, vol. 2, pp. 231–266. Editions du CNRS, Paris (in French)

Benveniste, A. (1982b) Méthodes d'orthogonalisation en treillis pour le problème de la réalisation stochastique. In: Outils et Modèles Mathématiques pour l'Automatique, l'Analyse de Systèmes, et le Traitement du Signal, vol. 2, pp. 267–308. Editions du CNRS, Paris (in French)

Benveniste, A. (1982c) Algorithmes simples d'estimation en treillis pour les séries longues. In: Outils et Modèles Mathématiques pour l'Automatique, l'Analyse de Systèmes, et le Traitement du Signal, vol. 2, pp. 309–330. Editions du CNRS, Paris (in French)

Benveniste, A. (1984) Design of One-Step and Multistep Adaptive Algorithms for the Tracking of Time-Varying Systems. INRIA Report no. 340

Benveniste, A. (1986) Advanced methods of change detection: an overview. In: (Basseville and Benveniste 1986), pp. 77–102, q.e.

Benveniste, A., Chauré, C. (1981) AR and ARMA identification algorithms of Levinson type: an innovations approach. IEEE Trans. on Automatic Control $\underline{AC-26}$ 6 (1981) 1243–1261

Benveniste, A., Goursat, M. (1984) Blind equalizers. IEEE Trans. on Communications $\underline{Com-32}$ 8 (1984) 871–883

Benveniste, A., Ruget, G. (1982) A measure of the tracking capability of recursive stochastic algorithms with constant gains. IEEE Trans. on Automatic Control $\underline{AC-27}$ 3 (1982) 639–649

Benveniste, A., Basseville, M., Moustakides, G. (1987) The asymptotic local approach to change detection and model validation. IEEE Trans. on Automatic Control $\underline{AC-32}$ (1987) 583–592

Benveniste, A., Bonnet, M., Goursat, M., Macchi, C., Ruget, G. (1978) Identification d'un Système à non Minimum de Phase par Approximation Stochastique. Rapport LABORIA no. 325 (in French)

Benveniste, A., Goursat, M., Ruget, G. (1980a) Robust identification of a non minimum phase system: blind adjustment of a linear equalizer in data communications. IEEE Trans. on Automatic Control $\underline{AC-25}$ 3 (1980) 385–399

Benveniste, A., Goursat, M., Ruget, G. (1980b) Analysis of stochastic approximations with discontinuous and dependent forcing terms, with applications to data communication algorithms. IEEE Trans. on Automatic Control $\underline{AC-25}$ 6 (1980) 1042–1058

Benveniste, A., Vandamme, P., Joindot, M. (1979) Analyse Théorique des Boucles de Phase Numériques en Présence de Canaux Dispersifs. Rapport CNET NT/MER/TSF/1 (in French)

Berger, E. (1986) Asymptotic behaviour of a class of stochastic approximation procedures. Probab. Th. Rel. Fields 71 (1986) 517–522

Billingsley, P. (1968) Convergence of Probability Measures. Wiley, London New York

Bitmead, R.R. (1984) Convergence properties of LMS adaptive estimators with unbounded dependent input. IEEE Trans. on Automatic Control AC-29 5 (1984) 477–479

Bitmead, R.R., Anderson, B.D.O. (1980) Performance of adaptive estimation algorithms in dependent random environments. IEEE Trans. on Automatic Control AC-25 4 (1980) 788–793

Blum, J. (1954) Multidimensional stochastic approximation methods. Ann. Math. Stat. 25 (1954) 735–744

Bogolyubov, N.N., Metropol'skii, Yu.A. (1961) Asymptotic Methods in the Theory of Nonlinear Observations. Gordon and Breach Science Publishers, New York

Bohlin, T. (1976) Four cases of identification of changing systems. In: Mehra, R.K., Lainiotis, D. (eds.) System Identification, Advances and Case Studies. Academic Press, New York

Borodin, A.N. (1977) A limit theorem for solutions of differential equations with random right-hand side. Theory of Proba. and its Applications 22, 3 (1977) 482–497

Bouton, C. (1985) Approximation Gaussienne d'algorithmes stochastiques à dynamique Markovienne. Thesis, Paris VI (in French)

Box, G.E.P., Jenkins, G.M. (1970) Time Series Analysis, Forecasting and Control. Holden-Day, San Francisco

Chung, K.L. (1954) On a stochastic approximation method. Ann. Math. Stat. 25 (1954) 463–483

Cottrell, M., Fort, J-C. Malgouyres, G. (1983) Large deviations and rare events in the study of stochastic algorithms. IEEE Trans. on Automatic Control AC-28 (1983) 907–920

Dacunha-Castelle, D., Duflo, M. (1982) Probabilités et Statistiques, Problèmes à Temps Fixe. Masson, Paris (in French)

Dacunha-Castelle, D., Duflo, M. (1983) Probabilités et Statistiques, Problèmes à Temps Mobile. Masson, Paris (in French)

Davies. (1973) Asymptotic inference in stationary Gaussian time series. Adv. Appl. Prob. 5 3 (1973) 469–497

Davis, M.H.A., Vinter, R.B. (1985) Stochastic Modelling and Control. Chapman and Hall, London

Deheuvels, P. (1973) Sur l'estimation séquentielle de la densité. C.R. Acad. Sc. Paris série A 276 (1973) 1119–1121 (in French)

Deheuvels, P. (1974) Conditions nécessaires et suffisantes de convergence ponctuelle presque sûre et uniforme presque sûre des estimateurs de la densité. C.R. Acad. Sc. Paris série A 278 (1974) 1217–1220 (in French)

Delyon, B. (1986) Un Théorème de Limite Centrale pour Certaines Équations Différentielles Aléatoires. Thesis, Paris VI (in French)

Derevitskii, D.P., Fradkov, A.L. (1974) Two models for analysing the dynamics of adaptation algorithms. Automation and Remote Control 35 1 (1974) 59–67

Deshayes, J., Picard, D. (1986) Off-line analysis of change-point models using non-parametric and likelihood methods. In (Basseville and Benveniste 1986), pp. 103–168, q.e.

Dvoretsky, A. (1956) On Stochastic Approximation. Proc. Third Berkeley Symp. on Math. Stat. and Prob. vol. 1, pp. 39–45

Ermoliev, Yu. (1983) Stochastic quasi-gradient methods and their application to system optimization. Stochastics 9 (1983) 1–36

Eweda, E., Macchi, O. (1983) Quadratic mean and almost sure convergence of unbounded stochastic approximation algorithms with correlated observations. Ann. Institut Henri Poincaré 19 1 (1983)

Eweda, E., Macchi, O. (1984a) Convergence of an adaptive linear estimation algorithm. IEEE Trans. on Automatic Control AC–29 2 (1984) 119–127

Eweda, E., Macchi, O. (1984b) Convergence analysis of self-adaptive equalizers. IEEE Trans. on Information Theory IT–30 2 (1984) 161–176

Eweda, E., Macchi, O. (1985) Tracking error bounds of adaptive non-stationary filtering. Automatica 21 3 (1985) 293–302

Eweda, E., Macchi, O. (1986) Bases théoriques pour l'égalisation adaptive en mode autodidacte. Annales des Télécom. 41 5–6 (1986) 280–294 (in French)

Eykhoff, P. (1974) System Identification. Wiley, London New York

Fabian, V. (1968) On asymptotic normality in stochastic approximation. Ann. Math. Stat. 39 (1968) 1327–1332

Fabian, V. (1971) Stochastic approximation. In: Rustagi, J. (ed.) Optimizing Methods in Statistics, pp. 439–470. Academic Press, New York

Fabian, V. (1978) On asymptotically efficient recursive estimation. Ann. Stat. 6 (1978) 854–866

Bibliography

Fabian, V. (1983) A local asymptotic minimax optimality of an adaptive Robbins–Monro stochastic approximation procedure. In: Herenkrath, U., Kalin, D., Vogel, W. (eds.) Mathematical Learning Models–Theory and Algorithms, pp. 43–49. Springer Verlag, Berlin Heidelberg New York

Falconer, D.D. (1976) Jointly adaptive equalization and carrier recovery in two dimensional digital communication systems. Bell Syst. Tech. J. 55 3 (1976) 317–334

Farden, D. (1981) Stochastic approximation with correlated data. IEEE Trans. on Information Theory IT-27 1 (1981) 105–113

Farden, D., Sayood, K. (1980) Tracking properties of adaptive signal processing algorithms. Proc. IEEE ICASSP, Denver, 1980, 466–469

Fogelman-Soulié, F., Robert, Y., Tchuente, M. (eds.) (1987) Automata Networks in Computer Science. Nonlinear Science, Theory and Applications. Manchester Univ. Press

Gardner, F.M. (1979) Phaselock Techniques. Wiley, New York

Gelfand, S.B. (1987) Analysis of Simulated Annealing Type Algorithms. PhD thesis, Report MIT-LIDS-TH-1668, May 1987

Gersh, W., Kitagawa, G. (1985) A smoothness priors time-varying AR coefficient modeling of non-stationary covariance time series. IEEE Trans. on Automatic Control AC-30 1 (1985) 48–57

Gladyshev, E.G. (1965) On stochastic approximation. Theory of Proba. and its Applications 10 (1965) 275–278

Goodwin, G.C., Sin, K. (1984) Adaptive Filtering, Prediction, and Control. Prentice Hall, Englewood Cliffs, New Jersey

Goodwin, G.C., Ramadge, P.J., Caines, P.E. (1980) Discrete time multi-variable adaptive control. IEEE Trans. on Automatic Control AC-25 (1980) 449–456

Goursat, M. (1984) Numerical results of stochastic gradient techniques for deconvolution in seismology. Geoexploration 23 (1984) 103–119

Gray, R.M. (1984) Vector quantization. IEEE ASSP Magazine 1 2 (1984) 4–29

Györfi, L. (1980) Stochastic approximation from ergodic sample for linear regression. Z. Wahrscheinlichtskeitstheorie verw. Gebiete 54 (1980) 47–55

Hajek, B. (1985) Tutorial Survey of Theory and Applications of Simulated Annealing. Proc. IEEE Decision and Control Conference 1985

Hall, D., Heyde, C.C. (1980) Martingale Limit Theory and Applications. Academic Press, New York

Henrici, P. (1963) Error Propagation for Difference Methods. Wiley, London New York

Himmelblau, D.M. (1978) Fault Detection and Diagnosis in Chemical and Petrochemical Systems. Elsevier, Amsterdam

Hinkley, D.V. (1971) Inference about the change point in a sequence of random variables. Biometrika 57 1 (1971) 1-17

Hiriart-Urruty, J.B. (1977) Algorithms of penalization type and of dual type for the solution of stochastic optimization problems with stochastic constraints. In: Barra, J.F., Brodeau, F., Romier, G., Van Cutsem, B. (eds.) Recent Developments in Statistics, pp. 183-219. North Holland Publ. Co., Amsterdam New York Oxford

Hoppenstaedt, F. (1971) Properties of solutions of ordinary differential equations with small parameters. Comm. on Pure and Applied Math. 24 (1971) 807-840

Ibragimov, I.A., Khas'minskii, R.Z. (1981) Statistical Estimation, Asymptotic Theory. Applications of Math., vol. 16, Springer Verlag, Berlin Heidelberg New York

Isermann, R. (1984) Process fault detection based on modeling and estimation methods, a survey. Automatica 20 (1984) 387-404

Kailath, T. (1980) Linear Systems. Prentice Hall, Englewood Cliffs, New Jersey

Kersting, G. (1977) Almost sure approximation of the Robbins-Monro process by sums of independent random variables. Ann. Probab. 5 (1977) 954-965

Khas'minskii, R.Z. (1966) On stochastic processes defined by differential equations with a small parameter. Theory of Proba. and its Applications 11 2 (1966) 211-228

Kiefer, J., Wolfowitz, J. (1952) Stochastic estimation of the modulus of a regression function. Ann. Math. Stat. 23 (1952) 462-466

Kindermann, R., Snell, J.L. (1980) Markov Random Fields and their Applications. Contemporary Mathematics vol.1, American Mathematical Society

Kligene, N.I., Tel'ksnis, L.A. (1983) Methods of detecting instants of changes of random process properties. Automation and Remote Control 44 (1983) 1241-1283

Korostelev (1981) Multistep procedures of stochastic optimization. Avtomatikha i Telemekhanika 5 (1981) 82-90

Krasovskii (1963) Stability of Motion. Stanford University Press

Kushner, H.J. (1981) Stochastic approximation with discontinuous dynamics and state dependent noise with w.p.1 and weak convergence. J. of Math. Anal. and Appl. 82 (1981) 527-542

Kushner, H.J. (1984) Approximation and Weak Convergence Methods for Random Processes, with Applications to Stochastic System Theory. MIT Press, Cambridge

Kushner, H.J., Clark, D.S. (1978) Stochastic Approximation for Constrained and Unconstrained Systems. Applied Math. Sci. no. 26, Springer Verlag, Berlin Heidelberg New York

Kushner, H.J., Huang, H. (1979) Rates of convergence for stochastic approximation type algorithms. SIAM J. Control and Opt. 17 1 (1979) 607–617

Kushner, H.J., Huang, H. (1981) Asymptotic properties of stochastic approximations with constant coefficients. SIAM J. Control and Opt. 19 1 (1981) 87–105

Kushner, H.J., Sanvicente, E. (1975) Stochastic approximation for constrained systems with observation noise on the system and constraints. Automatica 11 (1975) 375–380

Kushner, H.J., Shwartz, A. (1984) An invariant measure approach to the convergence of stochastic approximations with state dependent noise. SIAM J. Control and Opt. 22 1 (1984) 13–27

Lai, T.L., Robbins, H. (1978) Limit theorems for weighted sums and stochastic approximation processes. Proc. Nat. Acad. Sci. U.S.A. 75 (1978) 1068–1070

Lindsey, W.C., Simon, M.K. (1973) Telecommunication System Engineering. Prentice Hall, Englewood Cliffs, New Jersey

Ljung, L. (1977a) On positive real transfer functions and the convergence of some recursions. IEEE Trans. on Automatic Control AC–22 (1977) 539–551

Ljung, L. (1977b) Analysis of recursive stochastic algorithms. IEEE Trans. on Automatic Control AC–22 4 (1977) 551–575

Ljung, L. (1978) Convergence of an adaptive filter algorithm. Int. J. Control 27 5 (1978) 673–693

Ljung, L. (1984) Analysis of stochastic gradient algorithms for linear regression problems. IEEE Trans. on Information Theory IT–30 (1984) 151–160

Ljung, L., Caines, P.E. (1979) Asymptotic normality of prediction error estimation for approximate system models. Stochastics 3 (1979) 29–46

Ljung, L., Soderström, T. (1983) Theory and Practice of Recursive Identification. MIT Press, Cambridge

Lorden, G. (1971) Procedures for reacting to a change in distribution. Ann. Math. Stat. 42 2 (1971) 1897–1908

Marti, K. (1979) Approximationen stochasticher Optimierungsprobleme. Verlag Anton Hain, Meisenheim (in German)

McLeish, D.L. (1975a) A maximal inequality and dependent strong laws. Ann. Probab. 3 5 (1975) 829–839

McLeish, D.L. (1975b) Invariance principles for dependent variables. Z. Wahrscheinlichtskeitstheorie verw. Geb. 32 (1975) 165–178

McLeish, D.L. (1976) Functional and random central limit theorems for the Robbins–Monro process. J. Appl. Probab. 13 (1976) 148–154

McLeish, D.L. (1977) On the invariance principle for non-stationary mixingales. Ann. Probab. 5 4 (1977) 616–621

Métivier, M. (1983) Semimartingales. Walter de Gruyter, Berlin

Métivier, M. (1988) On the mathematical analysis of stochastic algorithms. Maryland Lecture Notes, System Research Center, College of Engineering, University of Maryland, College Park

Métivier, M., Priouret, P. (1984) Application of a Kushner and Clark lemma to general classes of stochastic algorithms. IEEE Trans. on Information Theory IT–30 (1984) 140–150

Métivier, M., Priouret, P. (1986) Convergence avec probabilité $1 - \varepsilon$ d'algorithmes stochastiques. Annales des Télécom. 41 5-6 (1986) (in French)

Métivier, M., Priouret, P. (1987) Théorèmes de convergence presque-sûre pour une classe d'algorithmes stochastiques à pas décroissant. Prob. Th. Rel. Fields 74 (1987) 403–28 (in French)

Mironovski, L.A. (1980) Functional diagnosis of dynamic systems, a survey. Automation and Remote Control 41 (1980) 1122–1143

Monnez, J.M. (1982) Etude d'un processus général multidimensionnel d'approximation stochastique sous contraintes convexes, application à l'estimation statistique. State doctoral thesis, University of Nancy (in French)

Moustakides, G. (1986) Optimal procedures for detecting changes in distribution. Ann. Stat. 14 (1986) 1379–1387

Moustakides, G., Benveniste, A. (1986) Detecting changes in the AR parameters of a non-stationary ARMA process. Stochastics 16 (1986) 137–155

Nevel'son, M.B., Khas'minskii, R.Z. (1976) Stochastic Approximation and Recursive Estimation. American Mathematical Society Translations of Math. Monographs, vol. 47

Neveu, J. (1972) Martingales à Temps Discrets. Masson, Paris (in French)

Nikiforov, I.V. (1983) Sequential Detection of Abrupt Changes in Time Series Properties. Nauka, Moscow (in Russian)

Nikiforov. I.V. (1986) Sequential detection of changes in stochastic systems. In (Basseville and Benveniste 1986), pp. 216–258, q.e.

Page, E.S. (1954) Continuous inspection schemes. Biometrika 41 1 (1954) 100-115

Pflug, G.Ch. (1981a) On the convergence of a penalty-type stochastic optimization procedure. J. Inf. Optimization Sci. 2 (1981) 249–258

Pflug, G.Ch. (1981b) Nichtregulare Familien von Dichten und rekursive Schätzung. Sitzungsber., Abt.II, Österr. Akad. Wiss., Math.-Naturwiss.Kl., 190 (1981) 347–383 (in German)

Pflug, G.Ch. (1986) Stochastic minimization with constant step-size: asymptotic laws. SIAM J. Control Optim. 24 (1986) 655–666

Pflug, G.Ch. (1987) Step-size rules, stopping times and their implementation in stochastic quasi-gradient algorithms. In: Wets, R. (ed.), Numerical Methods of Optimization. Springer Verlag, Berlin Heidelberg New York

Priouret, P. (1973) Processus de diffusion et équations différentielles stochastiques. Ecole d'été de probabilités de Saint-Flour, Lect. Notes in Math., vol. 390. Springer Verlag, Berlin Heidelberg New York (in French)

Prum, B. (1986) Processus sur un Réseau et Mesures de Gibbs. Techniques Stochastiques, Masson, Paris (in French)

Reinhard, H. (1982) Equations Différentielles. Gauthier Villars (in French)

Revesz, P. (1973) Robbins–Monro procedure in a Hilbert space, and its application in the theory of learning processes I. Studia Sci. Math. Hungar. 8 (1973) 391–398

Revuz, D. (1975) Markov Chains. North Holland

Robbins, H., Monro, S. (1951) A stochastic approximation method. Ann. Mat. Stat. 22 (1951) 400–407

Robbins, H., Siegmund, D. (1971) A convergence theorem for non-negative almost surmartingales and some applications. In: Rustagi, J. (ed.) (1971) Optimizing Methods in Statistics, pp. 235–257. Academic Press, New York

Roussas, G.G. (1972) Contiguity of Probability Measures, some Applications in Statistics. Cambridge University Press

Ruppert, D. (1982) Almost sure approximations to the Robbins–Monro and Kiefer–Wolfowitz processes with dependent noise. Ann. Probab. 10 1 (1982) 128–187

Ruppert, D. (1983) Convergence of stochastic approximation algorithms with non-additive dependent disturbances and applications. In: Herenkrath, U., Kalin, D., Vogel, W. (eds.) Mathematical Learning Models—Theory and Algorithms, pp. 182-190. Springer Verlag, Berlin Heidelberg New York

Ruszczynski, J., Sysksi, W. (1983) Stochastic approximation method with gradient averaging for unconstrained problems. IEEE Trans. on Automatic Control AC-28 12 (1983)

Sacks, J. (1958) Asymptotic distributions of stochastic approximation procedures. Ann. Math. Stat. 29 (1958) 373-405

Schmetterer, L. (1969) Multidimensional stochastic approximation. In: Krisnaiah (ed.) Multivariate Analysis II. Academic Press, New York

Schmetterer, L. (1980) Uber ein rekursives Verfahren von Herrn Hiriart-Urruty. Sitzungsber., Abt. II, Österr. Akad. Wiss., Math.-Naturwiss. Kl. 189 (1980) 139-147 (in German)

Shil'man, S.V., Yastrebov, A.I. (1976) Convergence of a class of multistep stochastic adaptation algorithms. Avtomatikha i Telemekhanika 8 (1976) 111-118

Shil'man, S.V., Yastrebov, A.I. (1978) Properties of a class of multistep gradient and pseudo-gradient algorithms of adaptation and learning. Avtomatikha i Telemekhanika 4 (1978) 95-104

Shiryaev, A.N. (1961) The problem of the most rapid detection of a disturbance in a stationary process. Soviet Math. Dokl. 2 (1961) 795-799

Shiryaev, A.N. (1978) Optimal Stopping Rules. Springer Verlag, Berlin Heidelberg New York

Soderström, T., Stoica, P. (1989) System Identification. Series in Systems and Control Engineering, Prentice Hall, Englewood Cliffs, New Jersey

Solo, V. (1979) The convergence of AML. IEEE Trans. on Automatic Control AC-24 (1979) 958-963

Solo, V. (1981) The second order properties of a time series recursion. Ann. Stat. 9 (1981) 307-317

Stoica, P., Soderström, T., Friedlander, B. (1985) Optimal instrumental variable estimates of the AR parameters of an ARMA process. IEEE Trans. on Automatic Control AC-30 11 (1985) 1066-1074

Stroock, D., Varadhan, S.R.S. (1969) Diffusion processes with continuous coefficients. Comm. on Pure and Appl. Math. 22 (1969) 345-400

Sunyach, C. (1975) Une classe de chaînes récurrentes sur un espace métrique complet. Ann. Institut Henri Poincaré 11 4 (1975) 325-343 (in French)

Tsypkin, Ya.Z. (1971) Adaptation and Learning in Automatic Systems. Academic Press, New York

Tsypkin, Ya.Z. (1973) Foundation of the Theory of Learning Systems. Academic Press, New York

Verdu, S. (1984) On the selection of memoryless adaptive laws for blind equalization in binary communication. In: Proc. of Sixth Int. Conf. on Analysis and Optimiz. of Syst., Lect. Notes in Control and Information Sc., vol. 62, pp. 239–249. Springer Verlag, Berlin Heidelberg New York

Wald, A. (1949) Sequential Analysis. Wiley, London New York

Walk, H. (1977) An invariance principle for the Robbins–Monro process in a Hilbert Space, Z. Wahrscheinlichkeitstheorie verw. Gebiete 39 (1977) 135–150

Walk, H. (1980) A functional central limit theorem for martingales in $C(K)$ and its application to sequential estimation. J. reine angew. Math. 314 (1980) 117–135

Walk, H. (1983-4) Stochastic iteration for a constrained optimization problem. Commun. Statist. Sequential Analysis 2 (1983-4) 369–385

Walk, H. (1985) Almost sure convergence of stochastic approximation processes. Statistics and Decisions, Supplement Issue 2 (1985) 137–141

Walk, H. (1988) Limit behaviour of stochastic approximation processes. Statistics and Decisions 6 (1988)

Walk, H., Zsido, G. (1989) Convergence of the Robbins–Monro method for linear problems in a Banach space. J. Math. Anal. Appl. 139 1 (1989) 152–177

Wasan, M.T. (1969) Stochastic Approximation. Cambridge University Press

Widrow, B., Walach (1984) On the statistical efficiency of the LMS algorithm with non-stationary inputs. IEEE Trans. on Information Theory IT-30 3 (1984)

Widrow, B., McCool, J., Larimore, M.G., Johnson, C.R. (1976) Stationary and non-stationary learning characteristics of the LMS adaptive filter. Proc. IEEE 64 8 (1976) 1151–1161

Willsky, A.S. (1976) A survey of design methods for failure detection in dynamic systems. Automatica 12 (1976) 601–611

Willsky, A.S. (1986) Detection of abrupt changes in dynamic systems. In: (Basseville and Benveniste 1986), pp. 27–49, q.e.

Willsky, A.S., Jones, H.L. (1976) A generalized likelihood ratio approach to detection and estimation of jumps in linear systems. IEEE Trans. on Automatic Control AC-21 (1976) 108–112

Younes, L. (1988a) Estimation and Annealing for Gibbsian Fields. Ann. Institut Henri Poincaré, Séries Probabilités et Statistiques, 24 2 (1988) 269–294

Younes, L. (1988b) Estimation for Gibbsian Fields, Applications and Numerical Results. Report of University of Paris–Sud Orsay

Younes, L. (1988c) Problèmes d'estimation pour des champs de Gibbs Markoviens. Application au traitement d'images. Thesis, University of Paris-Sud Orsay (in French)

Younes, L. (1989) Parametric influence for imperfectly observed Gibbsian fields. Prob. Th. Rel. Fields 82 (1989) 625–645

Subject Index to Part I

Adaptive
    control 35, 76
    forgetting factor 160
    gain 159
Admissible
    filter 146
    gain 112, 126
Algorithm analysis (guide) 48
Algorithm design
    guide 55, 137, 155
    optimal, constant gain 131, 134, 137, 140, 142, 150
    optimal, decreasing gain 110
ALOHA 90
AR, ARMA, ARMAX 71, 78, 166, 182, 192, 193, 205
Assumptions (A) 31
Asymptotic local method 176, 178
Average excess mean square error 158
Averaging 101

Back propagation 91

Chi-squared (χ^2) test 187
Conditionally linear dynamics 32
Constraints (algorithm with) 63, 65
Control, adaptive 35, 76
Convergence
    heuristics 41
    in finite horizon 42
    in infinite horizon 43, 44, 45
Cumulative Sum 172, 178
cov_θ 105

Detection delay 169
Discontinuities 53
Δ 113, 134
Δ_r^n 181
$D_{n,m}(\theta_0, \theta), D_n(\theta_0, \theta)$ 178

Echo cancellation 83
Equalisation 10
    blind 60
    in learning phase 17, 49
    self-adaptive 18, 53, 77, 85, 162
Exponential forgetting factor 140
$\varepsilon_n(\theta, X)$ 9
E_θ 33
$E_{\theta,z}$ 124

Figure of merit
    algorithms with constant gain 125, 147
    algorithms with decreasing gain 111
    off-line detection 170
    sequential detection 170
Filter 199
Fisher (information matrix) 113
Forgetting factor 140
Functional 55, 77

Gaussian approximations 107, 127
Generalised Likelihood Ratio test 175
Gibbs field 96
Gibbs sampler 97
Gradient, stochastic gradient method 55, 59, 141
γ_n 9
$\Gamma(\sigma)$ 143
$[\Gamma]$ 143

Hankel matrix 201
Hessian 73
Hoppenstaedt's method 132
Hypermodel 122, 136, 144
$H(\theta, X)$ 9
$H(\theta, z; X)$ 122
$[H]$ 145
$h(\theta)$ 28
h_θ 106, 124
$h(\theta, z)$ 124

Instrumental variables method 79, 183
Interaction system/algorithm 122, 145
Intrinsic quality criterion 113, 116, 134, 150

Kalman filter 139, 151, 157, 162, 208
$K(z, \zeta)$ 122
$k(z)$ 124
$[k]$ 145

Large deviations 31, 90
Lattice algorithm 36, 77, 118
Least squares 18, 50, 72, 73, 114, 121, 139, 157
   extended (ELS) 74, 78, 192
Level (of a test) 170
Likelihood method 56, 57, 71
Lloyd's algorithm 89
Local test 180
   instrumental test 183
   likelihood test 183

Markov chain (controlled) 25, 32, 167
Matrix inversion lemma 141
McLeish's theorem 189
Mean time between false alarms 169
Mixingale 188
Modelling 14, 21
Multistep algorithms 142, 161, 162
μ_θ 26
μ 122

Neural networks 91
Newtonian, quasi-Newtonian methods 55, 73
Noise suppression 80
Nominal model 177, 196
Non-stationarity 120, 124
Nuisance parameters 1951
∇ 49
ν 168

ODE 28, 33, 40, 55

Page–Hinkley stopping rule 171
Phase-locked loop 19, 58, 78, 85, 113, 116, 120, 138, 156, 196
Potential 49, 55
Power (of a test) 171
P_θ 33
π_θ 26
$\pi_{\theta,z}$ 122

Quantisation 53, 87
   Vector 89, 197
$Q(z)$ 126

Rare events 31, 90
Rate of convergence, algorithms with constant gain
   heuristics 104
   in finite horizon 107
   in infinite horizon 107
Rate of convergence, algorithms with decreasing gain
   heuristics 108
   results 110
Rational transfer function 200
Recursive Least Squares (RLS) 72, 115
Recursive Maximum Likelihood (RML) 74, 193
Robbins–Monro algorithm 38
$R(\theta)$ 106
$R(\theta, z)$ 126
Re λ 107
r 168

Search direction (choice of) 115, 131
Second order process 205
Sensitivity methods 193
Separation 159
Sequential detection 169
Singular perturbations 132
Sliding window algorithm 118, 161
Smith–McMillan degree 148
Spectral factorisation 208
Spectrum, spectral measure, spectral density 206
State space 202
Stopping rule 168

Transfer function 199
  Rational 200
Transient 75
θ_n 9

Θ_n 143
$[\theta]_n$ 143
$\tilde{\theta}_t$ 107
$\tilde{\theta}_n^\gamma$ 104

Validation 31, 185
Variable state vector 24, 32, 199
Vector field 33
Vector quantisation 89, 197

X_n 9
ξ_n 26

$Y_k(\theta_0)$ 180

z-transform 199
z_n 122
$[z]_n$ 145

Subject Index to Part II

(A) 334
(A.1) 213
(A.2) 213
(A.3) 216
(A.4) 216
(A.4-iii)' 236
(A.5) 220
(A'.5) 290
(A.6) 233
(A'.6) 301
(A.7) 233
(A'.7) 305
(A.8) 321

(B) 335
Burkholder inequalities 294
$[.]_p$ 252
$[.]_q$ 290

Canonical (process, filtration) 309
Conditionally linear dynamics 215, 290

Decision-feedback phase-locked loop 274
Diffusion 312
   Gaussian 313

F_n 213
$\widetilde{F}_t$ 309

γ_n 213
Γ_θ 217

$H(\theta, X)$ 217
H_θ 220
$h(\theta)$ 220

$I(\ldots)$, characteristic function 214

Least squares algorithm 272
$Li(p)$ 252
$Li(Q, L_1, L_2, p_1, p_2)$ 259
$Li(Q)$ 259
$Li^1(Q, L_1, L_2, p_1, p_2)$ 262
$\overline{Li}(\mathbb{R}^d, L_1, L_2, p_1, p_2)$ 265
$\overline{Li}(\mathbb{R}^d)$ 265
(L) 317
L_t 322, 328

$m(n, T)$ 214

$N_p(g)$ 253
$\|\cdot\|_{\infty,p}$ 252
ν_θ 216

ODE 230
$\widetilde{\Omega}_T$ 308

Poisson equation 217, 252
Process with (conditionally) linear dynamics 265
$\Pi_\theta, \Pi_\theta(x, A)$ 214
$P_{x,a}, P_{x,a}^{\gamma;\rho}$ 214

Recursive decision-feedback equaliser 215, 276
Robbins–Monro algorithm 215, 219, 229, 244, 343
$\rho_n(\theta, x)$ 213
$R(\theta)$, $R^{ij}(\theta)$ 321

Skorokhod space 308

Theorem 9, Chapter 1 232
Theorem 13, Chapter 1 236
Theorem 14, Chapter 1 237
Theorem 15, Chapter 1 238
Theorem 17, Chapter 1 239
Theorem 22, Chapter 1 244
Theorem 24, Chapter 1 246
Theorem 5, Chapter 2 259
Theorem 6, Chapter 2 262
Theorem 7, Chapter 2 265
Theorem 13, Chapter 2 278
Theorem 12, Chapter 3 301
Theorem 17, Chapter 3 304
Theorem 20, Chapter 3 305
Theorem 7, Chapter 4 322
Theorem 12, Chapter 4 328
Theorem 13, Chapter 4 332
Theorem 15, Chapter 4 335

Tight 310
Transversal equaliser, learning phase 271
θ_n 213
θ_n^γ 307
$\theta(t)$ 214
$\theta^\gamma(t)$ 308
$\bar{\theta}(t)$ 230
θ_n^N 327
t_n 214

$U(\theta)$ 239
$U^\gamma(t)$ 314

Weak compactness 310
Weak convergence of processes 310
w_θ^{ij} 321

$(X.1),(X.2),(X.3),(X.4)$ 256